T0260703

PROCESS-AWARE
INFORMATION SYSTEMS

PROCESS-AWARE INFORMATION SYSTEMS
Bridging People and Software Through Process Technology

Edited by

MARLON DUMAS
Queensland University of Technology

WIL van der AALST
Eindhoven University of Technology

ARTHUR H. M. ter HOFSTEDE
Queensland University of Technology

WILEY-INTERSCIENCE

A JOHN WILEY & SONS, INC., PUBLICATION

Library of Congress Cataloging-in-Publication Data:

Process-aware information systems : bridging people and software through
    process technology / Marlon Dumas, Wil van der Aalst, Arthur ter Hofstede
    (editors).
        p. cm.
    Includes bibliographical references.
    ISBN-13 978-0-471-66306-5
    ISBN-10 0-471-66306-9 (cloth : alk. paper)
    1. Computer-aided software engineering. 2. Human-computer interaction. I.
Dumas, Marlon. II. Aalst, Wil van der. III. Ter Hofstede, Arthur, 1966–
    QA76.758.P757 2005
    005.1'0285—dc22

                                                            2005001369

10 9 8 7 6 5 4 3 2 1

*To Inga and her admirable ability to marry
reason with emotion—Marlon*

*To Willem for showing that you do not have
to be smart to enjoy life—Wil*

Contents

▰▰▰ Preface

Process-aware information systems are at the heart of an ongoing "silent revolution." From the late 1970s to the early 1990s, the lion's share of attention in the area of information systems went to data. The focus was mainly on storing and retrieving information and, hence, data models were often the starting point for designing information systems, whereas database management systems were considered to be the heart of the run time infrastructure. During the 1990s, a number of parallel trends shifted the focus to processes. As a result, an increasing number of business processes are now conducted under the supervision of information systems driven by explicit process models. This shift of focus has resulted in a myriad of approaches to process engineering, modeling, and implementation, ranging from those supported by groupware and project management products to those supported by document, imaging, and workflow management systems, which are now finding their way into enterprise application-integration tools. The plethora of (sometimes subtly different) technologies in this area illustrates the relevance of the topic but also its complexity, and despite a number of discontinued and ongoing standardization efforts, there is still a lack of an overarching framework for designing and implementing process-aware information systems. Instead, process-awareness in information systems manifests itself in various forms, with similar concepts appearing under different names, in different combinations, and with varying levels of tool support.

The goal of this book is to provide a unifying and comprehensive overview of the technological underpinnings of the emerging field of process-aware information systems engineering. While primarily intended as a textbook, the book is also a manifesto for process-aware information systems, insofar as it puts forward the resemblances (and differences) between a number of technologies that up to now have evolved somewhat independently of one another. In this respect, it is hoped that the book will raise awareness of the need to look at new trends in the area in light of a broader perspective than has been employed up to now and to draw on the large body of existing theoretical and practical knowledge. In terms of scope, it should be mentioned that the focus of the book is on technical aspects, as opposed to strategic and managerial aspects, which are covered in a number of other publications (many of which are referenced throughout the book).

The book is intended to be used both as a textbook for advanced undergraduate and postgraduate courses and as reference material for practitioners and academics. Consistent with the former purpose, the book contains exercises, ranging from simple questions to projects and possible assignment subjects. Sample solutions for many of these exercises will be made available at a companion site, http://www.wiley.com/WileyCDA/WileyTitle/productCd-0471663069.html. Further information and material related to the book will be posted at: http://www.bpmcenter.org.

The book gathers contributions from a number of international experts and teams from both academia and industry. We acknowledge the contributors for their engagement and dedication in the preparation of their chapters and for their prompt help in peer-reviewing each others' chapters. It should be recognized that many of the topics covered in the book are still emerging or even groundbreaking, and authors had to put considerable effort into presenting them in a way that is accessible to the broadest possible audience. We also acknowledge the financial support of the Australian Research Council through its Discovery Projects scheme. Finally, we thank Wiley's editorial team, especially Val Moliere, for their support and patience that contributed to turning the original book project into a reality.

MARLON DUMAS
WIL VAN DER AALST
ARTHUR H. M. TER HOFSTEDE

Brisbane, Australia,
August 2005

Contributors

Otmar Adam, Institute for Information Systems (IWi), German Research Center for Artificial Intelligence (DFKI), Saarbrücken, Germany

Gustavo Alonso, Department of Computer Science, ETH Zentrum, Zürich, Switzerland

Paulo Barthelmess, Department of Computer Science, University of Colorado, Boulder, Colorado

Paul J. S. Berens, Pallas Athena, Apeldoorn, The Netherlands

Charles Brown, Logica CMG, Milton, Australia

Christoph Bussler, Digital Enterprise Research Institute, National University of Ireland, Galway, Ireland

Jun Chen, Department of Computer Science, University of Colorado, Boulder, Colorado

Francisco Curbera, Component Systems Group, IBM T.J. Watson Research Center, Hawthorne, New York

Jörg Desel, Catholic University, Faculty of Mathematics and Geography, Eichstätt, Germany

Marlon Dumas, Centre for Information Technology Innovation, Queensland University of Technology, Brisbane, Australia

Clarence A. Ellis, Department of Computer Science, University of Colorado Boulder, Colorado

Gregor Engels, University of Paderborn, Faculty of Computer Science, Electrical Engineering and Mathematics, Paderborn, Germany

Alexander Förster, University of Paderborn, Faculty of Computer Science, Electrical Engineering and Mathematics, Paderborn, Germany

Reiko Heckel, University of Paderborn, Faculty of Computer Science, Electrical Engineering and Mathematics, Paderborn, Germany

Rania Khalaf, Component Systems Group, IBM T.J. Watson Research Center, Hawthorne, New York

Jan Mendling, Vienna University of Economics, BA Department of Information Systems New Media Lab, Wien, Austria

Greg Meredith, Microsoft, Seattle, Washington

Nirmal Mukhi, Component Systems Group, IBM T.J. Watson Research Center, Hawthorne, New York

Andreas Oberweis, AIFB, University of Karlsruhe, Karlsruhe, Germany

Adrian Price, Versata, Inc., Oakland, California

Hajo A. Reijers, Eindhoven University of Technology, Department of Technology Management, Eindhoven, The Netherlands

Michael Rosemann, Centre for Information Technology Innovation, Brisbane, Australia

August-Wilhelm Scheer, Institute for Information Systems (IWi), German Research Center for Artificial Intelligence (DFKI), Saarbrücken, Germany

Arthur H. M. ter Hofstede, Centre for Information Technology Innovation, Queensland University of Technology, Brisbane, Australia

Oliver Thomas, Institute for Information Systems (IWi), German Research Center for Artificial Intelligence (DFKI), Saarbrücken, Germany

Sebastian Thöne, University of Paderborn, Department of Computer Science, Paderborn, Germany

Wil van der Aalst, Department of Technology Management, Eindhoven University of Technology, Eindhoven, The Netherlands

Alexander Verbraeck, Delft University of Technology, Faculty of Technology, Policy, and Management, Systems Engineering Group, Delft, The Netherlands

Jacques Wainer, Instituto de Computação, Universidade Estadual de Campinas, Caixa, Campinas, Sao Paulo, Brazil

Sanjiva Weerawarana, Component Systems Group, IBM T.J. Watson Research Center, Hawthorne, New York

A. J. M. M. Weijters, Department of Technology Management, Eindhoven University of Technology, Eindhoven, The Netherlands

Michael zur Muehlen, Stevens Institute of Technology, Wesley J. Howe School of Technology Management, Castle Point on Hudson, Hoboken, New Jersey

PART I

CONCEPTS

CHAPTER 1

Introduction

MARLON DUMAS, WIL van der AALST,
and ARTHUR H. M. ter HOFSTEDE

1.1 FROM PROGRAMS AND DATA TO PROCESSES

A major challenge faced by organizations in today's environment is to transform ideas and concepts into products and services at an ever-increasing pace. At the same time and following the development and adoption of Internet technologies, organizations distributed by space, time, and capabilities are increasingly pushed to exploit synergies by integrating their processes in the setting of virtual organizations. These forces triggered a number of trends that have progressively changed the landscape and nature of enabling technologies for information systems development.

Figure 1.1 illustrates some of the ongoing trends in information systems [2]. This figure shows that information systems consist of a number of layers. The center is formed by the system infrastructure, consisting of hardware and the operating system(s) that make the hardware work. The second layer consists of generic applications that can be used in a wide range of enterprises. These applications are typically used in multiple departments within the same organization. Examples of such generic applications are a database management system (DBMS), a text editor, and a spreadsheet editing tool. The third layer consists of domain-specific applications. These applications are only used within specific types of organizations or departments. Examples are decision support systems for vehicle routing, computer-aided design tools, accounting packages, and call center software. The fourth layer consists of tailor-made applications developed for specific organizations.

In the 1960s, the second and third layers were practically missing. Information systems were built on top of a small operating system with limited functionality. Since no generic or domain-specific software was available, these systems mainly consisted of tailor-made applications. Since then, the second and third layers have developed and the ongoing trend is that the four circles are increasing in size, that is, they are moving to the outside while absorbing new functionality. Today's operating systems offer much more functionality, especially in the area of networking.

Process-Aware Information Systems. Edited by Dumas, van der Aalst, and ter Hofstede
Copyright © 2005 John Wiley & Sons, Inc.

system
infrastructure

generic
applications

domain-
specific
applications

tailor-made
applications

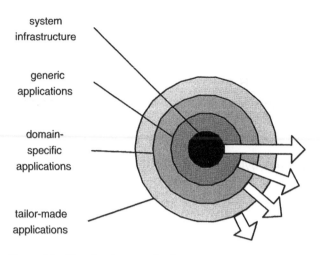

Figure 1.1   Trends relevant to business process management.

DBMSs that reside in the second layer offer functionality that used to be encoded in domain-specific and tailor-made applications. Also, the number and complexity of domain-specific and tailor-made applications has increased, driven by the need to support more types of tasks and users. In addition, the advent of the Web has resulted in these applications being made accessible directly to customers and business partners. The resulting proliferation of applications supporting various tasks and users has engendered a need for a global view on the operation of information systems. Accordingly, the emphasis has shifted from application programming to application integration. The challenge is no longer the coding of individual modules but rather the seamless interconnection and orchestration of pieces of software from all four layers.

In parallel with the trend "from programming to assembling," another trend changed the way information systems were developed. This trend is the shift "from data orientation to process orientation." The 1970s and 1980s were dominated by data-driven approaches. The focus of information technology (IT) was on storing, retrieving, and presenting information primarily seen as data. Accordingly, data modeling was the starting point for building an information system. This led to scalable and robust techniques and tools for developing data-centric information systems. The modeling of business processes, however, was often neglected. As a result, the logic of business processes was spread across multiple software applications and manual procedures, thereby hindering their optimization and their adaptation to changes. In addition, processes were sometimes structured to fit the constraints of the underlying information system, thus introducing inefficiencies such as manual resource allocation and work routing, poor separation of responsibilities, inability to detect work overflows and trigger escalation procedures, unnecessarily batched operations, and redundant data entry steps. Management trends in the early 1990s such as business process reengineering (see Section 1.3.1) brought about an increased emphasis on processes. As a result, system engineers are resorting to more process-driven approaches.

The last trend we would like to mention is the shift from carefully planned designs to redesign and organic growth. Due to the widespread adoption of Internet standards and the connectivity that this engendered, information systems are now required to change within tight deadlines in response to changes in the organization's environment; for example changes in the business focus or the business partners. As a result, fewer systems are built from scratch. Instead, existing applications are partly reused in the new system. Consequently, there is a continuous trend toward software componentization and dynamic and reuse-oriented software engineering approaches—approaches aimed at rapidly and reliably adapting existing software in response to changes in requirements. One of the most recent of these approaches, model-driven architecture (MDA), exploits automated code generation, code refactoring, model transformation, and model execution techniques to achieve a faster turnaround for propagating changes in the design into changes in the implementation.

The confluence of these trends, which are summarized in Figure 1.1, has set the scene for the emergence of an increasing number of *process-aware information systems* (PAISs). PAISs are built on top of a technological infrastructure that can take the form of separate applications residing in the second layer or integrated components in the third layer. Notable examples of PAIS infrastructure residing in the second layer are workflow management systems, process-aware groupware, and some enterprise application integration (EAI) platforms (see discussion in Section 1.3). The idea of isolating the management of processes in a separate component is consistent with the three trends discussed above. PAIS infrastructures can be used to avoid hard-coding the processes into tailor-made applications and thus support the shift from programming to assembling. Moreover, process awareness in both manual and automated tasks is supported in a way that allows organizations to efficiently manage their resources. Finally, pulling away the process logic from application programs and capturing this logic in high-level models facilitates redesign and organic growth. For example, today's workflow management systems and EAI platforms enable designers and developers to implement process change by working on diagrammatic representations of process models, a practice consistent with MDA. In addition, isolating the management of processes in a separate component is consistent with recent developments in the domain of intra- and interorganizational application integration (e.g., emergence of Web services and service-oriented architectures).

1.2 PAIS: DEFINITION AND RATIONALE

As illustrated by Figure 1.1, there has been a shift from data orientation to process orientation, triggering the development of PAISs. Since PAISs can be seen as special kinds of information system, we first discuss the term *information system*. Alter [6] provides the following definition of the term information system: "An *informa-*

tion system is a particular type of *work system* that uses *information technology* to capture, transmit, store, retrieve, manipulate, or display information, thereby supporting one or more other work systems." This definition uses two key terms: *information technology* and *work system*. Alter defines information technology as "the hardware and software used to [store, retrieve, and transfer] information," and a work system as "a system in which human participants perform a business process using information, technology, and other resources to produce products for internal customers."

Figure 1.2 depicts Alter's framework for information systems [6]. It shows an integrated view of an information system encompassing six types of entities: customers, products, business process, participants, information, and technology. The customers are the actors that interact with the information system through the exchange of products (or services). These products are being manufactured/assembled in a business process that uses participants, information, and technology. Participants are the people that do the work. Information may range from information on customers to information about the process. Technology is used in the business process to enable new ways of doing work. Diagrams like the one shown in Figure 1.2 always trigger a discussion on the scope of an information system. Some will argue that all six elements constitute an information system, whereas others will argue that only a selected subset (e.g., just business process, information, and technology) constitute an information system. In this chapter, we do not decide on a single definition of "information system" but use the term in different (although related) senses depending on the context. This book considers a specific type of information systems, that is, information systems that are process aware, and therefore link information technology to business processes. By process, we mean a way for an organizational entity to "organize work and resources (people, equipment, in-

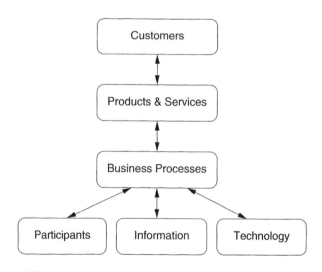

Figure 1.2   An integrated view of an information system.

formation, and so forth) to accomplish its aims" [23]. Sometimes, processes within an organization are hidden—they only manifest themselves in the way people and application programs interact with each other, without being driven by an a priori conception of the way work should be conducted. Other times, processes are captured as a priori defined (i.e., explicit) process models that are used to guide them or even to automate them.

Given these considerations, this book adopts the following definition of a PAIS: *a software system that manages and executes operational processes involving people, applications, and/or information sources on the basis of process models*. Although not part of the adopted definition, it can be noted that these process models are usually represented in a visual language, for example, a Petri net-like notation (Chapter 7). The models are typically instantiated multiple times (e.g., for every customer order) and every instance is handled in a predefined way (possibly with variations).

Given this definition, one can see that a text editor is not "process aware" insofar as it is used to facilitate the execution of specific tasks without any knowledge of the process of which these tasks are part. A similar comment can be made of an e-mail client. A task in a process may result in an e-mail being sent, but the e-mail client is unaware of the process it is used in. At any point in time, one can send an e-mail to any person without being supported or restricted by the e-mail client. Text editors and e-mail clients (at least contemporary ones) are applications supporting tasks, not processes. The same applies to a large number of applications used in the context of information systems.

The shift from task-driven to process-aware information systems brings a number of advantages:

- The use of explicit process models provides a means for communication between managers and business analysts who determine the structure of the business process, and the IT architects, software developers, and system administrators who design, implement, and operate the technical infrastructure supporting these processes.
- The fact that PAISs are driven by models rather than code allows for changing business processes without recoding parts of the systems, that is, if an information system is driven by process models, only the models need to be changed to support evolving or emerging business processes [3].
- The explicit representation of the processes supported by an organization allows their automated enactment [1, 17, 20]. This, in turn, can lead to increased efficiencies by automatically routing information to the appropriate applications and human actors, prioritizing tasks according to given policies, optimizing the time and resources required to deliver services to users, and so on. Also, providing a global view on the operations supported by an information system enables the reduction of redundant data entry tasks and provides opportunities for interconnecting otherwise separate transactions.
- The explicit representation of processes enables management support at the (re)design level, that is, explicit process models support (re)design efforts

[22]. For example, verification tools such as Woflan[1] allow for the verification of workflow models exported from tools such as Staffware[2] (see Chapter 14), ARIS,[3] and Protos.[4] Other tools allow for the simulation of process models. Simulation is a useful tool for predicting the performance of new processes and evaluating improvements to existing processes.

- The explicit representation of processes also enables management support at the control level. Generic process monitoring facilities provide useful information about the process as it unfolds. This information can be used to improve the control of the process, for example, moving resources to the bottleneck in the process. Recently, process monitoring has become one of the focal points of BPM vendors, as reflected by product offerings such as ARIS Process Performance Monitor (PPM) of IDS Scheer[5] and OpenView Business Process Insight (BPI) of HP.[6] This trend has also triggered research into workflow mining (Chapter 10) and process execution analysis and control [8, 25].

1.3 TECHNIQUES AND TOOLS

1.3.1 A Historic View of PAISs

To better understand the emergence and adoption of PAISs and their associated techniques and tools, it is insightful to take a quick historic overview. An interesting starting point, at least from a scientific perspective, is the early work on process modeling in office information systems by Skip Ellis [10], Anatol Holt [16], and Michael Zisman [24]. These three pioneers of the field independently applied variants of Petri net formalism (see Chapter 7) to model office procedures. During the 1970s and 1980s, there was great optimism in the IT community about the applicability of office information systems. Unfortunately, few applications succeeded, in great part due to the lack of maturity of the technology, as discussed below, but also due to the existing structure of organizations, which was primarily centered around individual tasks rather than global processes. As a result of these early negative experiences, both the application of this technology and related research almost stopped for nearly a decade. Hardly any advances were made after the mid-1980s. Toward the mid-1990s, however, there was a renewed interest in these systems. Instrumental in this revival of PAISs was the popularity gained (at least in managerial spheres) by the concept of *business process reengineering* (BPR) advocated by Michael Hammer [14, 15] and Thomas Davenport [9], among others. The idea promoted by BPR is that overspecialized tasks carried across different organizational

---

[1]http://www.tmis.tue.nl/research/woflan
[2]http://www.tibco.com/company/staffware.jsp
[3]http://www.ids-scheer.com
[4]http://www.pallas-athena.com
[5]http://www.ids-scheer.com
[6]http://www.hp.com

units need to be (re)unified into coherent and globally visible processes. In particular, IT should not only support the automation of individual tasks, but should also be seen as an instrument for coordinating and interconnecting tasks and resources (e.g., people, physical assets, software applications).

In the aftermath of the BPR wave, and despite some (sometimes well-founded) criticisms and early failures in the implementation of the underlying concepts, the importance of PAISs grew steadily. The early and mid-1990s saw the advent of business process modeling tools such as Protos and ARIS, as well as workflow management systems such as FlowMark [19][7] and Staffware. The number of PAIS-related tools that have been developed in the past decade and the continuously increasing body of professional and academic literature in this field of technology is overwhelming. Today's off-the-shelf workflow management systems and business process modeling tools are readily available. However, their application is still limited to specific industries such as banking and insurance. As pointed out by Skip Ellis [11], it is important to learn from the ups and downs of PAIS-related technologies. The failures in the 1980s can be explained by both technical and conceptual factors. In the 1980s, networks were slow, expensive, or not present at all; the development of suitable graphical interfaces was hindered by hardware limitations; and application developers were concentrated on addressing other problems such as scalable data storage and retrieval. At the same time, there were also more conceptual problems such as: (i) a lack of a unified way of modeling processes, (ii) a lack of methods for seamlessly propagating changes in the requirements into changes in the design and then into changes in the implementation, and (iii) the systems were too rigid to be used by people in the workplace. Most of the technical limitations have been more or less satisfactorily resolved by now. However, the more conceptual problems remain. In particular, widely adopted and unambiguous standards for business process modeling are still missing, and even today's workflow management systems enforce unnecessary constraints on the process logic (e.g., processes are made more sequential than they need to be). This book will discuss some of the traditional process models (e.g., Petri nets) and some of the emerging standards (e.g., BPEL). However, there is no consensus on which models and standards to use. New paradigms such as case handling (see Chapter 15) and associated products such as FLOWer offer more flexibility but still only provide a partial solution to the many problems related to the alignment of people, processes, and systems.

1.3.2 PAIS Development Tools

There are basically two ways to develop a PAIS: (i) develop a specific process support system, or (ii) configure a generic system. In the first case, an organization builds its own process support system "from scratch" with the specific aim of supporting its processes. This organization-specific system can be as simple as a soft-

---

[7]FlowMark was later integrated into the message-oriented middleware platform MQSeries to become MQSeries Workflow. Subsequently, this platform was renamed WebSphere MQ, so that the workflow system is currently known as "Websphere MQ Workflow."

ware library providing routines for incorporating process awareness into applications, or it can take the form of a process execution platform providing facilities for defining, testing, deploying, executing, and monitoring a large class of processes. This ad hoc approach ensures that the resulting system fits the needs of the organization and the specificities of its processes. However, the initial investment cost of this approach may be too high for some organizations, and the resulting system may not be scalable. As new processes are introduced, existing processes become more sophisticated, and users develop higher expectations, it becomes difficult to adapt the process support system to meet new demands.

Generic process support systems, on the other hand, are generally not developed by organizations actually using a PAIS (although there are cases in which an organization-specific system has subsequently evolved into a system comparable to a generic software product). A typical example of a generic software product is a workflow management system (WFMS) such as Staffware. WFMSs are generic in that they do not incorporate information about the structure and processes of any particular organization. Instead, to use such a generic system, an organization needs to configure it by specifying processes, applications, organizational entities, and so on. These specifications are then executed by the generic system. In the case of a WFMS, when certain types of events occur (e.g., arrival of a purchase order), an instance of the relevant process (called a *workflow*) is triggered, and this results in one or several tasks being enabled. Enabled tasks are then routed to people or applications who/which complete them. As tasks are completed, the WFMS proceeds by dispatching more tasks as per the process specification, until the process instance is completed.

At present, there are more than one hundred WFMSs. A typical workflow management system is composed of a design tool, an execution engine, a worklist management system, adapters for invoking various types of applications, and, in a few cases, modules for monitoring, auditing, and analyzing existing workflow models.

Although the classical apparatus for developing PAISs is workflow technology, "pure WFMSs" are far from being the only type of tool used for developing PAISs. Process awareness is also supported in different ways by the following types of tools:

- Process-aware collaboration tools such as Caramba (see Chapter 2).
- Project management tools such as AMS Realtime[8] and Microsoft[9] Project.
- Tracking tools (e.g., for job, issue, or call tracking) such as JobPro Central.[10]
- Enterprise resource planning (ERP) and customer relationship management (CRM) systems such as SAP[11] and Peoplesoft,[12] which incorporate a workflow management system within a broader enterprise system management solution.

---

[8]http://www.amsrealtime.com
[9]http://www.microsoft.com
[10]http://www.jobprocentral.com
[11]http://www.sap.com
[12]http://www.peoplesoft.com

- Case handling systems such as FLOWer (see Chapter 15).
- Business process design and engineering tools such as ARIS and Protos.
- Enterprise Application Integration (EAI) suites such as TIBCO[13] ActiveEnterprise and Microsoft BizTalk.
- Extended Web application servers (also called Web integration servers) such as BEA[14] WebLogic Integration and IBM[15] Websphere MQ.

Furthermore, process support may be found in various forms outside the realm of information systems. For instance, the emergence of process-centered software engineering environments (PSEEs) [13] illustrates that process awareness can be beneficial in other domains where people and applications need to interact in a coordinated manner.

The plethora of similar but subtly different enabling technologies for process-aware information systems is overwhelming. On the one hand, this demonstrates the practical relevance of process support. On the other hand, it illustrates that process support is far from trivial. At present, there is a "Babel of approaches" to deal with process awareness in information systems. This is hindering the emergence and general understanding of the common principles underlying these approaches.

1.4 CLASSIFICATIONS

A starting point from which to build a structured view on the landscape of supporting techniques, technologies, and tools for PAISs is to classify them according to orthogonal dimensions. The following subsections introduce and illustrate some of these dimensions.

1.4.1 Design-Oriented Versus Implementation-Oriented

Figure 1.3 summarizes the phases of a typical PAIS life cycle. In the design phase, processes are designed (or redesigned) based on a requirements analysis, leading to process models. In the implementation (or configuration) phase, process models are refined into operational processes supported by a software system. This is typically achieved by configuring a generic infrastructure for process-aware information systems (e.g., a WFMS, a tracking system, a case handing system, or an EAI platform). After the process implementation phase (which encompasses testing and deployment), the enactment phase starts—the operational processes are executed using the configured system. In the diagnosis phase, the operational processes are analyzed to identify problems and to find aspects that can be improved.

Different phases of the PAIS life cycle call for different techniques and types of tools. For example, the focus of traditional WFMSs is on the lower half of the PAIS

---

[13]http://www.tibco.com
[14]http://www.bea.com
[15]http://www.ibm.com

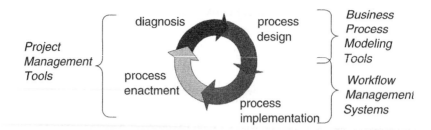

Figure 1.3   The PAIS life cycle.

life cyle. They are mainly aimed at supporting process implementation and execution and provide little support for the design and diagnosis phases. Indeed, although WFMSs are able to log process-related data, they rarely provide tools for real-time and offline interpretation of these data. There are some research proposals in the area of process-related data analysis (e.g., the Process Data Warehouse [7] and the Business Process Cockpit [8]) but these have made their way into commercial products only in a limited way (e.g., ARIS PPM and HP Openview BPI mentioned above). Moreover, support for the design phase is limited to providing a graphical editor, whereas model analysis (e.g., through simulation and static verification) and methodological support are missing.

At the other end of the spectrum, business process modeling tools are design-oriented, focusing on the top half of the PAIS lifecycle. For instance, ARIS (Chapter 6) supports a reuse-oriented design methodology by providing libraries of reference models that may be adapted to meet the needs of specific organizations.

Other types of PAIS-related tools (e.g., project management tools) are hybrid in the sense that they support both design (e.g., PERT and resource allocation analysis) and execution (e.g., Web-based project tracking). However, these hybrid tools tend to focus on very specific types of processes (e.g., projects, job handling in IT help desks, customer call handling). In a way, these tools may be seen as "vertical PAIS development tools," in that they cover a large section of the PAIS development life cycle, but do so by restricting their scope to specific problem domains.

1.4.2   People Versus Software Applications

Another way of classifying PAISs is in terms of the nature of the participants (or resources) they involve and, in particular, whether these participants are humans or software applications. In this respect, PAISs can be classified into human-oriented and system-oriented [12] or, more precisely, into person-to-person (P2P), person-to-application (P2A), and application-to-application (A2A) processes.

In P2P processes, the participants involved are primarily people, that is, the processes primarily involve tasks that require human intervention. Job tracking, project management, and groupware tools are designed to support P2P processes. Indeed, the processes supported by these tools usually do not involve entirely automated tasks carried out by applications. Also, the applications that participate in these processes (e.g., project tracking servers, e-mail clients, video-conferencing tools, etc.) are primarily oriented toward supporting computer-mediated interactions.

At the other end of the spectrum, A2A processes are those that only involve tasks performed by software systems. Such processes are typical in the area of distributed computing and, in particular, distributed application integration. Transaction processing systems, EAI platforms, and Web-based integration servers are designed to support A2A processes. It should be noted that sometimes the logic of these processes is captured by explicit process models, and other times it is implicitly coded into the programs that participate in the process. As the resources participating in A2A processes are applications, and these may share common databases, an important aspect that arises in this type of process is ensuring certain transactional properties as defined in the realm of database management systems (DBMSs). Techniques relevant to this aspect are presented in Chapter 11.

Finally, P2A processes are those that involve both human tasks and interactions between people, and tasks and interactions involving applications that act without human intervention. Workflow systems fall in the P2A category since they primarily aim at making people and applications work in an integrated manner. Note that since workflow systems support both people and applications, they can also be used to support interactions between people only, as well as interactions between applications only. A workflow system can, in principle, be used as a platform to implement A2A processes, although it may be preferable in these situations to use a platform specifically designed for this purpose. On the other hand, pure manufacturing workflow may be considered to be P2P rather than P2A. However, most workflow products nowadays support interactions between both people and applications and, therefore, we consider workflow technology as a whole to be P2A.

The boundaries between P2P, P2A, and A2A are not crisp. Instead, there is a continuum of techniques and tools from P2P (i.e., manual, human-driven) to A2A (automated, application-driven). In particular, ad hoc process and case-handling systems (see Chapters 2 and 15) can be placed in between the P2P and P2A categories. On the other hand, some tools target both A2A and P2A systems. For example, the IBM Websphere MQ family supports both application integration and workflow management.

1.4.3 Structure and Predictability of Processes

The degree of structure of the process to be automated (which is strongly linked to its predictability) is frequently used as a dimension to classify PAISs. In this respect, a traditional distinction is that between *ad hoc, administrative,* and *production processes* [21, 12]. An ad hoc process is one in which there is no a priori identifiable pattern for moving information and routing tasks among people; for example, a product documentation process or a process for preparing a response to a complex tender. Administrative processes, on the other hand, involve predictable processes with relatively simple task coordination rules. These rules may be revised with some frequency or may be adapted to fit exceptional cases, but, in any case, they capture the core of the process. Finally, production processes involve repetitive and predictable tasks with more or less complex but highly stable task coordination rules.

The above classification mixes the predictability of the process with its complexity. As process modeling has matured, it has become evident that some administrative processes can be relatively complex. A slightly different classification that considers only the predictability aspect is that between *unframed, ad hoc framed, loosely framed,* and *tightly framed* processes [4]. A process is said to be unframed if there is no explicit process model associated with it. This is the case for collaborative processes supported by groupware systems that do not offer the possibility of defining process models. Unframed processes are out of the scope of this book, although they are referenced in some parts (e.g., Chapter 2) insofar as unframed processes can lead to framed ones, and there is no clear-cut boundary between these categories.

A process is said to be ad hoc framed if a process model is defined a priori but only executed once or a small number of times before being discarded or changed. This is the case in project management environments in which a process model (i.e., a project chart) is often only executed once. It is also the case in grid computing environments in which a scientist may define a process model corresponding to a computation involving a number of datasets and computing resources, and then run this process only once (a type of process also known as *scientific workflows* or *grid workflows*). Chapter 2 provides an overview of a system designed to support ad hoc processes (Caramba).

A loosely framed process is one for which there is an a priori defined process model and a set of constraints, such that the predefined model describes the "normal way of doing things" while allowing the actual executions of the process to deviate from this model within certain limits. In other words, the trajectory of a process instance is restricted by some upper and lower bound. Case handling systems such as FLOWer support loosely framed processes by allowing implicitly specified routes. Ad hoc workflow systems such as TIBCO InConcert allow for adaptations of a process template or emerging processes such that every execution can be seen as corresponding to a different process model. In other words, the a priori defined process model is implicitly adapted to suit the requirements of each case.

Finally, a tightly framed process is one that consistently follows an a priori defined process model. This is the case of traditional workflow systems, of which Staffware (Chapter 18) is an example.

As with P2P, P2A, and A2A processes, the boundaries between unframed, ad hoc framed, loosely framed, and tightly framed processes are not crisp. In particular, there is a continuum between loosely and tightly framed processes. For instance, during its operational life a process considered to be tightly framed can start deviating from its model so often and so unpredictably that at some point in time it may be considered to have become loosely framed. Conversely, after a large number of cases of a loosely framed process have been executed, a common structure may become apparent, which may then be used to frame the process in a tighter way. Process mining techniques (see Chapter 12) provide a means for discovering such a "common structure" in a large number of process cases.

Figure 1.4 plots different types of PAISs and PAIS-related tools with respect to the degree of framing of the underlying processes (unframed, ad hoc, loosely, or tightly framed), and the nature of the process participants (P2P, P2A, and A2A).

1.4.4 Intraorganizational Versus Interorganizational

Initially, process-aware information systems were mainly oriented towards intraorganizational settings. Focus was on the use of process support technologies (e.g., workflow systems) to automate operational processes involving people and applications inside an organization (or even within an organizational unit). Over the last few years, there has been a push toward processes that cross organizational barriers. Such interorganizational processes can be one-to-one (i.e., bilateral relations), one-to-many (i.e., an organization interacting with several others), or many-to-many (i.e., a number of partners interacting with each other to achieve a common goal).

The trend toward interorganizational PAISs is marked by the emergence of *business-to-business (B2B) integration standards* that define collections of common B2B integration processes (e.g., for procurement) or support the definition of such processes (see Chapter 4). It is also apparent in the emergence of the notion of *(Web) service composition,* whereby applications are exported as services and composed by means of process models [5]. This notion is embodied in standards such as (WS-)BPEL (see Chapter 13) and WS-CDL [18]. A number of tools implementing these standards (or subsets thereof) are now emerging, and established tools for intraorganizational application and process integration are being extended to support these standards.

The modeling of collaborative interactions as explicit process models is a central issue in B2B integration (see Chapter 4). In this area, processes appear in two

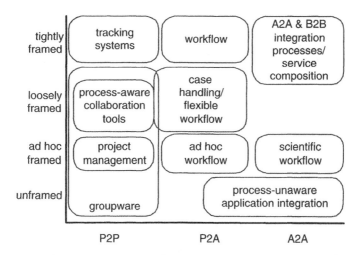

Figure 1.4 Types of PAISs and associated development tools.

forms: public and private. Public processes are those whose definitions are visible to parties outside the organization which implements the process. On the other hand, the definition of a private process is only visible to the organization that owns it. The rationale behind this distinction is twofold. On the one hand, organizations do not wish to expose the full details of their processes to other organizations. Instead, they would only expose the parts of the process that are relevant for establishing a given collaboration. On the other hand, it allows for partners to be replaced. An organization A partnering with an organization B in the context of an interorganizational process is able to substitute B for another organization C, so long as C provides a public process compatible with the requirements of A.

1.5   ABOUT THE BOOK

1.5.1   Goal and Intended Audience

The goal of this book is to provide a unifying and comprehensive overview of the technological underpinnings of the emerging field of *process-aware information systems engineering*. To achieve this goal, the book brings together contributions from leading experts in related fields. These contributions have been selected because they complement each other and cover some of the most salient aspects of the overall picture of process-aware information systems.

Building, deploying, and running a process-aware information system, especially in a mid- or large-scale environment, is a daunting task. It often involves a considerable number of stakeholders. These range from the chief technology officers, chief process officers, and/or managers who set the strategic directions for process automation, (re-)deployment, change, or continuous improvement projects, through the business analysts and IT architects who define the requirements and high-level specifications of the system, down to the process designers and application developers who refine the higher-level specifications into a deployable system. To this list should be added the most important actors: the users who interact with the system in their everyday conduct of business, as well as the operations managers, system administrators, and IT helpdesk assistants who ensure the day-to-day running and ad hoc troubleshooting of the system. The variety of involved stakeholders gives an idea of the multidisciplinarily nature of process-aware information systems engineering. This book does not intend to cover all aspects of this field. Instead, it focuses on technological aspects. Business and social aspects are only addressed when required to illustrate possible uses of certain techniques, technologies, or tools. Furthermore, the book does not directly address methodological issues although it refers to best practices in applying specific techniques.

The book is primarily intended for advanced students specializing in information systems technologies. It is designed to be used as a textbook for a one-semester, topic-oriented course on business process management, business process engineering, or workflow. It may equally well serve as a reference book for a course on enterprise systems. The book is also targeted at professionals involved in projects related to process-aware information systems, including business process modeling, workflow, groupware and teamwork, enterprise application integration, and business-to-business integration. In addition, since the book covers both practical and theoretical approaches to process support, it should also be of great interest to researchers and research students.

To support its pedagogic goal, chapters are structured in the style of tutorials. They present general aspects before zooming into specific technical issues. In addition, the book contains numerous examples and graphical illustrations, and each chapter includes a collection of thought-provoking questions and exercises of varying degrees of difficulty, allowing the reader to review major concepts and techniques. Solutions to most of these exercises are provided on the book's companion website. Finally, the book contains a list of resources including suggested readings as well as pointers to relevant portals, standardization bodies, initiatives, and consortia. These references complement those provided at the end of each chapter.

1.5.2 Overview of Contents

The book is divided into four parts. Part I exposes and illustrates some foundational concepts of PAISs. It also provides an overview of languages, techniques, standards and tools, but without entering into the level of detail of subsequent parts. In addition to the present chapter, this part includes three other chapters corresponding to the classification of PAISs according to the nature of the participants (i.e., P2A, P2P, A2A) as discussed in Section 1.4.2. Chapter 2 opens with an overview of P2A processes as embodied in WFMSs. This discussion of "mainstream" technology lays the ground for the discussions on P2P processes (Chapter 3) and A2A and B2B (Business-to-Business) processes (Chapter 4), which cover more "avant-garde" technology, reflecting the fact that for a long time process-awareness in information systems has been considered mainly in the setting of systems that intertwine human and automated tasks and the focus is now progressively expanding into more human-centric and system-centric processes.

Part II is dedicated to process modeling languages. Chapter 5 shows how UML, a widely adopted object-oriented modeling standard, can be applied to (business) process modeling. The authors demonstrate that the various types of diagrams included in the UML standard provide the building blocks for modeling processes, but that in order to apply them to the domain of process modeling, it is important to understand their overlap and how they complement each other. Chapter 6 presents the extended event-driven process chains (eEPCs) notation. In contrast to UML which is general-purpose, eEPCs are specifically designed to support business process modeling. They are supported by a well-known tool called ARIS, which provides a range of functionality for designing and analyzing business processes. To complement the overviews of UML and EPCs, two modeling languages widely used in practice, Chapter 7 looks at a formal notation for process modeling, namely Petri nets. This formal notation has been applied to a wide variety of domains such as concurrent systems analysis, communication networks design, critical systems verification, and workflow modeling. Several business process modeling and execution languages (or subsets thereof) have been given semantics in terms of Petri nets, including UML activity diagrams (Chapter 5) and BPEL (Chapter 13). There are also several products that directly support Petri nets, for example, workflow systems such as COSA[16] and Promatis[17] INCOME as well as business process modeling tools such as Protos. Part II closes with Chapter 8, which presents a set of patterns that have been used to evaluate the capabilities and limitations of a number of workflow specification languages (their original scope) but also process modeling and service composition languages.

Part III presents techniques relevant to the development of PAISs. As with the rest of the book, the intention is not to be exhaustive in terms of coverage. Instead, an in-depth presentation of techniques in selected areas is provided, namely process design, process mining, and transactional process development. Chapter 9 deals with issues at the frontier between the managerial and the technological views of PAISs. The methods and techniques introduced in this chapter are notably relevant in the design phase of the PAIS development lifecycle (Figure 1.3). When starting from scratch, business requirements can be mapped into process models. For existing process models, their alignment with the requirements could be improved with these techniques, in particular in terms of performance. Chapter 10 presents techniques that are relevant to the diagnosis phase of the PAIS life cycle. Specifically, it presents a set of techniques for automatically unveiling knowledge about the structure of process executions by analyzing event logs gathered during these executions. These techniques make it possible to identify discrepancies between the way processes are expected to execute (as captured in the corresponding process models) and the way they actually execute. Part III closes with Chapter 11, which deals with transaction management, and discusses how this aspect emerges in the context of business process execution.

To close the book, Part IV focuses on the application of the concepts, modeling approaches, and techniques presented in the previous parts by showing how some of them are embodied in specific standards and tools. Chapter 12 provides an overview of standards developed by the Workflow Management Coalition. These standards consolidate a number of concepts, language constructs, and interfaces supported by WFMSs. Chapter 13 presents a more recent standardization effort in the area of A2A processes, namely the Business Process Execution Language for Web Services (WS-BPEL or BPEL for short). Finally, Chapters 14 and 15 present two PAIS development tools. The first one, Staffware, is a representative of tightly framed P2A process development tools, whereas the second one, FLOWer, is intended to support loosely framed P2A processes, with some features relevant for P2P processes (in particular regarding work authorization and distribution). In line with the spirit of the book, these closing chapters do not focus on how to use the presented tools, but rather on how these tools provide realizations of general concepts and principles, as well as how they may be used to address PAIS development challenges.

---

[16]http://www.cosa.nl
[17]http://www.promatis.de/english

REFERENCES

1. W. M. P. van der Aalst and K. M. van Hee. *Workflow Management: Models, Methods, and Systems.* MIT Press, Cambridge, MA, 2002.

2. W. M. P. van der Aalst, A. H. M. ter Hofstede, and M. Weske. Business Process Management: A Survey. In W. M. P. van der Aalst, A. H. M. ter Hofstede, and M. Weske, editors, *International Conference on Business Process Management (BPM 2003),* volume 2678 of *Lecture Notes in Computer Science,* pp. 1–12. Springer-Verlag, Berlin, 2003.

3. W. M. P. van der Aalst and S. Jablonski. Dealing with Workflow Change: Identification of Issues and Solutions. *International Journal of Computer Systems, Science, and Engineering, 15*(5):267–276, 2000.

4. W. M. P. van der Aalst, M. Stoffele, and J. W. F. Wamelink. Case Handling in Construction. *Automation in Construction, 12*(3):303–320, 2003.

5. G. Alonso, F. Casati, H. Kuno, and V. Machiraju. *Web Services: Concepts, Architectures and Applications.* Springer-Verlag, Berlin, 2003.

6. S. Alter. *Information Systems: A Management Perspective.* Addison-Wesley, Reading, MA, 1999.

7. F. Casati. Intelligent Process Data Warehouse for HPPM 5.0. Technical Report HPL-2002-120, HP Labs, 2002.

8. M. Castellanos, F. Casati, U. Dayal, and M.-C. Shan. A comprehensive and automated approach to intelligent business processes execution analysis. *Distributed and Parallel Databases, 16*(3):239–274, 2004.

9. T. H. Davenport. *Process innovation: Reengineering Work through Information Technology.* Harvard Business School Press, Boston, 1992.

10. C. A. Ellis. Information Control Nets: A Mathematical Model of Office Information Flow. In *Proceedings of the Conference on Simulation, Measurement and Modeling of Computer Systems,* pp. 225–240, Boulder, CO, 1979. ACM Press.

11. C. A. Ellis and G. Nutt. Workflow: The Process Spectrum. In A. Sheth, editor, *Proceedings of the NSF Workshop on Workflow and Process Automation in Information Systems,* pp. 140–145, Athens, GA, May 1996.

12. D. Georgakopoulos, M. Hornick, and A. Sheth. An Overview of Workflow Management: From Process Modeling to Workflow Automation Infrastructure. *Distributed and Parallel Databases, 3*(2):119–153, 1995.

13. V. Gruhn. Process-Centered Software Engineering Environments, A Brief History and Future Challenges. *Annals of Software Engineering, 14*(1–4):363–382, 2002.

14. M. Hammer. Reengineering Work: Don't automate, Obliterate. *Harvard Business Review,* pp. 104–112, July/August 1990.

15. M. Hammer and J. Champy. *Reengineering the Corporation: A Manifesto for Business Revolution.* Nicolas Brealey Publishing, London, 1993.

16. A. W. Holt. Coordination Technology and Petri Nets. In G. Rozenberg, editor, *Advances in Petri Nets 1985,* volume 222 of *Lecture Notes in Computer Science,* pp. 278–296. Springer-Verlag, Berlin, 1985.

17. S. Jablonski and C. Bussler. *Workflow Management: Modeling Concepts, Architecture, and Implementation.* International Thomson Computer Press, London, UK, 1996.

18. N. Kavantzas, D. Burdett, and G. Ritzinger. Web Services Choreography Description Language Version 1.0. W3C Working Draft, http://www.w3.org/TR/wscdl-10, 2004.

19. F. Leymann and W. Altenhuber. Managing Business Processes as an Information Resource. *IBM Systems Journal,* *34*(2):326–348, 1994.

20. F. Leymann and D. Roller. *Production Workflow: Concepts and Techniques.* Prentice-Hall PTR, Upper Saddle River, NJ, 1999.

21. S. McCready. There is More Than one Kind of Workflow Software. *Computerworld,* *2:*86–90, November, 1992.

22. H. A. Reijers. *Design and Control of Workflow Processes: Business Process Management for the Service Industry,* volume 2617 of *Lecture Notes in Computer Science.* Springer-Verlag, Berlin, 2003.

23. A Sharp and P. McDermott. *Workflow Modeling: Tools for Process Improvement and Application Development.* Artech House, Norwood, MA, 2001.

24. M. D. Zisman. *Representation, Specification and Automation of Office Procedures.* Ph.D. thesis, University of Pennsylvania, Wharton School of Business, 1977.

25. M. zur Muehlen. *Workflow-based Process Controlling: Foundation, Design and Application of Workflow-driven Process Information Systems.* Logos Verlag, Berlin, 2004.

CHAPTER 2

Person-to-Application Processes: Workflow Management

ANDREAS OBERWEIS

2.1 INTRODUCTION

Information systems provide users in organizations with computer support to accomplish certain tasks. The functionality of an information system includes the allocation of resources such as data, communication services, or hardware devices to the users. Single-user tasks usually belong to (business) processes that are executed to realize a certain business objective. Typical examples of business processes include processing insurance claims, mortgage request handling, processing tax forms, order fulfilment, or recruitment of employees.

An information system is called *process-aware* if it supports process enactment by scheduling the activities according to the specified rules of the respective process type (for a more precise definition and discussion of the term *process-aware information system,* see Chapter 1). In this type of system users are expected to perform tasks in a certain pre-defined order. In traditional information systems, process support is either not available at all or is hard-coded in the programs. Maintenance of software systems in which application code is mixed with process logic is expensive and prone to errors. During their lifetime, processes require adaptation to changing organizational, technical, and environmental parameters. Changes in a process that is to be supported by a specific software system require modifying the source code, then compiling it, and, finally, reinstalling the software system. Each modification of existing source code may lead to programming errors or unexpected results. Therefore, the central new principle of workflow management is the separation of process logic and application functionality. Changes in processes can be made by using comfortable workflow tools without having to rewrite the source code of the software system [11]. The same principle of removing generic functionality from application programs has been successfully applied in the field of database management systems, in which data management functionality (such as query processing, integrity control, or concurrency control) is taken out of application programs.

Process-Aware Information Systems. Edited by Dumas, van der Aalst, and ter Hofstede
Copyright © 2005 John Wiley & Sons, Inc.

Figure 2.1 shows the simplified example process of order fulfilment in a company. In this figure an incoming order is checked for availability of the ordered products. If there is enough quantity on hand, then the order is confirmed by e-mail. Otherwise, the customer is informed that the ordered products are out of stock. After the order is confirmed, the products are delivered and the invoice is sent to the customer. Finally, the customer's payment is received and booked by the accounting system.

Tasks (or activities, represented as rectangles in Figure 2.1) involve interactions with software applications or tools (represented as grey colored hexagons) such as inventory control systems, billing software, or e-mail systems. In Figure 2.1, the node inscribed with the symbol "$\vee$" denotes a choice between two tasks (EXCLU-SIVE OR), and the nodes inscribed with "$\wedge$" denote parallel execution of the tasks in between (AND).[1] Arcs between hexagons and rectangles indicate that the respective software system is needed to perform the respective task.

A process-aware information system assigns users and other resources (e.g., application programs) to tasks or, from another point of view, tasks and applications are allocated to users. Furthermore, the information system controls the required routing of tasks. This functionality can be provided by a workflow management system that is integrated into or coupled to the information system.

This chapter presents some basic concepts of workflow management. The following section introduces the terminology. Section 2.3 considers aspects of workflow modeling. Section 2.4 surveys the functionality of workflow management systems. A reference architecture for workflow management systems is presented. The chapter ends with a brief outlook on some important current and future developments in the field of workflow management.

Workflow management is introduced here as a generic concept. Concrete commercial workflow management systems will not be considered in this chapter. For a detailed description of a representative workflow system (namely Staffware), refer to Chapter 14 of this book.

2.2   WORKFLOW TERMINOLOGY

Workflow is usually regarded as "the computerized facilitation or automation of a business process, in whole or in part" [14]. It consists of a coordinated set of activities that are executed to achieve a predefined goal. Workflow management aims at supporting the routing of activities (i.e., the flow of work) in an organization such that the work is efficiently done at the right time by the right person with the right software tool. It focuses on the structure of work processes, not on the content of individual tasks. Individual tasks are supported by specific application programs. Workflow management links persons (end users, workflow participants, workflow agents) to these applications in order to accomplish the required tasks (see Figure

---

[1]In Figure 2.1, a general notation for process modeling is used. There exist many specific graphical languages for business process modeling. Some of them (UML, EPCs, Petri nets) are described in detail in Chapters 5–7 of this book. Some elements of the notation used here were inspired by the EPCs notation described in Chapter 6.

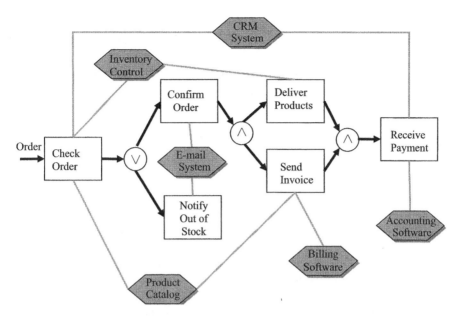

Figure 2.1   Order fulfillment example.

2.2). A workflow system is an information system based on a workflow management system (WFMS) that supports a specific set of business processes through the
execution of a process specification. A process specification (or workflow schema)
describes a type of process that can be interpreted as a template for the execution of
concrete workflow instances.

There must be an organizational instance that is responsible for the specification
of the workflow schema and makes decisions about modifications of a given

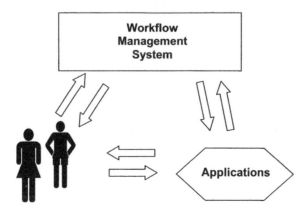

Figure 2.2   A workflow management system as a link between users and applications.

schema. For each workflow instance, one organizational instance must be responsible. This instance initiates a workflow execution and makes decisions about deviations from the given workflow schema and about termination.

Structural changes in process types can be made using comfortable workflow tools without having to rewrite the application program. The use of workflow management systems simplifies application development because application components can be reused and process enactment functionality, which is common to many applications, is provided by the underlying workflow management system.

A workflow engine interprets the workflow description, which is represented in a workflow programming (or modeling) language. Changes to the workflow schema are directly reflected by the workflow execution.

Workflows are typically *case-based,* that is, every piece of work is executed for a specific case. Examples of cases are mortgage applications, insurance claims, tax forms, orders, or requests for information. A case defines an instance of a workflow type and a workflow type is designed to handle similar cases.

A WFMS is a software package that provides support for the definition, management, and execution of workflows, together with certain interfaces to its environment (e.g., to an application system or to another WFMS) and to its users. Workflow technology originates from earlier technologies such as office automation, document management, database management, and electronic messaging systems.

2.3 WORKFLOW MODELING

2.3.1 Different Perspectives on Workflows

Workflow modeling is a prerequisite for planning workflow execution, and for analyzing, training, executing, modifying, and archiving workflows. In order to cope with the complexity of workflow models, it is useful to look at them from different perspectives [1, 16]. Furthermore, different persons in an organization might have different views on a workflow. The different perspectives described in the following are partially overlapping; for example, with respect to workflow management, organizational units can also be regarded as resources.

Resources and Resource Management. Resources include all kinds of objects that are necessary to perform a workflow or a task. For each workflow, it must be decided which resources are required for execution. At the schema level, resource classes are specified. At the instance level, concrete instances of resources must be allocated to task instances. Some resources are shareable between different tasks; others are for exclusive use only. There must exist rules according to which conflicts are resolved if several task instances compete for the same resource at the same time.

Organizational Units. Organizational units (such as departments, groups, and roles in a company) might also be regarded as resources for a workflow. The description of the organizational units includes the organizational structure (i.e., the set of relationships between the units) and the individuals that belong to it, together with their main attributes (competencies, responsibilities). For more information on organizational issues of workflow management see [23].

Tasks and Task Management. The task view restricts itself to a description of the tasks to be performed within a certain workflow and to the relationships between these tasks. Relationships between tasks mainly concern the routing of tasks. This aspect is considered in more detail in Section 2.3.2.

Data and Data Flow. Data is a specific kind of resource. Several types of data are relevant for workflow management and must be stored in respective databases:

- Application data is data that is specific to applications and is usually not accessible by the workflow management system. Documents such as invoices, orders, delivery notes, and protocols, which might be stored in document management systems, also belong to this type of data.
- Workflow schema data is data that specifies the structure of a workflow. This type of data may be versioned, which means that there may exist different versions of the same workflow type. Schemas of workflow types that are no longer in use are archived.
- Historical data is data about past executions of workflows that is stored in order to analyze and reorganize workflow types in the future.
- Internal data is administrative data which is only used internally by a workflow engine. This kind of data usually concerns the current status of active workflow instances.

Temporal Aspects. This view includes information about deadlines and durations of activities, temporal distances between activities, availability times for resources, working hours of employees, and other temporal restrictions that are to be obeyed by workflow executions (e.g., "an activity A2 must not start earlier than 3 days after another activity A1 has finished"). A given set of temporal restrictions may be analyzed for consistency. Information about expected durations of activities and about availability times of resources may be used to compute minimum/maximum durations of complete workflow executions. For a more detailed discussion of temporal aspects related to workflow management see [7].

Applications. The applications view focuses on the application programs that are used in order to perform certain tasks. Requirements concerning the invocation of the applications, the format for input and output data, and behavior in exceptional situations are documented.

Business Rules. Business rules represent policies or principles that are to be obeyed in organizations independently of certain processes. In order to simplify the maintenance of workflow models, it is useful to separate business rules from process descriptions. Business rules usually have the form "If a certain event occurs and a certain condition holds, then do something" (ECA-rules, which stands for event, condition, action). An overview about modeling concepts for business rules in the area of workflow management is given in [19].

Exception Handling. An important feature of workflow systems is that they should support end users in case of exceptional situations. A workflow schema usually describes the regular process execution. However, exceptional situations may occur, for example, if resources are not available, if a deadline is missed, if applications fail, or in case of technical faults. A workflow management system should provide means for systematic recovery from these faulty situations. In some cases, there exist predefined procedures for the handling of exceptional situations. Exceptional situations are defined by exceptional conditions (e.g., "a resource R1 is not available") at design time of a workflow schema. An exception is indicated for a given workflow state if the respective condition is fulfilled in this state. For each exceptional condition, an exception handling mechanism must be specified. A taxonomy of failures and exceptions together with an overview about exception handling mechanisms in the area of workflow management is given in [9, 13, 22].

Interorganizational Cooperation. Workflows in the fields of supply chain management and e-business cross the borders of organizations. One part of the related tasks is executed in one organization and the other part is executed in another organization or organizations. This view concentrates on the interfaces between the involved organizations, the respective data exchange formats, the fragmentation of a workflow schema, and the rules for task allocation. For a detailed discussion of these issues see [2, 8, 20].

2.3.2 Routing of Tasks

The routing definition for a workflow type specifies which tasks need to be executed and in what order. Basic routing patterns include *sequential execution, choice, iteration,* and *parallel execution* [1]:

- Sequential execution of tasks. Tasks are executed one after the other. Two tasks A and B must be executed in sequential order if, for example, B consumes an output object of A.
- Exclusive choice between tasks. A subset of a given set of tasks is selected to be executed. Exclusive choice is required if, for example, two activities need to access the same resource that can only serve one activity at the same time. An exclusive choice between tasks also occurs if different cases require different treatment (e.g., "enough stock at hand" and "out of stock" in an order fulfilment process). In Figure 2.3 exclusive choice between activities A1 and B1 is expressed by a so-called XOR-split. The process is split into two alter-

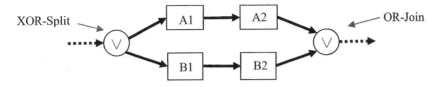

Figure 2.3   Choice between tasks.

native branches. The two alternative branches of a process are again integrated by a so-called OR-join.

- Iteration of tasks. A particular task is executed several times, either for a predefined number of iterations or repeated until a certain termination condition is fulfilled.
- Parallel execution of tasks. Two or more tasks are executed independently of each other "in parallel." This does not necessarily mean that the tasks are executed at the same time. Parallel branching of processes is expressed by a so-called AND-split, whereas an AND-join synchronizes parallel branches of a process (see Figure 2.4). Usually, this type of relationship between activities is called "parallelism." The term "parallel," in the sense of "at the same time," is not distinguished from "parallel" in the sense of "independently of each other."

A survey of other routing patterns relevant for process and workflow modeling is given in Chapter 8 of this book.

2.3.3   Workflow Model

The workflow model (or workflow schema) specifies all aspects of a workflow that are relevant for workflow execution. Each workflow schema defines a workflow type. For each workflow type, many workflow instances (workflow executions) may be created. A workflow schema contains at least information about tasks that are to be executed, relationships between tasks, and conditions that enable tasks, and resources and resource management rules.

Workflow modeling usually starts with a model of the underlying business process, which is then adapted and refined to the specific requirements of workflow

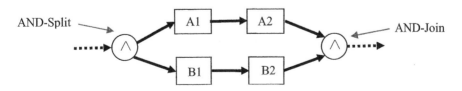

Figure 2.4   Parallel execution of tasks.

management. However, usually it is not appropriate to translate a business process model directly into a workflow model. In general, it is more effective to reorganize the complete process in order to exploit the full potential of process automation and computer support. Recent approaches propose so-called workflow or process mining to discover process models from available data about past process executions (see [3, 4] and Chapter 10 of this book).

Several languages have been proposed for workflow modeling; some of them are vendor proprietary and some are standardized. The spectrum of languages ranges from textual languages in a programming or algebraic style to graphical notations, enriched by iconographic representations of activities and resources for better understanding (see Chapters 5, 6, and 7). Some specific languages for workflow modeling are described in [1, 16]. Reference [26] discusses quality issues for workflow models and proposes to adapt metrics such as cohesion and coupling from the field of software programming for workflow models.

2.3.4 Workflow Life Cycle

Each workflow type has a life cycle that covers different stages (see Figure 2.5). The life cycle starts with the recognition of a need for a new workflow type in an organization. The new workflow type is planned, its boundaries are clarified, and its schema is modeled and evaluated. For each workflow type, an event that initiates the workflow execution and an event that defines the end of the workflow execution are defined. Formal analysis methods as well as simulation techniques may be applied at this stage of workflow design (activities related to process design and redesign are considered in Chapter 9 of this book; for a detailed description about concepts for process simulation see [10]). Future workflow participants are trained to use the new workflow type and then the workflow type is deployed in an organization. At this stage, workflow instances are created and executed according to the workflow schema. The deployment stage is either followed by a stage in which a

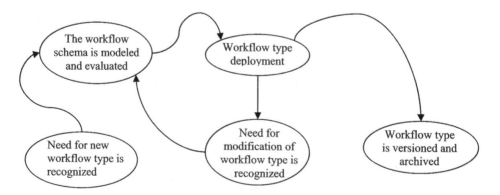

Figure 2.5 Life cycle of a workflow type.

need for modification is recognized or by a terminating stage in which the work-flow type is archived (e.g., for legal reasons). If a need for modification is recognized, a new phase of modeling and evaluating starts. The modified workflow type is deployed and the old schema is finally stored in a workflow schema repository.

Similarly, each workflow instance has a life cycle of its own that follows certain templates of the underlying workflow type (see Figure 2.6; in [16] the term "execution model" is used instead of "life cycle" for a similar state transition diagram representation). After a preparation stage, the life cycle starts with the initiation of the workflow. Then the workflow is executed according to the specification of the underlying workflow schema. Under certain conditions, the running workflow might deviate from the schema. The workflow execution ends in a predefined termination state, representing either successful termination or termination in an exceptional manner. The execution is documented in a workflow instance archive (for later analysis and for legal reasons). Usually, many instances of the same workflow type might be active at the same time.

Reference [12] proposes methods and tools for comprehensive business process life cycle management and also considers the impact of workflow technology on the process life cycle.

2.4 WORKFLOW MANAGEMENT SYSTEMS

2.4.1 Basic Functionality

Workflow management systems support the definition and administration of work-flow types (at design time) as well as the execution and monitoring of workflow instances (at run time). We distinguish between autonomous and embedded workflow management. This is again similar to database management systems, which might also operate as (autonomous) standalone systems or as (embedded) components of information systems. Autonomous workflow management is based on a standalone workflow management system. Embedded workflow management means that a

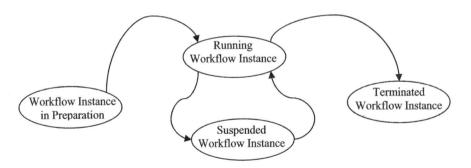

Figure 2.6 Life cycle of a workflow instance.

workflow engine and other system components are integrated in an information system, for example, in an ERP system such as SAP R/3.[2]

Traditional groupware systems may also be used to support collaborative group work (see Chapter 3). However groupware is usually not process-aware and process enactment is not supported by groupware. Workflow management systems are process-aware but in comparison to groupware they are also rather restrictive and leave little room for individual user decisions and adjustments to individual preferences. There is usually a trade-off between flexibility and process support as well as between flexibility and performance.

2.4.2   The WfMC Reference Model

The need for standardization in the field of workflow management is obvious, since workflow management systems have to interoperate with their environment. The Workflow Management Coalition (WfMC) is a nonprofit, international organization whose role it is to standardize workflow management terminology and related technology. Founded in 1993, the WfMC had more than 250 members in 2004, including vendors and users of workflow products, analysts, and research organizations in the field of workflow management. The Web site of the WfMC[3] provides a huge reservoir of material on the topic of workflow management, especially concerning all relevant aspects of standardization in the field.

In the following, we survey the central concepts of the WfMC workflow reference model as introduced in [14]. This reference model includes the main components of a generic workflow management system architecture (see Figure 2.7) together with a set of major standardized interfaces to its environment. More details on some of the WfMC reference model components are given in Chapter 12.

2.4.3   Workflow Engine

Central to the reference architecture is the workflow enactment service, which includes one or more workflow engines. The workflow enactment service generates new instances of a workflow type. A workflow engine is responsible for control and execution of the workflow instances. It creates work items, matches capabilities (skills, knowledge, and experience) of workflow participants with requirements of tasks, und allocates work items to workflow users. The workflow engine also records data about task and workflow instances.

2.4.4   Interfaces

A workflow management system provides five interfaces to components in its environment. One of the Workflow Management Coalition's objectives is to standardize these interfaces.

---

[2]http://www.sap.com
[3]www.wfmc.org

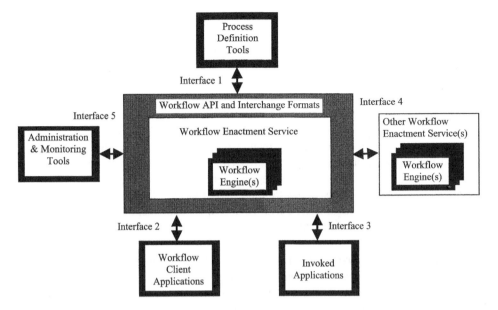

Figure 2.7   WfMC workflow reference model—components and interfaces.

Process Definition Tools.   The underlying workflow is defined in the workflow schema, which is edited by a process definition tool. This class of tools also includes components for the analysis and evaluation of workflows. Simulation is an important method used to evaluate a workflow schema with respect to users' requirements. Formal analysis methods might be applied, for example, to check whether a given workflow schema contains tasks that can never be executed, whether deadlocks might occur in a given workflow under certain circumstances, or whether there exist bottlenecks with respect to certain resources.

Each workflow editor usually supports one specific workflow modeling language. The process definition tool must also include components for storage and versioning of workflow schemas. The process definition tool is used by the workflow designer.

Workflow Client Applications.   Workflow client applications provide an interface to the workflow users who take part in the workflow execution. The basic concept is the so-called *worklist,* containing the work items that are to be executed [1]. A worklist manager is a software component that manages the interaction between workflow participants and a scheduling component. The basic functionality of a worklist is to present tasks that may be executed to a workflow participant. It provides users with instance-specific information on the respective cases. The worklist allows users to select tasks and to invoke application systems to execute the tasks.

A worklist might be based either on the *push* or on the *pull* principles. In push-based worklists, the workflow participant is given the next task to perform; the workflow participant does not have a choice. In pull-based worklists, a list of open tasks to perform is presented to the workflow participant. The workflow participant then has to select a task to execute. In push-based worklists, the workflow engine is responsible for the scheduling of tasks; in pull-based worklists the workflow participant himself does the scheduling.

Invoked Applications. Executing a task usually involves invoking an external application. These software systems may be standardized office tools such as an editor or a spreadsheet program. They may also be software systems that are developed for a specific business function. Invoked applications include interactive applications as well as fully automated applications. The former type of application includes interaction with users; for example, a user has to fill out a form. The latter type of application automatically performs a task; for example, the calculation of a bill or the search for data in a database.

Workflow Interoperability. This interface defines requirements for interoperability with other workflow systems. For reasons of scalability or to connect systems of different organizations, a workflow system may be coupled to other workflow enactment services. Problems of heterogeneity, for example, concerning the workflow languages or different enactment policies of the involved enactment services, have to be solved.

Administration and Monitoring. This interface is for administrators and persons who manage the execution of workflow instances. According to certain process parameters, modifications of the underlying process schema might be proposed.

Administration of running workflows includes the addition or removal of resources, for example, staff, and the modification of properties of resources. This component also provides managers with information about the current status of an ongoing process or of tasks being executed.

Process analysts can use the provided historical data as the starting point for a reorganization of workflow schemas. Throughput times, for example, might serve as a performance indicator for workflow executions. Service times, transport times, and waiting times can be influenced and reduced. Other metrics may include the percentage of rework, the percentage of errors, the number of defects, the number of customer complaints, and so on.

2.5 OUTLOOK

This chapter has given a survey on workflow management. Workflow management is a technology that links processes to persons and resources. However, several specific aspects on workflow management have not been considered. Due to space limitations, we restrict ourselves here to referring the reader to other work on these topics.

- *Flexibility* in workflows [17]. Usually, a distinction is made between two levels of flexibility in workflow systems; flexibility at the workflow schema level und flexibility at the workflow instance level. A workflow schema must be modified if the underlying process is reengineered, for example, due to changing requirements of the environment. A workflow instance, for example, a running workflow, must be modified if there occurs an exceptional situation in the organization (for example, a resource is not available) such that the workflow cannot be executed as planned.

- Workflow management in *distributed and mobile environments* [5, 6, 15, 24]. Distribution of workflow execution means that collections of tasks, that is, workflow partitions, are executed at different sites in a network. Resources (such as data) are also geographically distributed. Allocation of workflow partitions to nodes in a network depends on the availability of required resources and on the workload at the respective node. There exist several strategies for workflow allocation, ranging from *static allocation* of a workflow schema at design time to *dynamic allocation* at execution time of a process [6].

  Mobility of workflow agents during workflow execution causes additional problems due to the limited computational power and the restricted resources (display size, battery capacity, memory) of the mobile enduser devices (such as PDAs or mobile phones) and due to unreliable wireless network connections. Reference [24] describes a prototypical architecture with which mobile workflow users are connected to a central workflow engine via mobile devices.

- *Security aspects* in workflows, such as *authentication* (ensures the identity of a person performing a task, typically involving use of a password, a PIN, or other information that can be used to validate the identity of a person), *authorization* (the process of deciding whether a person is allowed to perform a certain task), *data privacy* (a person is only allowed to access sensitive documents if this is required by a task he or she has to perform). Security constraints are specific rules that state the security requirements to be enforced during the execution of a task [18]. Security constraints are guaranteed by respective security control mechanisms. Violations of security constraints can be treated as specific exceptional situations.

- *Scheduling* workflows under resource allocation constraints [27]. Resources in workflow environments are typically not limitless and not sharable between different activities that are executed at the same time. Therefore, scheduling workflow execution involves decisions as to which resources to use and when. Scheduling workflows is a problem of finding a correct execution for the workflow tasks, such that the constraints of the underlying workflow type are fulfilled. Reference [27] proposes techniques for workflow scheduling.

From a conceptual point of view, traditional workflow management systems neglect the product that is to be created for the (internal or external) customer of the business process. Due to the process focus of workflow management, products are only implicitly considered. Future systems will, therefore, have to integrate product management and workflow management. So-called *product machines* will allow customization and personalization of digital products to meet the specific needs of a customer. From the product model and a generic workflow model, the respective customized workflow model will automatically be generated.

From a technical point of view, future workflow systems will not only be linked to traditional application programs but also to so-called *Web services* that are provided via the Internet. Web services provide a specific functionality to support certain tasks that might be used in order to compose a complete workflow. For a discussion on this issue see [21] and [25].

EXERCISES

Exercise 1
(a) For the order fulfilment process presented in Section 2.1, describe which tasks are to be performed manually, semi-automatically, and fully automatically.
(b) Give a list of resources that are needed for the order fulfilment process. Try to classify the resources with respect to similar characteristics.
(c) What exceptional situations might occur during the order fulfilment process? Describe exception handling mechanisms.

Exercise 2
(a) Give general requirements for a workflow modeling language. Think of different groups of people that have to use this language.
(b) What aspects of a workflow should be expressible by a workflow modeling language?
(c) Describe characteristics of a "good" workflow schema.

Exercise 3
(a) Compare workflow management systems and database management systems. How are both types of systems related to each other? How does the availability of workflow and database management systems influence the development of information systems?
(b) Describe the functionality of the five interfaces provided by a workflow management system according to the Workflow Management Coalition.

Exercise 4
(a) Discuss aspects of flexibility for a workflow instance. Describe situations in which a running workflow is to be modified. What kind of problems might occur?
(b) Describe situations in which a workflow schema is to be modified. Which problems have to be solved? Give examples.

REFERENCES

1. W. M. P. van der Aalst and K. van Hee. *Workflow Management. Models, Methods, and Systems.* MIT Press, Cambridge, MA, 2002.

2. W. M. P. van der Aalst. Inheritance of Interorganizational Workflows to Enable Business-to-Business E-Commerce. *Electronic Commerce Research, 2,* 195–231, 2002.

3. W. M. P. van der Aalst and A. Weijters. Workflow Mining—Discovering Workflow Models from Event-Based Data. In *Proceedings of ECAI Workshop "Knowledge Discovery from Temporal and Spatial Data (ECAI),* Lyon, 2002, pp. 78–84.

4. R. Agrawal, D. Gunopulos, and F. Leymann. Mining Process Models from Workflow Logs. In *Proceedings of the 6th International Conference on Extending Database Technology (EDBT),* 1998, pp. 469–483.

5. G. Alonso, B. Reinwald, and C. Mohan. Distributed Data Management in Workflow Environments. In *Proceedings of the 7th International Workshop on Research Issues in Data Engineering (RIDE'97),* 1997, pp. 82–90.

6. T. Bauer, M. Reichert, and P. Dadam. Intra-Subnet Load Balancing in Distributed Workflow Management Systems. *International Journal on Cooperative Information Systems, 12,* 3, 295–323, 2003.

7. C. Bettini, X. S. Wang, and S. Jajodia. Temporal Reasoning in Workflow Systems. *Distributed and Parallel Databases, 11,* 269–306, 2002.

8. C. Bussler. The Application of Workflow Technology in Semantic B2B Integration. *Distributed and Parallel Databases, 11,* 163–191, 2002.

9. F. Casati and G. Cugola. Error Handling in Process Support Systems. A. Romanovsky, C. Dony, J. Lindskov Knudsen, and A. Tripathi (Eds.), *Advances in Exception Handling Techniques,* pp. 251–270, Springer-Verlag, 2001.

10. J. Desel and T. Erwin. Modeling, Simulation and Analysis of Business Processes. W. van der Aalst, J. Desel, and A. Oberweis (Eds.), *Business Process Management,* pp. 129–141, Springer-Verlag, 2000.

11. A. Elmagarmid and W. Du. Workflow Management: State of the Art Versus State of the Products. In A. Dogac, L. Kalinichenko, M. Tamer Özsu, and A. Sheth (Eds.), *Workflow Management Systems and Interoperability,* pp. 1–17, Springer-Verlag, 1998.

12. D. Georgakopoulos and A. Tsalgatidou. Technology and Tools for Comprehensive Business Process Lifecycle Management. In A. Dogac, L. Kalinichenko, M. Tamer Özsu, A. Sheth (Eds.), *Workflow Management Systems and Interoperability,* pp. 356–395, Springer-Verlag, 1998.

13. C. Hagen and G. Alonso. Exception-Handling in Workflow Management Systems. *IEEE Transactions on Software Engineering, 26,* 10, 943–958, 2000.

14. D. Hollingsworth. The Workflow Reference Model. Workflow Management Coalition, Document Number TC00-1003, Winchester, 1995.

15. S.-Y. Hwang and C.-T. Yang. Component and Data Distribution in a Distributed Workflow Management System. In *Proceedings of 5th Asia-Pacific Software Engineering Conference (APSEC '98),* pp. 244–253, Taipei, Taiwan. IEEE Computer Society, 1998.

16. S. Jablonski and C. Bussler. *Workflow Management: Modeling Concepts, Architecture and Implementation.* International Thomson Computer Press, 1996.

17. G. Joeris and O. Herzog. Towards Flexible and High-Level Modeling and Enacting of

Processes. In M. Jarke and A. Oberweis (Eds.), *Proceedings of Advanced Information Systems Engineering (CAiSE'99),* pp. 88–102, Springer-Verlag, 1999.

18. K. Karlapalem and P.C.K. Hung. Security Enforcement in Activity Management Systems. In A. Dogac, L. Kalinichenko, M. Tamer Özsu, and A. Sheth (Eds.), *Workflow Management Systems and Interoperability,* pp. 165–194, Springer-Verlag, 1998.

19. G. Knolmayer, R. Endl, and M. Pfahrer. Modeling Processes and Workflows by Business Rules. In W. van der Aalst, J. Desel, and A. Oberweis (Eds.), *Business Process Management,* pp. 16–29, Springer-Verlag, 2000.

20. K. Lenz, A. Oberweis. Interorganizational Business Process Management with XML Nets. In H. Ehrig, W. Reisig, G. Rozenberg, and H. Weber (Eds.), Petri Net Technology for Communication Based Systems, pp. 243–263, Springer-Verlag, 2003.

21. F. Leymann and K. Güntzel. The Business Grid: Providing Transactional Business Processes via GRID Services. In M.E. Orlowska et al. (Eds.), *Proceedings of the International Conference on Service Oriented Computing (ISCOC),* pp. 256–270, Springer-Verlag, 2003.

22. Z. Luo, A. Sheth, K. Kochut, and B. Arpinar. Exception Handling for Conflict Resolution in Cross-Organizational Workflows. *Distributed and Parallel Databases, 13,* 271–306, 2003.

23. M. zur Mühlen. Organizational Management in Workflow Applications—Issues and Perspectives. *Information Technology and Management Journal, 5,* 3, 2004.

24. S. Müller-Wilken, F. Wienberg, and W. Lamersdorf. On Integrating Mobile Devices into a Workflow Management Scenario. In *Proceedings of the 11th International Workshop on Database and Expert Systems Applications (DEXA'00),* pp. 186–192, 2000.

25. G. Piccinelli, W. Emmerich, S.L. Williams, and M. Stearns. A Model-Driven Architecture for Electronic Service Management Systems. In M.E. Orlowska et al. (Eds.), *Proceedings of the International Conference on Service Oriented Computing (ISCOC),* pp. 241–255, Springer-Verlag, 2003.

26. H. A. Reijers and I.T.P. Vanderfeesten. Cohesion and Coupling Metrics for Workflow Process Design. In J. Desel, B. Pernici, and M. Weske (Eds.), *BPM 2004,* pp. 290–305, Springer-Verlag, 2004.

27. P. Senkul, M. Kifer, and I. H. Toroslu. A Logical Framework for Scheduling Workflows Under Resource Allocation Constraints. In *Proceedings of the 28th VLDB Conference,* pp. 694–705, Hong Kong, 2002.

CHAPTER 3

Person-to-Person Processes: Computer-Supported Collaborative Work

CLARENCE A. ELLIS, PAULO BARTHELMESS, JUN CHEN, and
JACQUES WAINER

3.1 INTRODUCTION

Computer-based systems that support communication and collaboration among people are an important category of tools for assisting computer-supported collaborative work (CSCW). This chapter addresses some of the issues concerning these systems and their underlying human communication processes. These systems and processes are important because they focus on sociotechnical issues within information systems. An increasingly large population of diverse humans is interacting with technology these days; it is, therefore, increasingly important for systems to address issues of human collaboration. The incorporation of these processes into computerized systems is particularly challenging because people processes are complex, semistructured, and dynamically changing. In this chapter, we argue that these systems are important and intriguing because there is frequently a need for exception handling and dynamic change performed by humans. It is, thus, necessary to take into account factors that impinge upon the organizational structures, the social context, and the cultural setting.

In this chapter, we explain and characterize person-to-person (P2P) processes and systems. We discuss a class of systems, called *process-aware systems,* which embed explicit representations of group processes. First, we discuss collective work, its implications for collective technology, and its problems and pitfalls. We then develop some concepts and language for describing P2P systems. We employ this language to characterize and contrast five exemplary P2P systems. The chapter ends with a summary, conclusions, exercises, and references.

3.2 CHARACTERIZATION OF PERSON-TO-PERSON INTERACTIONS

This section discusses issues related to collective work in a broad sense. It introduces the complexity and variability of collective group interactions through examples, and discusses how structure can be identified in these interactions. Finally, it explores the implications of technology and its problems and pitfalls.

3.2.1 People Working Together

It remains true that "how people work is one of the best kept secrets" (David Wellman, cited by Suchman [44]). Collective work is characterized by its fluidity and complex weaving of organizational, social, political, cultural, and emotional aspects. Interaction at work takes a wide variety of forms. Consider, for instance, the following examples:

- **Extreme Collaboration.** Mark [28] describes a "war room" environment employed by the NASA's Jet Propulsion Laboratory (JPL) to develop complex space mission designs in a very short time—nine hours over a single week for a complete and detailed mission plan. During these interaction sessions, sixteen specialists are physically colocated in a room that contains a network of workstations and public displays.

  Collaboration is prompted by a complex combination of physical awareness, by monitoring of the parallel conversations in the noisy environment, and in response to data that is published through customized networked spreadsheets that allow team members to publish data they produce and subscribe to data published by others. Team members move around the room to consult other specialists, or flock to the public display to discuss problems of their interest. Their movements impart important awareness information to others in the room, who may choose to join a group based on the perceived dependencies of their own work on what a specific set of specialists is discussing. While working, team members are peripherally aware of multiple simultaneous conversations and react to keywords that concern their part of the job by giving short answers from their workplace or moving to join a group. Finally, data that is made available through the computerized spreadsheet system may also trigger collaboration.

- **Congressional Sessions.** These are highly formalized interactions based on the *Robert's Rules of Order* [33]. Participants have very specific roles and duties; for example, the Speaker of the House is the presiding officer, responsible for maintaining proper order of events, and the Chief Clerk is responsible for day-to-day operation of the house. The structure of each session is predefined—the allowable items of discussion are known to all in advance. There are strict rules that determine how issues may be debated, including the order of speakers (for and against), the time they are allowed to speak, and, to some degree, the content of their addresses. Deliberation is based on voting, which is regulated as well, for example, by rules that specify when a vote can be called or waived, and the proportion of voters needed for approval in many different situations.

- **Policy Making and Design.** Rittel and Webber [32] discusses the inherently intractable nature of design and planning problems, which he names "wicked problems." This class of problems is characterized by their ill-defined nature—in many cases it is not possible to separate the understanding of the problem from the solution, as the formulation of the problem is equated to statements of potential solutions [31]. Multiple solutions are, in general, possible, and it might be even hard to determine which solutions are superior to others. Sometimes, these problems emerge as a result of conflicting interests, for example, when deciding how to allocate a limited number of rooms to different individuals who might have coinciding preferences. Possible solutions for these problems involve compromise—ideal solutions are replaced by acceptable ones.

Rittel and Kunz [31] propose tackling wicked problems through an argumentative method in which questions are continually raised, and advantages and disadvantages of multiple possible responses are discussed. The method, called issue-based information systems (IBIS), is based on documenting and relating issues, positions about issues, and arguments that either support a position or object to it. Each separate issue is the root of a (possibly empty) tree, with the children of the issue being positions and the children of the positions being arguments [10]. Links among these three basic elements are labeled; for example, issues and positions are connected by *responds-to* links, and arguments are connected to positions either by *supports* or *objects-to* links.

An IBIS discussion starts with the elicitation of one or more (abstract) issues, to which participants respond with positions and arguments and refine them into more concrete subissues. Contradictory positions are resolved by consensus or voting. The end result is a forest of linked elements that represent the evolution of the discussion, alternatives that were considered, and the rationale for decisions.

In all the above examples, the actual interactions represent a small fraction of a much larger interaction over time and space. The members of the extreme collaboration team at NASA have been working together for many years, and have detailed knowledge of each others' peculiarities and expertise. They also share common engineering knowledge of their field, as well as the common approaches and problem solving strategies that are part of the cultural heritage of their field. Thus, a war room interaction session succeeds because the group has a larger interaction context lasting over many years. Likewise, for the congressional sessions, these formal meetings represent just the visible tip of the political iceberg. Complex backstage negotiations shape the performance at the session and result from economic and political pressures of a multitude of stakeholders. Finally, the policy making and design based on IBIS is guided by a deep understanding of the issues in discussion that the participants bring to the interaction based on a lifetime of experiences, shared or not, and by expert opinion and supporting documentation that is sought as part of the process. All these interaction contexts are in turn embedded in larger societal settings, as parts of organizations, a government, a nation, and so on.

Although these examples are all extracted from work life, clearly, the complexity and variability exhibited here also extends to human interaction beyond work environments.

3.2.2   In Search of Structure

Although complete details of interactions and the intricate factors that govern them are usually beyond what can be understood, constituting implicit processes that are mostly inaccessible, certain emergent regularities and patterns of group behavior can be observed. Rather than being unconstrained, interactions usually follow a structure that is repetitively reproduced by participants at each new instance [26, 27]. This structure is a result of shared belief and value systems, and is frequently learned from previous experiences of participants in similar situations. This structure reinforces the enacted behaviors, helping to shape future interactions. More than repeating patterns, participants make implicit or explicit statements about/ through their actions, as they go about their activities. Participants exert "reflective self-monitoring" [26] so as to *act accountably,* that is, in a manner that is "observable and reportable" [23]. Acting accountably means acting explicitly (even if unconsciously) according to values and rules shared by a social group, that get, at the same time, instantiated and reinforced by actions of individuals [36].

Participation in interactions may be constrained by organizational rules, goals, and norms. Participants are able to make sense of each other's actions (and reorient their own accordingly) because individual actions are recognizable by the group as being one of the meaningful actions that are sensible within a context. Bittner [4] suggests that "a good deal of the sense we make of things happening in our presence depends on our ability to assign them to the phenomenal sphere of influence of some rule" (cited in [15]). Participation in interactions is further constrained to specific sets of behaviors that are associated with the *roles* played by participants (e.g., teacher, student, meeting chair, meeting participant). Roles to some extent determine the behaviors of any person occupying a certain position within a context, independently of personal characteristics [2, 3]. Some of these roles may be noninstitutionalized and sometimes even pathological, for example, *the devil's advocate* and the *scapegoat,* respectively.

The linguistic interchanges among participants of an interaction can be seen as forming an elaborate game as well, in which each speech act [39] constrains and directs subsequent acts. Intuitively, the act of asking a question is bound to elicit some response related to the nature of this question, even if indirectly. Searle [39] and others associated with the *language/action* perspective (e.g., [21, 46, 13]) identify a set of illocutionary points that constitute the essential components of conversations for actions. Individual acts are combined into acceptable "move sequences," so that, for example, a *request* by a participant can be *accepted, declined,* or *counteroffered.* Each of these, in turn, has its possible continuations; for example, a counteroffer can be accepted, the original request might be canceled, or a new counteroffer might be generated [47]. Collective discourses thus display structure and can be equated to an evolving process. In practical terms, that means that interactions, even seemingly unstructured ones, are regulated by linguistic, social, and cultural norms that dictate to a large extent the way interactions are "played out." In other words, interactions constitute *social processes*. Such processes take place at many distinct levels, embedded within each other in a recursive structure. *Debate* and *voting*, for instance, can be considered subprocesses within a meeting in which they occur; meetings in turn are part of larger processes within organizations, which are embedded within yet larger organizational and societal settings.

This chapter focuses upon person-to-person (P2P) interactions. These interactions tend to have a high degree of human involvement, which poses special challenges to technological augmentation.

3.2.3 Formalized Interactions

Mature interactions develop into *routines*—"if social practice becomes reasonably stable over time and space, then routines—practices in which actors habitually engage—develop. Routines constitute the habitual, taken for granted character of the vast bulk of the activities of day-to-day social life" [26, p. 376]. The stability associated with routines brings about less need for clarification and amplification among participants [30]. Awareness of the expectations and norms of an interaction may eliminate the need for communication between interdependent parties and between superiors and subordinates [22]. Stability brings about a potential for enhanced efficiency by allowing participants to go about their work without engaging in redundant discussions and negotiations. The maturity of interactions offers an enhanced opportunity for reflection, as a clearer understanding of roles may emerge, and as practices become stable enough to grant better perception of their nature. This reflection may result in (1) the definition of roles, which might receive specific denominations, such as "session chair" and "facilitator," and be associated with explicit duties; and (2) the definition of formalized rules that regulate interactions.

Reified roles and norms may dictate the acceptable form of individual interactions, for example, by establishing specific forms of address or by requiring specific formal documents to be used (e.g., by requiring that all requests of a certain nature be presented in writing). These rules may also impose limitations on the acceptable actions that each role is entitled to take depending on context. The identification of specific contexts of interaction might result in the emergence of a partial ordering among activities. An interaction may contain potentially complex patterns in which structured, recursively nested subinteractions are embedded within interactions. This certainly is the case in the extreme collaboration example presented earlier. In extreme collaborations, collaborative subinteractions are initiated in response to participant's perceptions and intuitions with respect to the perceived state of the problem and the dependencies among its parts. Similarly, IBIS-based interactions are composed of nested subinteractions that explore issues by developing branches of the discussion trees. In congressional sessions, on the other hand, interactions are strictly choreographed and played out as participants make use of their (restricted) rights to call for action (e.g., call for voting or object to a motion).

It is important to note that whereas formalized norms crystallize rules of interaction, that does not imply that the whole of the interaction becomes mechanical (or "mechanizable"). In fact, the very act of abstracting that is involved in formalization implies omission of details. Formal norms just make explicit the structural backdrop against which complex and diverse interactions unravel. Consider, for instance, the example given previously of the formalized interaction that takes place in houses of representatives. A congressional session is not restricted to the superficiality of the turn-taking and formal actions exercised by the representatives. These are just the mechanisms through which very complex games are played. The reasons that motivate these formal actions are still deeply connected to intertwined social, political, economic, and personal interests of a large number of stakeholders.

3.2.4 Implications for Collaboration Technology

In a later section of this chapter, we will discuss collaboration technologies that have explicit embedded representations of group processes. Informally, we define these technologies as being "process aware." They are capable of capturing process information, interpreting, presenting, and utilizing this information in various ways to assist in the control, coordination, and execution of process activities. This technology can span a range from a passive information repository to a fully automated process control and enactment system. We offer a more rigorous definition of *process* and of *process awareness* in Section 3.3.

Process-aware collaboration technologies make statements about work, and operate on abstracted representations of work. Both the advantages and the risks associated with this kind of technology emerge from the explicit way the mechanisms of work are addressed. The fundamental property associated with process-aware technologies is therefore the *visibility* of work it affords.

Dourish [15] relates the usefulness of process-aware technologies to the mechanisms they provide for explaining work, in addition to coordinating it. Explaining work requires that the myriad of actions that comprise work be associated with identifiable abstract activities toward which these actions would have been applied. This would correspond to the *understandability* criterion needed within organizations, and also to the intrinsic *accountability* mechanisms that play an important role in group collaboration. Activities such as planning and control rely upon criteria such as understanding and accountability, that in turn can be enhanced by providing *visibility*. Representations of work associated with process-aware technologies serve, then, as a rational structure within which individual actions can be made sensible [15], making these actions visible in a meaningful way. In this sense, process-aware technologies are "technologies of understandability and accountability," imposing rational order for the purpose of practical reasoning and action [42]. This reasoning and action takes different forms and serves various functions as they are applied to representations of work made visible at different moments in time:

- **Visibility of Past Work.** Being able to examine representations of work performed in the past affords accountancy functions to be performed. These functions might encompass quality control measurements, auditing by tracing back to determine how certain outcomes were achieved, or to guaranteeing that outcomes did conform to organizational policies. These representations are sometimes also used as interorganizational artifacts, for example, to allow for financial compensation based on work performed in oil rigs shared by multiple organizations [6], or by a print shop to demonstrate to its clients the proper running of outsourced printing facilities [5, 15]. The articles referenced above discuss workflow management systems, but the arguments are applicable to process-aware systems and technologies in general.

- **Visibility of Current Work.** Visibility of work currently being/to be performed serves purposes of orientation and guidance, and is associated with the intrinsic function of *coordination* that is a concern of process-aware technologies. Making activities at hand explicit serves the purpose of making clear to participants the objectives toward which actions ought to be taken. Visibility of the current state of work has also consequences in terms of *overviewing* [34]. By being able to examine a concise representation of state of work, workers can identify potential problems (e.g., risk of missed deadlines) with further confidence.

- **Visibility of Future Work.** Process-aware technologies can maintain logs and histories. These histories can be analyzed using predictive models. Thus, past trends can be translated into future predictions. The ability to anticipate outcomes of recurrent situations is an essential aspect of human activity [1]. Reflecting and planning are intrinsic to humans, and are related to the mechanisms that make stable routines naturally emerge. Process-aware technology creates an opportunity for innovative *synthetic* processes to be developed and experimented with. Reflection may bring about new processes that would not normally emerge from routines created over an extended period of time. Consider, for instance, the processes implemented in group decision support systems (GDSS). These artificial processes may avoid common meeting pitfalls such as individual domination, sidetracking, and information overload [29], by carefully organizing work around a few activities during which participants are oriented (and constrained) towards some specific actions, for example, anonymous idea generation, followed by categorization, followed by voting.

3.2.5 Problems and Pitfalls

Because of the explicit nature in which work is addressed, process-aware technology can impact negatively the fluid nature of actual work practices. Critics of process-aware technology (e.g., Suchman [41, 43, 40] and Robinson [35]) point out that abstracted representations of work are in principle incompatible with the rich situated nature of work itself, in which variations are the rule. Abstracted representations of work, the essential ingredient of process-aware technologies, do not account for this rich variety and for the intertwined social and cultural aspects of work.

According to critics, action is situated, based on context, and often improvised. "Offices do not follow procedures as their main purpose. On the contrary, offices have certain goals to obtain and the procedures are only a way to reach their goals. The 'smooth flow' of office procedures is a result of how practitioners orient their work—it is not the work itself' [41]. Plans would, therefore, be resources for reflecting about work, or for explaining it after is has been completed, but would be a hindrance to the actual work performance, or would at least not provide necessary resources for the performance of work. "Systems adopting procedural work flow models only weakly support the cooperation, collaboration and coordination that would greatly aid in achieving the goals of the system" [37].

There is, in fact, evidence (e.g., [25, 24, 5, 38]) that process-aware technology might introduce obstacles to the fluid performance of work in certain situations. In practical terms, technological hindrances result in work being performed around the systems [25, 24], or through the use of "kludges" (e.g., inserting fictitious workers used by participants working on two jobs, described in [5]), or through a reinterpretation of the characteristics of technical objects (e.g., understanding zero hours as meaning "incomplete" in the time sheets described in [6]). These mechanisms, necessary for the actual performance of work, defeat the purpose of the technology, to at the same time coordinate and make work visible in sensible terms, given that workarounds by definition are invisible to the technology. The invisibility hinders the explanatory function of the technology because part of the work cannot be accounted for; the coordination function is crippled by its misfit with actual performance requirements.

An important body of research in collaboration technology steers clear of the process-related aspects of work, concentrating on offering collaboration tools whose collective use must be regulated by their users themselves. These tools are described as "process-unaware technologies." The processes are thus kept implicit from the perspective of the technology. This approach is indeed flexible, and allows for unanticipated use [34], making these technologies more adaptable to different peculiarities of work situations and, therefore, useful in a potentially larger number of occasions. The downside of this "process-oblivious" approach is, of course, that users of such technology have to bear the burden of structuring the interactions in a sensible way themselves. That may not be so straightforward, if the mechanisms offered are incompatible with the structuring that a group wishes to impose. That may happen, for example, if a laborious mapping of tools is required to support simple recurrent activities. Other criticism is related to the potential political role that representations of work play [44]. Representations serve specific interests, as they highlight (or hide) specific aspects of work. One concern is that if these politically biased representations are incorporated into technologies, only the specific interests served by the representation will be enforced, which might be at odds, again, with the actual practices of the stakeholders whose perspective was not considered. The main problem seems to be that stakeholders many times have an incomplete understanding of the complexity of work in which they are not directly involved. Work performed at a distance, particularly if it is performed well, tends not to be visible, except to those doing it. What appears to be menial "nobrains" work at a distance might in fact involve complex, elaborated work (e.g., the filing job reported by Suchman [44]).

The lack of proper understanding of the finer details of work and the consequent incompatibility of representations with actual work practices is exacerbated when the group responsible for the development of technology is detached from the users of this technology [30, 44] (as is many times the case). In this case, there is a danger that the technology will embody a vision of work that does not correspond to the perspectives of *any* stakeholder, resulting in potentially negative results. Technology-supported processes (automated or not) are by definition embedded in larger, more complex systems; technology that is at odds with the systems within which it is embedded becomes a source of potential disruption for these systems. Intuitively, a synthetic process that differs too much from the environment in which it is introduced is bound to cause problems.

3.3 CHARACTERIZATION OF PERSON-TO-PERSON SYSTEMS

3.3.1 CSCW

The field of CSCW (computer-supported collaborative work) is concerned with theories and technologies to help groups to accomplish work tasks. This field usefully integrates social sciences and technical sciences. Journals and conferences in this field [12] describe a plethora of computer-based tools to assist in group activity coordination. These tools range from e-mail systems to real-time group editors [e.g.: CoWord]. We refer to these computer-based tools and technologies as "person-to-person systems" or P2P systems. Very few of these P2P systems have an explicit embedded representation of the underlying group process. This section develops concepts and language to describe different P2P systems with a particular emphasis on P2P systems that are process-aware, that is, they do have explicit embedded representations of group processes.

3.3.2 Definitions

A *process* is a set of (greater than one) *activities* carried out to attain some *goals*. These activities are structured within a partial order such that some activities may be required to precede others. Each activity may have many attributes (e.g.: input data, persons performing activity, time constraints). Each activity may be composed of lower-level subactivities (a subprocess), and may span a spectrum from totally manual to totally automated execution. Decision-making activities and "branching nodes" can be included, but are not discussed here.

A *process-aware system* is a computer-based system that contains an explicit embedded representation of the underlying group process.

3.3.3 Examples

To illustrate the concepts of processes and process awareness, let us discuss two hypothetical P2P systems. In each case, we describe a linear sequence of activities for the process, and omit parallelism and decision activities.

The first example is a software release control system (SRCS) used by a large software company to release a new version of its software product every six months. It is both an information repository for code, documentation, and messages, as well as a process control system that automatically initiates activities, runs compilers, and distributes new versions of software. It is an asynchronous, always available system that can be used by the software engineers anytime during the day or night.

Activity 1 of the software release process is bug report activity, which lasts two months. During this time, anyone can electronically submit software error reports that are evaluated during this activity.

Activity 2 is bug repair activity, during which programmers fix errors in the software product. This activity lasts for exactly two months, so fixes must be submitted electronically before the fixed cutoff date.

Activity 3 is prerelease compilation and system testing, which lasts for one month. This activity also includes patching of any faulty software uncovered during testing.

Activity 4 is final release compilation and documentation, which occupies most of the sixth month. During the final days of this month, the software is distributed electronically to customers (activity 5).

Although this process control and management system (SRCS) exerts automatic control over all five activities of the release process, it is categorized as a P2P system because the work activities are labor intensive, demanding the collaboration of many highly skilled knowledge workers. These five activities are rather high-level activities. It is easy to envision that each of these activities has a subprocess of finer-grained, detailed activities embedded within it.

The second example system is a hypothetical meeting system for structured brainstorming and problem solving using the nominal meeting methodology [14]. This real-time distributed technology enables a group of people to have a decision-making meeting over the Internet from their homes using their workstation, voice connection, and a network.

This general category of systems, called a decision support system (DSS), has been successfully used for technology assessment meetings [16]. It works as follows. A group of experts on some topic are paid (or volunteer) to participate in a technology evaluation meeting. At a designated time, all participants connect to the meeting system from their home computers. The meeting is conducted and controlled by the meeting facilitator, a person trained in conducting meetings who is responsible for the smooth running of the meeting. He or she also controls availability and usage of the tools during all activities of the meeting. In the nominal meeting methodology, the facilitator leads the group through the following activities.

Activity 1 is issue exploration. The main question to be explored at the meeting is introduced and clarified by the principal investigator using a presentation tool.

Activity 2 is solution brainstorming. Participants are all encouraged to submit solutions. Others can request clarifications. The goal of this activity is to get a wide variety of creative possible solutions, so no criticisms are allowed until later activities. The meeting system has a brainstorming tool that allows all participants to concurrently type in solution ideas and clarification requests. The system then makes these typed texts conveniently available to all group members.

Activity 3 is solution convergence. All solutions are discussed, debated, categorized, and their feasibility is explored using a discussion tool.

Activity 4 is voting. Participants vote on which of the remaining solutions is most workable. This activity is a complex, computer-controlled subprocess. The meeting system provides a voting tool for this activity.

The meeting system provides the following set of tools:

1. A presentation tool for use during activity 1
2. A brainstorming tool for use during activity 2
3. A discussion tool for use during activity 3
4. A voting tool for use during activity 4
5. A voice conference tool

The facilitator decides on the appropriate time to begin and end each activity. The facilitator has control of all tools, and can thus enable or disable any tool at any time on all users' workstations. She will typically enable the correct tools for the particular activity. All tools, except the voice conferencing tool, enable communication via typed nonverbal means. This information forms the formal record of the session. When appropriate, the facilitator can enable the voice tool that allows anyone to speak to the group at any time. This constitutes the second-level informal group communication channel.

Meetings can be intensive people-interaction entities. Meeting systems are a classic example of P2P systems. Most meeting systems are not aware that there is any process. In this example, the meeting system is process-aware because the meeting system knows the activities, and knows which activities are currently ongoing throughout the meeting via tool usage.

3.3.4 P2P Systems Characteristics

With these two systems in mind, let us define some terminology that will allow us to describe and characterize P2P systems. One important attribute of P2P systems is *process awareness*. A P2P system is *process-aware* if and only if it has an explicit embedded representation of the process it supports. In the first example, the SRCS system is clearly process-aware because it changes access rights, notifies individuals, and so on, during each activity change. The five activities are explicitly embedded in the code composing the software control system. In other systems, the process may be represented as a table, a set of rules, or (e.g., workflow systems) an explicit internally stored model.

Note that workflow management systems, discussed extensively in other chapters of this book, are salient examples of process-aware systems. Some of these systems have been quite productive and successful. In this book, we (arguably) classify workflow management systems as P2A systems.

On the other hand, a P2P system such as a typical real-time group editor [11] has no knowledge of and no internal representation of the editing process. Any process (turn taking, section responsibility, or precedence of edits) is a process that is informally agreed upon by the group—not known to the system. In these cases, we say that the system is *process-unaware.*

The activities of the process may be explicitly known by a P2P system, but not used to control the system. In the case of the example meeting system, various tools correspond to various activities, so the system has process awareness. The sequence of tool invocations and usage times are recorded by the meeting tool, but not used to control the system. If the process had been described in more detail, including activities of reading the text and listening to the voice, then this meeting system would only have partial process awareness. Note that the system can detect some activities such as speaking and typing, but cannot detect other activities such as listening and reading. If the facilitator injects other activities using the voice tool, the system will never know about it. For many reasons, a system may be aware of some activities but not others. In all of these cases, we say that the system has partial process awareness.

Another important characteristic for categorizing P2P systems is *precedence enforcement.* If activity A is specified to be performed before activity B, then a process-aware P2P system may enforce this precedence constraint. Alternatively, the system may emphasize flexibility and require that precedence enforcement be done by users. There are also hybrid systems in which some enforcement actions are performed by the system and others are left to the users. The example software control system is a precedence-enforcement system, whereas the meeting management system is not precedence enforcing.

Related to enforcement are the concepts of *activity initiation* and *activity termination.* For each system, we can observe whether activities are forcibly terminated (and initiated) by the system or controlled by users. As with enforcement, the possibilities include system controlled, user controlled, or hybrid controlled. The software control system is system controlled, but the meeting system is user controlled.

Humans are adept at handling exceptional conditions and problem solving. Much human time and capital in the office is spent in exception handling. P2P systems have a high need for flexibility to handle exceptions and to encourage human problem solving (typically higher than A2A and P2A systems). Thus, when a P2P system has precedence enforcement, activity initiation, and activity termination all controlled by the system, it is important to have reasonable, thoughtful system-override mechanisms. In some systems, it simply requires a keystroke by a human. In others, it requires reprogramming by the vendor.

Organizations are constantly changing. Thus, another particular need of P2P systems is dynamic change. This takes many forms and can be quite complex. We investigate systems according to the flexibility of their binding times. Different systems allow users to design or change processes easily at design time, at instantiation time, and/or on the fly at execution time. The software control example system has activities and durations of activities coded in at design time. We say that a system, such as this, which buries the process in the tool, has an "embedded process." It is not very flexible for dynamic change. However, when a particular release cycle is begun (called the instantiation time), managers can set the exact dates that various activities must terminate. Thus, this system allows some dynamic change at instantiation time.

When considering binding times, we particularly investigate the times at which we can create new activities within a process, the time at which we can specify or change the precedence relations between activities, and the times at which we can specify or change the duration of activities. Activities have many other attributes (persons associated, information inputs, etc.), so we investigate the time at which these attributes can be changed.

Meeting facilitation is a demanding art and sometimes requires dynamic changes of agenda to make it successful. In the meeting system, the meeting facilitator has control of the tools that represent the activities. Thus, the duration of an activity, as well as the ordering of activities, can be dynamically decided at execution time. This is a very flexible system that allows binding at design time, at instantiation time (when the meeting begins), and at execution time (during the meeting). The facilitator has the flexibility to skip an activity or repeat an activity if desired. Also, the facilitator can use tools in unintended ways. Furthermore, the voice tool can be used at any time to inject an unanticipated activity such as "introduce yourself" or "question and answer."

Since P2P systems emphasize communication and collaboration among people, it is useful and frequent that tools are present allowing the users to efficiently perform activities. Thus, one parameter of characterization of P2P systems is concerned with *tool integration*. A system may be composed of different tools (called multitooled), or may be a single tool (called unitooled). Note that a single tool can be present on multiple users' screens simultaneously (e.g., a chat tool is on the screen of all communicators simultaneously). The software control system appears to the users as a single integrated tool; it is unitooled. However, the meeting system appears as a suite of tools; it is multitooled.

Other parameters of characterization of P2P systems include *activity execution mode* (defined as manual, automatic, or semiautomatic execution) and also process visibility (characterized as past, present, and/or future). Visibility was defined and discussed in Section 3.2. Other parameters are not related to process, and are thus beyond the scope of this chapter.

3.4 EXAMPLE SYSTEMS

This section uses the terminology introduced previously to characterize five exemplary P2P systems: CoWord, WebEx Meeting Center, IPMM, LeadLine, and Caramba. These five systems represent the state of the art of P2P systems. They cover different applications, including coauthoring, online meeting, CASE, chat, and virtual team. Processes in these five systems have different characteristics in terms of complexity and dynamics. They are from different developers, including Griffith University, WebEx, The Hong Kong Polytechnic University, Microsoft, and Caramba. All of them are recent and leading systems in their respective domains.

A comparison among these five systems is given in Table 3.1. For each system, we examine its *process awareness, precedence enforcement, binding time, activity execution mode,* and *process visibility.* Detailed explanation is given in the following sections.

3.4.1 CoWord (Griffith University, Australia)

CoWord (Collaborative Word) is a real-time, collaborative word processing system that allows multiple users to edit the same Microsoft Word document at the same time over the Internet [11]. From the user's point of view, CoWord is Word plus some new features that enable multiple users to perform and undo editing operations concurrently and consistently. For example, the semantics of CoWord undo commands is richer than that of Word undo commands. In CoWord, an undo can be applied to operations from a single user instead of those from all users. From the implementation perspective, CoWord is implemented by adding a collaboration engine on top of Microsoft Word to handle the concurrent behaviors and support new features.

CoWord is not aware of the editing process (e.g., turn taking, section responsibility) among multiple users. The specification and enactment of the editing process happen outside CoWord. CoWord is only aware of various editing actions, such as

Table 3.1   Comparison among five example systems

	CoWord	WebEx	IPMM	LeadLine	Caramba
Domain	Coauthoring	Online meeting	Project management	Chat	Virtual Team
Developer	Griffith Univ.	WebEx	Hong Kong Polytechnic Univ.	Microsoft	Caramba
Process awareness	No	No	Yes	Yes	Yes
Precedence enforcement	—	—	Hybrid	System	User
Binding time	—	—	Instantiation time, Execution time	Instantiation time	Instantiation time, Execution time
Activity execution mode	—	—	Semiautomatic	Manual	Manual
Process visibility	—	—	Past, present	Present	Past, present

inserting a new word, deleting an old line, changing the size of the title, and undoing the previous operation.

3.4.2   WebEx Meeting Center (WebEx™ Communications, Inc.)

WebEx Meeting Center is a general-purpose online meeting service supporting various types of meetings, such as departmental meetings, new customer calls, production demonstrations, and document review meetings. WebEx Meeting Center provides meeting support via a set of tools, including a presentation tool, document editing tool, application/desktop sharing tool, Web browser sharing tool, chat tool, polling tool, participant awareness tool, meeting recording tool, and multimedia support tool [45].

As explained above, a meeting often has a process. However, WebEx Meeting Center does not provide any process support and thus is process-unaware. The process resides within meeting participants' minds, and the enactment of the meeting process depends on the meeting facilitator who dynamically enables and disables tools for different activities. Since the meeting process is not explicitly specified in the system, WebEx Meeting Center could detect neither the current activity nor the change from one activity to another.

3.4.3   IPMM (Hong Kong Polytechnic University, Hong Kong)

IPMM (Integrated Process and Project Management) is an integrated process and project management tool for multisite software development [9]. Process management and project management are two important tasks in software development. Although processes and projects are related and share some characteristics in common, in the past, the management of them was separated. This leads to various problems, such as efficiency, consistency, and costly communication.

For example, in Microsoft Project, tasks are defined with dependencies such as A must finish before B starts. In a process management tool, there are definitions of process steps such as route from programmer A to software manager B. IPMM formally integrates these so that routing happens smoothly and automatically. The goal of IPMM is to integrate process management and project management in order to create a more effective software engineering environment.

IPMM is implemented on top of IBM's Lotus Notes/Domino and Microsoft Project. Users first define a process model, consisting of processes, activities, actors, forms and their relationships, in a graphical editor. This process model is then exported into Microsoft Project, a project management tool. According to the mapping between process and project shown in Figure 3.1, one project is created for each process object. Then users can manage projects in Microsoft Project. In the meanwhile, IPMM keeps track of process execution status and automatically updates project progress.

IPMM allows users to explicitly specify a process and thus is process-aware. Precedence constraints are enforced by both systems and users. A process is defined at instantiation time, but users can modify it at execution time. Although most tasks

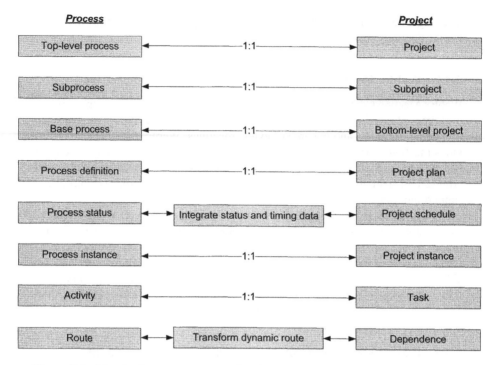

Figure 3.1   Mapping between process and project [9]. Reproduced with permission from Kluwer Academic Publishers.

are still executed manually, more and more tasks are being executed automatically. Therefore, IPMM has semiautomatic execution. The process/project definition and their status are visible from the user interface, supporting more effective process and project management.

3.4.4   LeadLine (Microsoft)

LeadLine is a process-aware text chat tool [20]. Most chat tools are not process-aware, under the assumption that chat is often informal and thus has no internal structure. However, chat tools today are also frequently used in organizations for business purposes, for which there often exists a process. LeadLine is designed to meet such needs.

Figure 3.2 shows an example hiring process, in which a committee is required to discuss three candidates for a technical writer position and pick one of them to hire within 20 minutes. This process has five activities: position review, candidate #1 discussion, candidate #2 discussion, candidate #3 discussion, consensus, and decision. Each activity has an assigned time.

Users specify a process in LeadLine via two major concepts: script and scene. The concept of script is similar to the concept of process, and the concept of scene

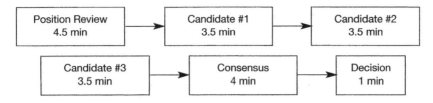

Figure 3.2   Hiring process.

is similar to the concept of activity. A script defines a set of roles and divides a session into a sequential set of scenes. For each role and scene combination, the script specifies instructions for users playing that role. Figure 3.3 shows the script for the hiring process given above.

LeadLine supports simple and static processes. The script is preauthored at the instantiation time and cannot be changed at the execution time. The system enforces

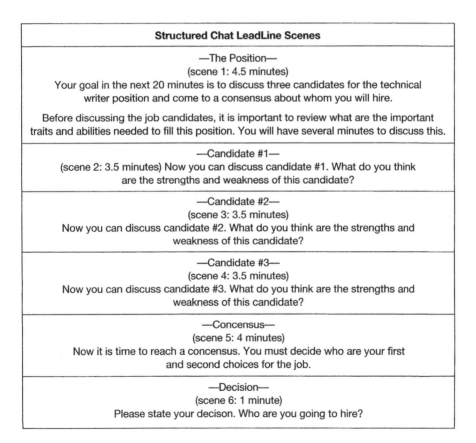

Figure 3.3   Chat script for the hiring process.

the precedence constraints according to the script. When the specified time of a scene passes, LeadLine automatically moves to the next scene. However, LeadLine cannot tell whether users are actually talking about the issues of this scene or not. In order to do this, advanced conversation-understanding technology would be needed. All activities in LeadLine are executed manually. In the user interface, LeadLine shows the status of the current scene, such as what the current scene is about and the time left for this scene. Users can also scroll up the chat window to see the chat history.

3.4.5   Caramba (Caramba Labs Software AG)

Caramba is a process-aware collaboration system for virtual teams [17, 8]. It is mainly concerned with the asynchronous collaboration among team members by providing a shared workspace for teams to communicate, cooperate, and coordinate their work. Processes exist in virtual teams. However, different from processes in many other working environments, processes in virtual teams are often ad hoc and dynamic. Caramba supports both predefined processes and ad hoc processes. Process awareness is enhanced by providing an identical process view to all team members.

Before introducing mechanisms used by Caramba for supporting processes, we first give an ad hoc process example in virtual teams. As shown in Figure 3.4,

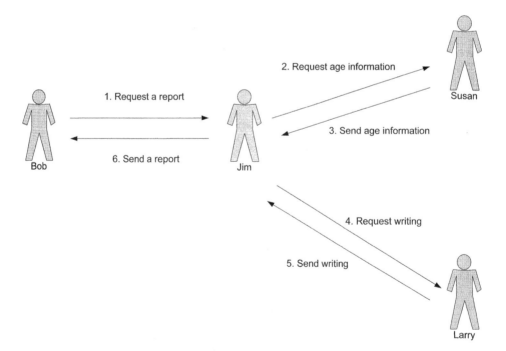

Figure 3.4   An ad hoc process example.

manager Bob initiates a process by requesting Jim to write a report about the aging situation in Colorado within one week. Jim does not have the age information about current Colorado residents and therefore contacts Susan. Susan collects the age information, puts it in a document and sends the document to Jim. One week is really too short for Jim to write such a nontrivial report that is expected to address how the aging situation might affect Colorado in the next 10 years. Jim knows that Larry is an expert about healthcare, so he asks Larry to help write the section about "impact of aging on healthcare." Larry agress to do it. Finally, Jim puts everything together in the final report and sends it to Bob. This process is not predefined. The decisions made by Jim and responses from Susan and Larry are all ad hoc.

Caramba designers define the process in new way according to their real-world experience. A Caramba process consists of a set of activities combined together via a sophisticated coordination model (for detailed information about the coordination model, please refer to [17]). An activity is composed of work items, which consist of one or several actions. In Caramba, a process can be predefined and then enacted; a process can also be started with no underlying process model. For a predefined process, Caramba enacts the process according to the process definition. For a nonpredefined process, Caramba tracks the relationships between activities, work items, and individual actions, and the relationships between activities, resources, and artifacts. Therefore, the model for a nonpredefined process is recorded at execution time.

Caramba allows users to explicitly specify a process and thus is process-aware. Precedence constraints are enforced by users. Caramba supports flexible binding time. For predefined processes, users design them at instantiation time and are allowed to change them at execution time. For nonpredefined processes, users design them at execution time. Activities are mostly executed manually. Process information and status are visualizable to all team members. Figure 3.5 shows a view of the ad hoc process given above.

Workcase:0001
Subject:Investigate Aging Situation
Started:2.9.2004
Sender:Bob
Addressee:Jim

Time	Sender	Addressee	Recipient	Subject	Status
5pm,2.13.2004	Jim	Bob	Bob	Re:Investigate Aging Situation	Done
2pm,2.13.2004	Larry	Jim	Jim	Re:Aging Impact on Healthcare	Done
3pm,2.11.2004	Jim	Larry	Larry	Aging Impact on Healthcare	Done
10am,2.10.2004	Susan	Jim	Jim	Re:Aging Information	Done
3pm,2.9.2004	Jim	Susan	Suaan	Aging Information	Done
2pm,2.9.2004	Bob	Jim	Jim	Investiate Aging Situation	Done

Figure 3.5   A process view.

3.5 SUMMARY AND CONCLUSIONS

In summary, P2P processes are important because they focus on sociotechnical issues within information systems. An increasingly large population of diverse humans is interacting with technology these days; it is, therefore, increasingly important for systems to address issues of person-to-person collaboration. The incorporation of P2P processes into computerized systems is particularly challenging because people processes are complex, semistructured, and dynamically changing. In this chapter, we have argued that P2P systems are important and intriguing because there is frequently a need for exception handling and dynamic change performed by humans. It is, thus, necessary to take into account factors that impinge upon the organizational structures, the social context, and the cultural setting.

P2P processes exist at many levels. Thus, a voting subprocess may be nested within in a meeting process, which may exist within a complex software engineering project, which itself is a higher-level process. Furthermore, the meeting, which may appear rather unstructured, may have very structured subdialog processes. For example, a debate and then a voting process within the meeting can both be considered as lower-level processes nested within the meeting. We argue that modern technology and modern social science can be combined to effectively facilitate interaction among people at many of these levels. However, most of the current CSCW technology is aimed toward specific task assistance, and is not process-aware.

In this chapter, we explained and characterized person-to-person (P2P) processes and systems. First, we discussed collective work, its implications for collective technology, and its problems and pitfalls. We then developed some concepts and language for describing P2P systems. We defined and examined concepts of process awareness, partial awareness, precedence enforcement, activity initiation and termination, binding time, activity execution modes, and process visibility. We categorized five salient P2P systems according to this ontology.

This leads to the question of "when should an organization move to process aware technology?" There is an issue of goodness of fit of technology to process. Examples have been observed of significant gains in efficiency and effectiveness by the incorporation of process-aware P2P systems [17]. In some of these cases, the ability of people to view the current state of the entire process has been extremely valuable for exception handling. Examples have also been observed of cumbersome inhibition of the work people do by overly inflexible information systems imposing strict process orderings. In some of these cases, the ability of people to get their work done in a timely fashion has been seriously impeded by unnecessary formality and complexity introduced by the system [18]. In general, P2P process-aware technology is most useful in situations of nontrivial P2P processes that are standardly followed within a structured, stable environment.

We envision many future benefits of P2P technology that are being investigated in the research domain today. Role-based information systems can help to partition complexity. Affective computing techniques enable virtual agents to actively participate in group communications in a fashion that is familiar and natural to humans, rather than requiring people to learn details of the technology's interface.

Multimedia and multimodal systems are becoming feasible, available, and useful. As businesses become more distributed and intertwined, we see an increasing need for intelligent process technology. We see exciting research progress and significantly enhanced technology in the future. A thoughtful marriage of information technology and social science is necessary to produce P2P systems that are organizationally aware, socially aware, culturally aware, and truly process-aware.

3.6 EXERCISES

Exercise 1
Read the following description of CA-ForComments (adapted from [19]) and classify it according to the criteria discussed in Section 3.3.4. Classify this system according to (1) *process awareness,* (2) *precedence enforcement,* (3) *binding time,* and (4) *activity execution mode,* similarly to what is done in Section 3.4.

CA-ForComments [7] is a document editing system that supports certain forms of collaborative use. It offers the textual objects "comment," "revision," and "dialog" (among others). *Comments* are supposed to contain the reviewer's comment on a sequence of lines of the original document. Thus, besides operations for creating and editing of a comment, ForComments offers an operation for attaching a comment to a sequence of lines in the original text. The *revision* is supposed to contain a replacement text for a sequence of lines and, besides the creation and attachment operations, the system provides an operation for replacing the text of the original document with the text of the revision object attached to it. *Dialog* is a comment on a comment and can only be attached to another comment.

CA-ForComments distinguishes a few different activities. In the activity of creating the document, for which the available objects are the components of the document (e.g., lines, characters, etc.) and the available operations are the usual edit operations on the document components, the goal is to create a draft of the document, and the performer of this activity is the author. There are also review activities, one for each reviewer. The many review activities, which follow the createdocument activity, can be performed in parallel or in sequence, depending on how the system is configured. The available objects for the review activities are the document, comments, revisions, and dialogs. The operations available are read (but not modify) the document components; create, modify, and attach comments and revisions to the document components; and, possibly, read other reviewers' comments and create, modify, and attach dialogs to them. Finally, there is the activity of incorporating the reviewers' comments into the document, performed by the author. The author may then choose to restart the review activities or declare that the document as finished.

Exercise 2
Read the description of the meeting example in Section 3.3.3. This example describes a meeting process composed of four activities: (1) issue exploration, (2) brainstorming, (3) convergence to solution, and (4) voting. Also read the description of the process-unaware system (WebEx) described in Section 3.4.2.

(a) Which of the tools available in WebEx would you activate to support each of the four activities of the meeting process if you were the meeting facilitator responsible for manually organizing a process as described in Section 3.3.3? Notice thatWebEx does not provide the exact same tools listed in the Section 3.3.3 example—there is no brainstorming or discussion. Your task is to decide how to map available tools to support each of the four phases.

(b) Which phase presents the most marked differences in terms of tool use, when you compare the description in Section 3.3.3 with what is available in WebEx?

Exercise 3

Consider three groups of people you have never met before that collaborate: (1) using an extreme collaboration style, (2) following *Robert's Rules of Order,* or (3) using IBIS.

(a) Consider that you are given one day to prepare before you join one of the three groups. Which group do you think you would be better prepared to join, given the short preparation time?

(b) Which one would require the longest time before you could participate fully in the collaboration?

Exercise 4

For the same groups described in Exercise 3, answer the following.

(a) If you were to develop a process-aware collaboration system, which of the three groups do you think would provide the most benefit? Which would provide the least?

(b) Assume that your system supports distributed collaboration, that is, it allows people that are dispersed to collaborate, for example, by communicating through the Internet from their homes or offices. Which of the three interaction styles would fit your system best? Answer this question based on your perception of how important it is for participants of each group to be in the same room, thus being able to observe each other's actions.

REFERENCES

1. J. Bardram. Plans as Situated Action: An Activity Theory Approach to Workflow Systems. In *European Conference on Computer-Supported Cooperative Work—ECSCW' 97,* pp. 3–14, Lancaster, UK, September 1997.

2. D. Berlo. *The Process of Communication.* Holt, Rinehart and Winston, New York, 1960.

3. B. Biddle and E. Thomas (Eds.). *Role Theory: Concepts and Research.* Wiley, New York, 1966.

4. E. Bittner. The Concept of Organisation. *Social Research, 32,* 1965. (Reprinted in Turner (Eds.), *Ethnomethodology.* Penguin, Harmondsworth, UK.

5. J. Bowers, G. Button, and W. Sharrock. Workflow from Within and Without: Technology for Cooperative Work on the Print Industry Shopfloor. In *Proceedings of the European Conference on Computer-Supported Cooperative Work (ECSCW),* 1995.

6. B. A. T. Brown. Unpacking a timesheet: Formalisation and representation. *Computer Supported Cooperative Work, 10*(3–4): 293–315, 2001.

7. Broderbund Software, San Rafael, California. *CA-ForComments 2.5 PC User Guide,* 1991.

8. Caramba, http://www.carambalabs.com/eng/index.html, 2004.

9. K. C. C. Chan and L. M. L. Chung. *Integrating Process and Project Management for MultiSite Software Development,* Volume 14 of *Annals of Software Engineering,* pp. 115–144. Kluwer Academic Publishers, 2002.

10. J. Conklin and M. L. Begeman. gIBIS: A Hypertext Tool for Exploratory Policy Discussion. *ACM Trans. Infrastructure Systems, 6*(4):303–331, 1988.

11. CoWord, http://reduce.qpsf.edu.au/coword/home content.html, 2004.

12. Computer Supported Cooperative Work, *The Journal of Collaborative Computing.* Kluwer Academic Publishers, 1992–2004.

13. G. De Michelis and M. A. Grasso. Situating Conversations within the Language/Action Perspective: The Milan Conversation Model. In *Proceedings of the Conference on Computer Supported Cooperative Work—CSCW,* pp. 89–100, 1994.

14. A.L. Delbecq and A.H. Van de Ven. *A Group Process Model for Identification and Program Planning,* Volume 7 of *Journal of Applied Behavioral Sciences,* pp. 466–492. 1971.

15. P. Dourish. Process Descriptions as Organisational Accounting Devices: The Dual Use of Workflow Technologies. In *Proceedings of the 2001 International ACM SIGGROUP Conference on Supporting Group Work,* pp. 52–60. ACM Press, 2001.

16. Decision Support Systems, htt://dssresources.com.

17. S. Dustdar. Caramba—A Process-Aware Collaboration System Supporting Ad Hoc and Collaborative Processes in Virtual Teams. *Distributed and Parallel Databases, 15*(1):45–66, 2004.

18. C. Ellis and G. Nutt. Multidimensional Workflow. In *Proceedings of the Second World Conference on International Design and Process Technology,* Austin, TX, December 1996.

19. C. Ellis and J. Wainer. A Conceptual Model of Groupware. In *Proceedings of the 1994 ACM Conference on Computer-Supported Cooperative Work,* 1994.

20. S. Farnham, H. R. Chesley, et al. Structured Online Interactions: Improving the Decision-making of Small Discussion Groups. In *Proceedings of CSCW 2000,* pp. 299–308, Philadelphia, 2000.

21. F. Flores and J. J. Ludlow. Doing and Speaking in the Office. In G. Fick and H. Spraque Jr., (Eds.), *Decision Support Systems: Issues and Challenges,* pp. 95–118. Pergamon Press, New York, 1980.

22. J. R. Galbraith. *Organization Design.* AddisonWesley, Reading, MA, 1977.

23. H. Garfinkel. *Studies in Ethnomethodology.* Prentice-Hall, Englewood Cliffs, NJ, 1967.

24. L. Gasser. The Intergration of Computing and Routine Work. *ACM Transactions on Office Information Systems, 4:*205–252, 1986.

25. E. M. Gerson and S. L. Star. Analyzing Due Process in the Workplace. *ACM Transactions on Information Systems, 4*(3):257, 1986.

26. A. Giddens. *The Constitution of Society: Outline of the Theory of Structuration.* Polity Press, 1984.

27. M. A. K. Halliday. *Language as Social Semiotic: The Social Interpretation of Language and Meaning.* University Park Press, Baltimore, MD, 1978.

28. G. Mark. Extreme Collaboration. *Communications of the ACM, 45*(6):89–93, June 2002.

29. J. F. Nunamaker, A. R. Dennis, J. S. Valacich, D. Vogel, and J. F. George. Electronic Meeting Systems. *Communications of ACM, 34*(7):40–61, 1991.

30. W. Orlikowski. The Duality of Technology: Rethinking the Concept of Technology in Organizations. *Organization Science, 3*(3):398–427, 1992.

31. H. Rittel and W. Kunz. Issues as Elements of Information Systems. Working paper 131, Institut fur Grundlagen der Planung, University of Stuttgart, 1979.

32. H. Rittel and M. Webber. Dilemmas in a General Theory of Planning. *Policy Sciences, 4:*155–169, 1973.

33. H. M. Robert. *Robert's Rules of Order Revised for Deliberative Assemblies.* Scott, Foresman, 1915. Online edition at http://www.bartleby.com/176/.

34. M. Robinson. Design for Unanticipated Use. In C. Simone, G. de Michelis, and K. Schmidt, (Eds.). *Proceedings of the Third European Conference on Computer-Supported Cooperative Work,* pp. 187–202, Milan, Italy, September 1993. Kluwer Academic Publishers.

35. M. Robinson and L. Bannon. Questioning Representations. In *Proceedings of 2nd European Conference on Computer Supported Cooperative Work,* pp. 219– 233, Amsterdam, Netherlands, September 1991.

36. J. Rose and R. H. Hackney. Towards a Structurational Theory of Information Systems: A Substantive Case Analysis. In *Proceedings of the Hawaii International Conference on Systems Science,* pp. 258–267, Honolulu, HI, 2003.

37. H. Saastamoinen. Exceptions: Three Views and a Taxonomy. Technical report, Department of Computer Science, University of Colorado at Boulder, 1994.

38. H. Saastamoinen. *On Handling Exceptions in Information Systems.* Jyväskylä studies in Computer Science, Economics and Statistics, University of Jyväskylä 1995.

39. J. Searle. *Speech Acts: An Essay in the Philosophy of Language.* Cambridge University Press, Cambridge, UK, 1969.

40. L. Suchman. Office Procedure as Practical Action: Models of Work and System Design. *ACM Transactions on Office Information Systems, 1*(4):320–328, 1983.

41. L. Suchman. *Plans and Situated Actions: The Problem of Human–Machine Communication.* Cambridge University Press, Cambridge, UK, 1987.

42. L. Suchman. Technologies of Accountability: Of Lizards and Aeroplanes. In G. Button (Ed.), *Technology in Working Order: Studies of Work, Interaction and Technology,* pp. 113–126, Routledge, London, 1993.

43. L. Suchman. Do Categories Have Politics? The Language/Action Perspective Reconsidered. *Computer Supported Cooperative Work, 2*(3):177–190, 1994.

44. L. Suchman. Making Work Visible. *Communications of ACM, 38*(9):56–61, 1995.

45. WebEx Meeting Center, http://www.webex.com/services onlinemeeting. html, 2004.

46. T. Winograd and F. Flores. *Understanding Computers and Cognition: A New Foundation for Design.* Ablex, Norwood, NJ, 1986.

47. T. Winograd. A Language/Action Perspective on the Design of Cooperative Work. In *Proceedings of the 1986 ACM Conference on Computer-Supported Cooperative Work,* pp. 203–220. ACM Press, 1986.

CHAPTER 4

Enterprise Application Integration and Business-to-Business Integration Processes

CHRISTOPH BUSSLER

4.1 INTRODUCTION

Organizations like businesses, governmental organizations, or others maintain their business data in so-called legacy applications. A long time ago, one of these legacy applications was enough in order to maintain all the data of an organization. At that time, everybody in this organization who accessed the computing infrastructure did so through this unique centralized application system, and did the data processing directly with it. Over time, however, an increasing number of users gained access to the organization's computing infrastructure and more and more specialized applications were introduced inside the organization, each one dealing with different business data. This need for specialization came from the fact that business data became more complex in nature, but also because specialized applications were offered from software vendors and not self-built anymore inside the organizations.

Due to the inherent relationship between the business data in the various legacy applications, they have to communicate with each other to ensure proper data exchange and data synchronization. This integration of legacy application systems, called enterprise application integration (EAI), will be discussed in more detail in Section 4.1.1.

Organizations strive for efficiency in order to compete in the global marketplace. One factor of efficiency is speed of operation within, but also between, organizations. Organizations found out quickly that utilizing computer networks for interorganization data exchange contributes significantly to the speed of operation. Standards like EDI [5] and SWIFT[1] were put in place in the manufacturing and banking industries over 30 years ago in order to accomplish efficiency by means of electron-

---

[1]www.swift.com

ic data interchange. Although these industries have been pioneering this mode of operation, virtually every industry today uses this approach. In recent times, this has been called business-to-business (B2B) integration. B2B integration uses standards for documents as well as conversational behavior to ensure interoperable data exchange. Section 4.1.2 will describe this type of integration between organizations in more detail.

EAI and B2B integration are really two sides of the same coin. Currently, one cannot exist without the other. Data that is maintained inside legacy applications must be communicated across organizational boundaries in the general case. Still, historically, EAI and B2B were considered different and this separation will be followed in this chapter for didactical reasons. In the software industry, however, the separation is disappearing.

Section 4.1 introduces EAI and B2B integration processes (or "B2B processes" for short). Both are introduced separately from the viewpoint of requirements as well as properties and functionality. In real integration scenarios, however, both are relevant and require each others' cooperation in order to not only integrate application systems, but also businesses. A separate discussion about the cooperation of EAI and B2B processes is provided. This allows showing not only their cooperation, but also their differences.

After reading Section 4.1, the reader is expected to understand the need for the two types of processes, their cooperation, and understand the business problems that can be solved with the two types of processes.

This chapter can only introduce the basic concepts and approaches of EAI and B2B integration. A comprehensive discussion, technical as well as historical, can be found in [3].

4.1.1 Enterprise Application Integration (EAI)

Enterprise application integration (EAI) is concerned with the integration of enterprise applications. Enterprise applications are information systems that organizations use for their internal management and processing of data. Depending on the specific domain, these can have many forms. A hospital will have patient record management systems as well as lab systems; a manufacturing company will have product planning, financial, supply ordering, and manufacturing systems; and all organizations have an enterprise resource planning (ERP) system in one form or another. All these systems together are synonymously called enterprise application systems, information systems, or legacy application systems.

The term "legacy application system" requires some explanation. Application systems or information systems are managing the data of an organization. The structure and the meaning of data changes over time; however, one has to start at some point in time with its implementation. In the past, organizations implemented their application systems according to their needs at some point in time. Due to the changing needs, application systems require modification. However, as it turns out, modifying application systems is not that easy and, in many cases, it entails high costs and unacceptable risks. Hence, they become legacy systems. Many organizations acknowledge this fact by not modifying them anymore but introducing new ones that complement the existing ones. This is one of the reasons why enterprises have more than one application system installed.

Why do the legacy application systems in an organization have to be integrated? If application systems manage different sets of data, it seems that there is no reason for integration. However, in reality, there is overlap of the data as well as references between them. The fact that data is managed by different application systems does not mean that the data are disjoint. For example, a patient record system manages patient data. Personal data, medical history, and so on, are part of the patient record management system. The lab system of a hospital manages the different lab tests and their results. It makes sure that no test gets lost and all tests are performed to their completion. The relationship between the patient record management system and the lab management systems is that lab test are performed for patients. A patient record contains a list of lab test that have been performed, are currently being performed, and will be performed. Once a lab test is done, the patient record will be updated with the results. If the patient record and the lab test data are in different systems, then these need to be synchronized so that the patient record contains the status of the currently ongoing lab tests.

The only way for an organization to avoid the integration of its application systems is to have all necessary data management implemented in one single application system. Although this is an appealing thought, given the complexity of integration, in reality, it is not feasible due to the data complexity as well as the complexity of the software itself.

Application systems are HAD (heterogeneous, autonomous, and distributed) systems. "Heterogeneous" means that their underlying data model implements the same and similar concepts in different schemas. For example, a patient address might be defined differently in the patient record system and in the lab test system. One might have one field for the patient's address in form of a string, whereas the other might have separate fields for street, city, and country.

"Autonomous" means that the state of the application systems can change independently of each other. For example, a patient record can be added while at the same time a lab test is marked invalid. None of the two systems involved have a direct dependency that requires their direct synchronization. The updating of a record and a lab test are completely independent of each other.

"Distributed" means that the state of the application systems is not shared between the application systems. Each application system has its own storage for managing its state. Usually, database management systems are used for implementing the particular storage an application system needs.

Heterogeneity is a problem that requires careful solutions if data are to be communicated from one application system to another one. A concept extracted from one is represented in its own schema. If this is passed on to another application system, this system expects the data to be represented in its schema. If the schemas do not match, then the representation of one has to be transformed into the representation of the other. This is called transformation or mediation. Of course, the transformation must not change the semantic meaning of the data. In the example above, the address represented as one single string has to be transformed into an address in which the different parts are called out separately in different fields. Whenever an address is communicated between the systems, then transformation has to ensure schema compliance. This is applicable for both directions of communication. In practical terms, this means that the string representation of an address needs to be split exactly in such a way that the separate fields can be filled in, and vice versa.

Autonomy is a problem that needs to be addressed, too, when integrating application systems. Data are not communicated randomly between application systems, but based on state changes of application systems. For example, when an address is updated, the update is communicated. Updates happen through state changes. Whenever an application system changes its state, then the question arises whether, due to the state change, data needs to be communicated to any other application system. Since this is usually the case, state changes need to be detected in application systems and data needs to be communicated accordingly. Data sent to an application system might cause a state change in it as a consequence. Autonomy means in this context that the state changes can happen at any time in any information system independent of each other.

Distribution is less of a problem since the application systems manage their state themselves. Since this state is internal to the application systems, the independence is not visible to the integration in general. The only specific situation when the distribution becomes relevant is that of distributed transactions. If transaction control is necessary, the distribution of the application systems' state requires distributed transaction execution. This, in turn, means that information systems in this case require data management systems that are capable of participating in distributed transaction protocols. Some of these protocols are discussed in Chapter 15.

Fundamentally, EAI synchronizes the state changes of application systems. In addition to data being heterogeneous, behavior is heterogeneous, too. One state change in one application is not necessarily equivalent to one state change in another application. A state change in one can cause many state changes in the other one. The synchronization of state changes in application systems is called behavior integration.

In summary, EAI means data integration and behavior integration. Data integration requires extracting data from an application system, transforming it, and inserting it into those application systems that require knowledge of the data. The data is extracted upon state changes to an application system, and the insertion of data might cause state changes in application systems. Section 4.3 will discuss in detail the concepts of integration that can accomplish this functionality.

4.1.2   Business-to-Business Integration (B2B)

Business-to-business (B2B) integration is concerned with the exchange of electronic documents between organizations of any type, commercial, governmental, public, and so on. Electronic documents are exchanged over computer networks like the Internet or value added networks. These networks are, in general, unreliable, unsecured, and do not guarantee a specific level of service quality. Whenever documents are communicated between organizations over networks, a clear agreement has to be in place between the communicating organizations as to how to make the transmission reliable and secure and how to achieve specific quality levels of transmission.

Reliability is achieved through elaborate exchange protocols that can detect if a document has been transmitted. Based on this, a single transmission can be achieved. Security requires a set of technologies. Confidentiality and authentication can be achieved through public and private keys for encrypting the document. Integrity can be achieved with sequence numbers to avoid the loss, replay, or injection of messages. Nonrepudiation mechanisms are put in place that allow an organization to verify the receipt of a document from another organization. With all these mechanisms in place, organizations can achieve confidentiality, nonrepudiation, integrity, and authentication [4].

The structure of documents has to be described in such a way that all organizations involved in a document exchange understand the structure unambiguously. These document definitions can be in any form as long as common understanding is achieved. Recent approaches are to use XML schemas [7, 9, 2], and future approaches might be to use ontology languages like OWL [6]. This common understanding is necessary for the organizations to interpret the structure and look for the right fields in the right places within a document.

In addition to the document structure itself, the content of the document in terms of values of business data has to be agreed upon. For example, a book identifier following the ISBN definition must be known to the communicating organizations to be useful, or country codes identifying the countries of the world must be agreed upon so that all organizations involved in a communication use the same values. It would not be useful if one organization uses "Ireland" and another one expects the value to be "EIR." Fundamentally, every element of a document as defined by its structure has to have a clear definition of its possible values so that the communicating organizations know how to construct the document contents and what to expect.

Finally, the semantics of the document content has to be agreed upon. Agreeing on possible values does not guarantee the same common understanding of what the values mean. For example, a "ship date" can be interpreted in many ways. It can be the estimated date of delivery on your doorstep or it can be the date when the goods are given to the shipping company (and some days or months later it might arrive at your doorstep). Another example is the price of goods on a purchase order. That might be the price net of VAT (value added tax) or including VAT. This type of interpretation of the value is called "semantics," and the communicating organizations have to have the exact same understanding of it to avoid any miscommunication and incorrect action upon receipt of a document.

Documents are, in general, not sent in isolation as singular transmissions. Instead, usually a set of document exchanges together form a conversation between organizations. For example, sending a purchase order to a supplier causes it to return a purchase order acknowledgment that either confirms that the goods are going to be shipped or that the goods cannot be provided. This example conversation consists of two documents that both carry business data and are necessary for businesses to agree on a deal.

However, in order to achieve the single transmission of each document, "helper documents" might be necessary, like transmission acknowledgements. For example, after a purchase order is sent, a transmission acknowledgement is returned acknowledging that the purchase order has been received. At this point, no statement is made about the fulfillment of the order. The only purpose of a transmission acknowledgement is to let the sender know about a successful electronic transmission. The same applies to the purchase order acknowledgement. This also will be acknowledged by a transmission acknowledgement message. This means that, in total, four messages are sent—two with business data (purchase order and purchase order acknowledgement) and two for stating that the low-level transmissions happened without any fault. If certain aspects of security need to be ensured, even more document exchanges may be necessary.

If two organizations plan to exchange electronic documents, they have to agree on everything that has been discussed so far: low-level messaging infrastructure ensuring single, reliable, secure, and predictable transmission of documents; document structure; content; semantics; as well as the set of documents that are sent back and forth as a conversation. This is quite an effort and there is no real benefit in defining all of this "from scratch" every single time organizations decide to engage in B2B integration.

Therefore, standards have been developed over the years that specify all of the above. This means that organizations only have to state which standard they will comply to for B2B integration; they do not have to define their own documents and protocols. Examples are EDI,[2] SWIFT,[3] and RosettaNet.[4] These standards are maintained by standards organizations or groups of interested organizations for their own benefit. Changes, adjustments, and mistakes are dealt with and, once incorporated, benefit the whole community. Also, establishing B2B integration is facilitated because standards also incorporate a huge set of past experience.

4.1.3 Cooperation of EAI and B2B Processes

At a first glance, EAI and B2B appear to be very different types of integrations. One is solely concerned with integrating organization–internal application systems with a variety of proprietary interfaces, whereas the other is concerned with integration of organizations through communication over public insecure and unreliable networks according to B2B standards. However, this difference disappears when looking more closely and turns into a complementary situation.

Internal application systems implement the business logic and business process-

[2]www.unece.org/trade/untdid/welcome.htm
[3]www.swift.com
[4]www.rosettanet.org

es of an enterprise. As introduced earlier, this means to facilitate the integration of application systems by means of exchange of data between them. In addition, some of the business processes require communication with other organizations. For example, purchasing a part requires sending a purchase order to a supplier. Exactly at this point, the data become the events sent between organizations. At this point, the data have to be passed over the public networks according to defined B2B standards and according to a defined communication protocol. When the data are transmitted, then in the receiving organization the data must be processed and that is going to happen again in internal application systems.

Fundamentally, EAI and B2B cooperate in order to ensure that data are communicated not only between application systems within enterprises, but also between enterprises. Without one, the other cannot fulfill its purpose.

Going one step further, one can observe that the integration concepts required to implement EAI and B2B integration do not differ at all. The same integration concepts can model and represent both types of integration. This is the reason why in Section 4.3 the integration concepts are introduced without being attributed to either EAI or B2B. In both EAI and B2B, the different communicating entities (application systems or organizations) require data in their specific schema. This means that in both environments transformation (mediation) is required. In both cases, networks are involved that, by nature, are unsecured, unreliable, and do not provide guaranteed service levels. Not only organizations require a specific pattern of sending and receiving data; application systems also do. There, too, data are sometimes acknowledged through acknowledgement messages.

In summary, the nature of organizations and application systems are very similar when viewed from the conceptual level of integration concepts. Although this is not immediately visible when viewed from a higher level, it is visible when implementing software systems providing integration functionality.

The next section will introduce some examples before the integration concepts are introduced in Section 4.3.

4.2 EXAMPLES OF EAI AND B2B PROCESSES

Integration can assume very different forms depending on the specific situation. The following examples introduce a variety of integration problems. The examples show some specific integration solutions that can be commonly found and give some basic insight into details.

4.2.1 Multiapplication System Business Processes

In general, organizations have several information systems that must be integrated in order to implement the organization's business processes. An example is an ordering process in which a customer orders a product that has to be custom manufactured. A series of information systems has to play well together to achieve this, as follows.

A salesman finds a new customer who wants to buy a product that needs to be custom built for him. The first activity of the salesman is to create a new customer in the customer management system. The entry of the new customer automatically triggers a credit check request that must be performed. Only upon a successful credit check can products be ordered by the new customer. Credit checking is not done in-house but requires contacting a credit bureau. Therefore, once the new customer is created, an event is sent to the credit bureau asking for a credit check. The credit bureau returns the result. The result is given to the customer management system and, depending on the outcome, the new customer is rejected or accepted. The credit bureau requires data in the form of a financial standard. This means that the organization has to transform its data into the standard's definition. Upon receiving the credit check result, the transformation process has to transform the data back into the format required by the customer management system.

Once the credit check result has been sent back by the credit bureau, the credit bureau sends an invoice to the organization asking for payment of the credit check service used. The organization has to acknowledge the invoice and initiate payment with its bank. This means that the credit bureau sends a message to the organization, which then sends a payment notice to its bank. The bank, in turn, acknowledges the payment. In both cases, financial standards are used as the underlying definition of the data sent.

If the new customer is confirmed and the credit check went well, the salesman talks to the new customer and gets the specifications of the product the customer would like to order. The salesman creates a new order for the new customer in the order management system by typing in the customer identification as well as the product specification. The order management system contacts the customer management system first to check for the existence of the customer. This requires a message sent to the customer management system and a message back acknowledging the existence of the customer. No external organizations are involved in this and so no specific standard has to be used for this communication. Once the existence of the customer is confirmed, the order management system checks the consistency of the product specification. Assuming it is consistent, the order management system has to be integrated with two other systems. One is an engineering system that can draw up the engineering order and the other is the planning system that plans the production and comes back with a production and delivery schedule. Again, the integration with these two systems is in-house and does not require any external communication with another organization. Once the engineering order is drawn up, the salesman gets a notification that it is possible to build the product. Once the planning system returns a result, the salesman is notified about estimated production and delivery dates that he can share with the customer.

From the viewpoint of the order system, no external communication took place (the engineering systems and the planning system are both internal application systems), but the planning system actually communicates with an external organization. Since the product is not built in-house, potential suppliers of the product have to be involved for planning purposes. So in order for the planning system to provide a good schedule to the order management system, it contacts potential suppliers to find out what production schedules they could entertain. In this case, the external integration was hidden from the order management system.

The example stops here although, in reality, the business process would continue with a confirmation by the customer that he wants the specified product. An advance payment of a certain amount may have to be deposited so production can start. The supplier of the product has to be selected, shipping arranged for, and final payments made by the customer and to the supplier.

4.2.2 Business Data Replication

The example in Section 4.2.1 has clearly shown that both information systems as well as external partner organizations provide functionality required to implement a rather complex business process. The integration between these is actually part of implementing the business process itself, making sure that the right application system or organization is contacted at the right time, and sending and receiving the correct data according to the agreed upon standards and protocols. From a data viewpoint, all systems and organizations exchange data for advancing their state and that of the business process.

A different type of integration is business data replication. In this case, the information systems and organizations are not integrated to implement a business process, but to synchronize replicated data that all of them have to have. The requirement is that all the data one has must also be stored at all the others locations. Furthermore, any change in one system's or organization's data must be propagated to all those that have the same data item.

The example in this section is that of customer data synchronization. This extends the example from Section 4.2.1. The salesman is responsible for a certain area; let's say EMEA (Europe, Middle East, and Asia). In addition, a second salesman is also responsible for the same area and both synchronize well. A third salesman is responsible for the United States alone. Both EMEA and the United States have separate customer management systems. Each of the salesmen uses the systems of its area. However, one specific situation is that multinational customers have offices located in both sales areas, EMEA and the United States. Based on sales volume, discounts apply. In order to know if one company orders in both sales areas, it was decided that each customer that is entered in one of the customer management systems has to be automatically entered in the other customer management system. This means that once a customer is entered in EMEA, it has to be automatically replicated in the U.S. system.

What happened "behind the scenes" is that once the salesman entered the new customer in the example of Section 4.2.1, and once the customer was accepted based on a successful credit check, the customer record was sent through the integration process to the U.S. system and automatically entered there, too.

The integration process has to fundamentally monitor both customer management systems. Whenever a new customer is confirmed or an existing customer changes, then the change in data has to be replicated in the other system. The same must happen with deletions, of course.

4.2.3 Request for Quotation with Incomplete Responses

The examples in Sections 4.2.1 and 4.2.2 were of point-to-point integrations. Two information systems or two organizations were directly communicating with each other and each event sent resulted in an event being sent back.

In specific circumstances, other integration patterns exist. This is best introduced by extending the example in Section 4.2.1. The planning system has to contact other supplier organizations and find out if they can produce the specified product and what the schedule for it will be. Later in the process, when the customer agrees to purchase the product, a product price has to be determined, too. Therefore, the suppliers have to be contacted and asked them for a quote in addition to a realistic production schedule. One possible way to accomplish this would be to contact each supplier in turn, waiting for a response from each of them. However, requesting a quote does not mean that a supplier must respond with one. A supplier might not respond at all.

So, instead of sending a request for quote separately to each supplier, a request for quote is broadcast to all possible suppliers. A deadline is set for accepting the quotes. Quotes coming in after that are rejected. Therefore, the organization waits up to the deadline and collects all quotes. Once the deadline is reached, the quotes are returned as a set of quotes and the selection process can continue.

From an integration perspective, a broadcast took place and there were an unknown number of responses. The maximum number is the number of potential suppliers. However, the exact number is not known until the deadline is passed. This pattern of integration is called "scatter–gather" in [8].

4.2.4 Purchase Order Update

Another property of the example so far was that it was "forward progressing" in the sense that the business process was constructively executed in order to achieve the goals of selling a product (and making money in the end). However, sometimes decisions have to be revised, causing parts of a business process to be repeated.

In terms of the example, it might be that the customer agreed on the product price and delivery date. Consequently, the downpayment was made and the supplier was formally dispatched a purchase order based on the quote he provided. The supplier scheduled the production of the product and things progressed as expected.

What can happen is that the customer requires a change in the product specification due to some late discovery of new or changed requirements. In this case, the customer contacts the salesman and he, in turn, accesses the order management system to organize the change. The interesting part in this case is that the impact of the changes is different depending on the progress of the overall order. The extreme cases are as follows. In the simplest case, the purchase order has not yet been dispatched to the supplier and the change is not going to change either price or schedule. In the worst case, the product has been manufactured and has already been shipped. The change requires the building of a new product and all negotiations have to start from the beginning. Of course, there can be cases in between these extremes. For example, the purchase order has been dispatched and the manufacturing has been scheduled but did not start yet. In this case, a change can be implemented relatively easily. The schedule will be cancelled and the modified product rescheduled.

The important issue here is that the part of the business process that needs to be executed again depends on its state. In the example, in one case the whole business process has to be repeated from the beginning, whereas in the other case only an updated engineering order has to be sent.

In summary, the examples introduced provided a glimpse of the integration problems that require solutions. In [3], many more cases and details are discussed and this reference is recommended for the reader who requires more insight.

4.3   CONCEPTS, ARCHITECTURES, AND TOOLS

Modeling integration is an involved task due to the complexity of the problems involved. Looking back to the example in Section 4.2.4 (purchase order update), it is apparent that expressive integration concepts have to be in place in order to be able to define the real situation completely. This section will introduce the most important concepts in detail; a complete set can be found in [3].

4.3.1   Integration Concepts

In the following, we introduce the individual integration concepts one by one.

Endpoint.   An endpoint is either an information system or an organization. A user usually interacts through a user environment. This is subsumed under information systems. In this regard, a user can easily be part of any integration, be it B2B or EAI. In summary, information systems, organizations, as well as users are all equally modeled in the different types of integration through the concept of the endpoint.

An endpoint has a unique identifier and it is known if an endpoint is an organization, user, or information system. With each of these three subclasses of endpoint comes more detailed meta-data like phone numbers, public keys, and so on. This meta-data is used at run time, for example, when data needs to be encrypted with the public key of the receiving endpoint.

Event.   An event carries data and is either sent to an endpoint or received from an endpoint. For example, "create purchase order" or "check customer credit" are events. In the example above, the customer management system sends a "check customer credit" event to the credit check bureau that receives it. The return event carrying the result is the "notify result customer credit check" and it is sent back from the credit check bureau to the customer management system.

For each event, the syntax and semantics as well as the possible value of the various fields have to be defined. For example, in the "check customer credit" event, the name and address of the new customer must be specified as well as the upper amount of credit to be checked.

Events have a source and target endpoint to specify from which endpoint they are sent and to which endpoint they have to go. The integration infrastructure accesses this meta-data in order to deliver the events as requested.

Interface Process (also called Public or Abstract Process). Endpoints expose a specific interface behavior. This means that they send and receive events in particular sequences. For example, the credit check bureau's interface behavior is that whenever they receive a "check customer credit" event they return a "notify result customer credit check" event. The interface process for this endpoint defines this pattern. Another interface process of the credit bureau is the payment process, in which they send a "pay invoice" event and expect back a "payment completed" event from their bank. Of course, the customer management system has its own interface processes. It sends out the "check customer credit" event and waits for the "notify result customer credit check" event. This interface process is the complementary interface process, so to speak.

Interface processes define the publicly visible behavior of an endpoint. They define which events an endpoint expects, which events an endpoint sends out, and the order in which they occur. Also, some event might be optional and only sent or received in a specific case. The internal behavior of an endpoint is defined in another type of process, discussed later on. This "private" process is not visible from outside the endpoint and implements the business logic.

Integration Process. An integration process defines the business logic of an endpoint, that is, the internal behavior of it. For example, once the credit bureau receives the "check customer credit" event, it retrieves the customer data and the credit limit from this event and starts determining the credit limit. The extraction as well as determining the credit limit is the internal process of the credit bureau. This represents the knowledge of this company and is private to it. From the outside, only the interface process is visible, not the integration process. This scope limits the visibility and allows endpoints to keep their internal logic hidden from the public.

Integration processes operate on events, as do interface processes. Whereas interface processes are concerned about the exchange sequence of events from an interface viewpoint, integration processes are concerned about implementing the business logic. Integration and interface processes, therefore, have "touch points" at which events are handed over. Once an interface process receives an event, it provides it to the appropriate integration process. Once an integration process requires the sending of an event, it hands it over to the appropriate interface process. In this sense, interface and integration processes are cooperating processes that are defined independently, but are synchronized during execution time.

BPEL (see Chapter 14) is an example of a language intended to describe both interface and integration processes (which in BPEL are called abstract and executable processes, respectively).

Transformation. Transformation is the definition of rules that transform one event into another event according to their syntax, values, and semantics. A transformation allows mediation between the event endpoints, send and receive, without modifying the endpoints themselves. An example would be the transformation of an address represented as a string to an address represented as individual fields.

Transformation can be used in either integration or interface processes for transforming events from one schema to another. However, the most appropriate place to use transformations is between the interface and integration processes. This allows an integration process to be defined based on a schema and an interface process to be defined based on another schema. The transformation between the events of the two schemas is between the integration process and the interface process.

Now that the main integration concepts have been introduced, the example in Section 4.2.1 can be represented in terms of these concepts. Figure 4.1 shows part of the example graphically. The UML notation is used (see Chapter 5 for details). The dashed arcs indicate the message flow, whereas the solid arcs indicate control flow between activities. Each back-end application system or each business partner has one or more interface processes defining its external behavior. The integration processes are between the interface processes relating individual messages of those processes with each other. A diamond-shaped activity denotes transformation between different syntactic as well as semantic message representations.

The integration in Figure 4.1 integrates the customer management system as a back-end application with the credit check bureau as a business partner. This means that an application and an external business partner are integrated. The flow of messages in this example is fairly simple—a request message is sent out and a response message is received. According to the example, the message sent to the business partner has to follow a particular financial standard. This requires a transformation from the internal representation as provided by the customer management system to the financial standard. The diamond shapes represent the transformation and the resulting messages are in the appropriate representation. Also, the figure indicates that the interface process of the customer management system and the integration process operate on the same message representation, as there is no transformation between them.

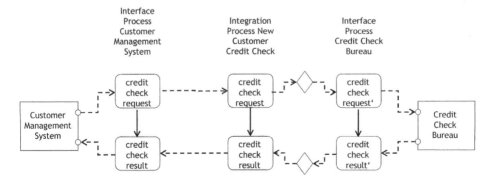

Figure 4.1   Example defined with integration concepts.

According to the example, once the credit check result is received by the customer management system, the salesman will see if the new customer is accepted or not, depending on his credit check.

Figure 4.2 shows the payment process for the customer credit check. Several additional features of the integration concepts introduced above can be observed. First, three systems are integrated: an ERP, the credit check bureau, and a bank. Second, each of these three systems has its own interface processes that are integrated by the integration process. As can be seen, there are three general communications going on. (1) is the invoice communication in which the invoice is communicated. (2) is the payment initiation and (3) is the payment notification. These three integrations are separate in themselves and only the integration process orders them appropriately.

Furthermore, for the payment initiation, the transformation takes place before the integration process, as in this situation the ERPs format must be transformed before the integration process.

4.3.2   Integration Architectures

In general, two types of integration architectures can be distinguished: component-based and holistic. The component-based architecture implements integration by

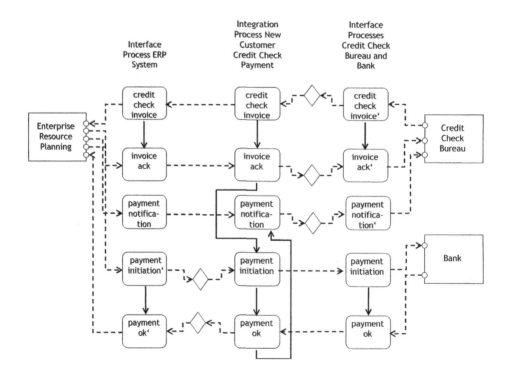

Figure 4.2   Example defined with integration concepts.

combining base technologies like databases, workflow management systems, queues, XML processors, XSLT processors, security systems, and so on. Therefore, the integration architecture internally is a mix of existing smaller components, each with its own architecture. Of course, these individual technology components have to be integrated themselves. Therefore, "glue" code in form of the integration logic has to be developed to define how these components work together in order to achieve integration. Figure 4.3 gives an example of a component-based integration architecture.

In Figure 4.3 the components are displayed individually. Events are produced and consumed by the integration architecture through the transport system component. From then on, the integration logic ensures that the individual components are invoked in the correct order. It picks up the events from the transport system and invokes the next component. This might involve putting the event in a queue to establish persistent storage and maintain the order of receipt. Then it checks with the workflow management system to see if the incoming event is expected from an existing interface process or if a new interface process has to be launched. This might involve the transformation system in order to transform the event before passing it to an integration process for internal processing. The sending out of an event by an integration process follows the opposite order of processing.

This approach of architecture has severe downsides. One is that each component has its own conceptual model and its own implementation style and persistent storage. This means that a constant rerepresentation takes place between the integration concepts and the concepts of the components. At the same time, none of the components implements the integration concepts directly. The integration logic layer has to implement the integration concepts by using the components to give the appearance that the integration architecture is genuine.

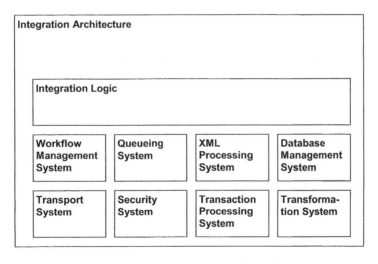

Figure 4.3 Component-based integration architecture.

The alternative architecture is the holistic style. In this type of architecture, all components are developed directly following the integration concepts. No integration logic layer is necessary since the components directly implement the integration concepts. Figure 4.4 shows a graphical view of this type of architecture.

In the holistic architecture style, two types of components can be distinguished. One set is the integration-specific set containing components like event management, interface process management, integration process management, endpoint management, and the transformation system. These components directly implement the integration concepts introduced earlier and, therefore, no rerepresentation is necessary as in the component-based architecture. Furthermore, it can be expected that an implementation is more efficient and correct since all these components can share the same representations.

The other set is a set of base technology components like security, transaction processing, or transport that are independent of the integration concepts and have to be in place for specific system-level functionality. This separation is indicated in Figure 4.4 as two different layers.

4.3.3 Integration Systems and Tools

A large number of commercial systems providing integration functionality are offered as products in this area. Each of these integration systems is based on a very different model of integration so that a direct comparison is not possible. Also, many commercial systems do not provide sufficient documentation publicly to analyze their concepts and architecture. Exceptions are systems from BEA, IBM, Mi-

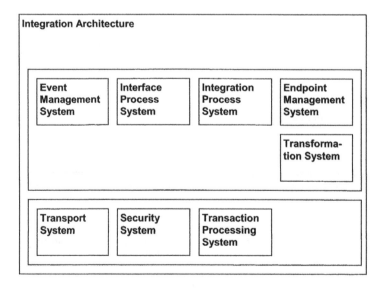

Figure 4.4 Holistic integration architecture.

crosoft, and Oracle. These companies provide the product documentation publicly on their Web sites so that it is possible to read about their functionality and architectures. BEA's product is called WebLogic Integration,[5] IBM's family of products is called WebSphere,[6] Microsoft's product is called BizTalk Server,[7] and Oracle's product is called Oracle Integration.[8]

The description of each of the products would require a chapter by itself since all would have to be discussed in sufficient detail to show the precise functionality and the differences. In [3] these and others are described in more detail.

4.4 FUTURE DEVELOPMENTS

Currently, integration solutions are either based purely on workflow management systems or are systems put together based on middleware components like message queues, databases, and XML processing. However, none of the integration solutions is built on a holistic integration meta-model that provides all modeling concepts necessary in order to model integration. Concepts as introduced in Section 4.3.1 and in more detail in [3] are not explicitly represented. The database management systems' schemas underlying the integration solutions are not based on integration concepts and holistic integration models but persistently stored for execution purposes. This situation leads to integration solutions that are difficult to operate and hard to understand. Defining and executing integration becomes very cumbersome in the absence of a conceptual integration model. Instead, different technologies are put together and the modeler has to deal with each of those separately.

Research efforts are ongoing that promise to address this situation by providing explicit representations of integration concepts. An effort ongoing for some time is DAML-S (DARPA Agent Modeling Language for Web Services).[9] It is based on the ontology language OWL [6]. A more recent effort was launched in Europe called WSMO (Web Service Modeling Language).[10] This effort has a wider scope than DAML-S. A specific language, the Web Service Modeling Language (WSML), was developed in order to define Semantic Web Services and integration.[11] Finally, an interpreter for WSML has been built, called Web Service Modeling Ontology Execution (WSMX).[12] WSMX allows one to compile WSML-defined Web Services and executes them according to their semantic definition.

These efforts not only take the approach of defining a complete set of integration concepts but also use ontology languages for the definition of the concepts them-

---

[5]http://bea.com/products/weblogic
[6]http://www.ibm.com/software/websphere
[7]http://www.microsoft.com/biztalk
[8]http://otn.oracle.com/products/integration
[9]www.daml.org/services
[10]www.wsmo.org
[11]www.wsmo.org/wsml
[12]www.wsmx.org

selves. This means that at least the concepts are defined in semantically well-defined languages. As underlying technology, both use Web Services [1].

Commercial implementations based on these new efforts are not yet in place, but can be expected soon. It is possible to see the advantages of the approach to implement integration solutions based on semantically well-defined holistic integration concepts.

4.5 EXERCISES

A few exercises are provided that allow the reader to become more familiar with the introduced concepts in EAI and B2B integration. The exercises' problems can be found often in real integration projects and are, therefore, as realistic as possible. The reader is encouraged to solve the problems and is invited to discuss possible solutions with the author.

4.5.1 Integration State Inconsistency Resolution

The first part of this exercise requires an integration process that synchronizes states of several different information systems.

- Assume three information systems, each with the same customer address data (system 1, system 2, and system 3).
- The systems are connected by a B2B integration system. Any communication or data exchange between the three systems is accomplished by the B2B integration system.
- Users can update customer addresses in each of the three information systems independently of each other. This means that a user can update one customer address in system 1, another user can update a customer address in system 2, and yet another user can do so in system 3. All these updates are independent of each other.
- Because of this possibility of independent updating, the list of customer addresses can start to differ between the three information systems.
- An integration process is needed (with appropriate interface processes) that is executed from time to time and that ensures that the customer address lists in the three information systems are identical. Whenever one list has a customer address that the others do not have, the address is added to the others. What does such an integration process look like?

The second part asks for a more elaborate integration process that takes deletions of customer addresses into consideration.

- Users can delete customer addresses at any time. If a user deletes a customer address in one of the information systems, the goal is to also have the address deleted from the other two information systems.

- The integration process from the first part of this exercise, however, will reintroduce the customer address (why?). If an address is missing in one information system (because of the deletion) it will be added again since the other two systems still have it in their list of customer addresses.
- An integration type is needed that detects that an address was deleted and will delete it from all the other information systems. What does such an integration process look like?

The third part of this exercise deals with updates of customer addresses.

- In addition to inserting and deleting customer addresses, an address can be updated. The integration process from the first part will not necessarily ensure that a change in a customer address will be propagated to the other two information systems (why not?).
- An integration process is needed that detects customer address updates and propagates the changes to all the other information systems. What does such an integration process look like?

The fourth and final part of this exercise deals with a varying number of information systems that require customer address synchronization.

- In enterprises, the number of information systems can increase and decrease over time. An additional information system should not require integration processes to be changed. Instead, integration processes should be modeled in such a way that adding or removing information systems will be independent of the particular business logic.
- An integration process is required that can support the synchronization of customer addresses between any number of information systems.

Hints:

- It is left to the reader to construct either one integration process for each of the four parts of the exercise or a single one that provides all functionality. However, if four different integration processes are defined, there might be considerable duplication in functionality.
- Detecting change and updating might be accomplished in many ways. An information system might send a customer address change event to the B2B integration system, it might maintain version numbers of addresses, or it might maintain time stamps or other mechanisms that allow the detection of modifications. In the worst case, it would not maintain any meta-data and the integration process have to find out itself if a deletion, update, or insert took place. A sophisticated integration process can deal with all mentioned forms of modification detection.
- What was not discussed so far is the situation in which customer addresses can change while the integration process is synchronizing the information

systems. In this situation, the update and the synchronization are parallel activities. If this concurrency happens, then the integration process might not operate correctly. Please check your designs to see if they work under this concurrency assumption.

4.5.2   Master/Slave Data Update

The previous exercise showed how difficult it is if two or more information systems update their internal data autonomously. The difficulty is that the integration process has to check the internal state of each of the information systems and act on it. This situation can be drastically improved if updates of data are restricted to one information system (acting as a master) only. Based on its state, all the other information systems are updated in turn (acting as slaves).

An integration process is required that will recognize one information system as the master. Changes in the master internal data will be picked up by the integration process, which will then update all other information systems as needed.

Hint: this integration process is very similar to those of the first exercise except that it listens only to one information systems instead of all of them.

4.5.3   Application Integration Extension

After the integration process has been implemented as in the previous exercise, the synchronization of the information systems works flawlessly. After some time, however, the company merges with another company. As it turns out, the newly acquired company also has successfully implemented a master/slave integration process internally. Now there are two environments that require integration. You are put in charge of combing the two environments. Fundamentally, you have two sets of information systems and each works perfectly with a master/slave integration process in operation. The task at hand is now to integrate the two with the goal of having a perfectly synchronized set of information systems again. Alternatively, you could implement a third master/slave integration process that replaces the two existing ones. The choice is up to you.

- One possibility is to redefine one of the two master/slave processes in such a way that the master becomes a slave. Then there will be only one master remaining.
- However, the acquired company is in a time zone 12 hours away and they have to update the customer addresses during their working hours. The solution has to be adjusted so that two masters exist, one in each time·zone. This requires the integration of the two separate mater/slave integration processes so that both are cooperating to synchronize the information systems.

Hint: for each master, all the other information systems are slaves. So, depending on the master, the other master takes the role of a slave.

4.5.4 B2B Process Termination

Imagine that your company and one of its suppliers exchange events for quite some time successfully. An integration process as well as the interface processes are set up and run successfully for a while. However, recently problems emerge with the termination of the integration processes. For some unknown reason, it happens more and more often that the processes with your supplier do not terminate any more but stop processing in the middle of the execution. You suspect a deadlock situation, but you cannot confirm this yet. Whenever you start the processes again, they execute fine for a while, even on the same datasets. However, terminating the stopped processes is quite difficult and restarting them is not easy either. You need an easier way to do this until you find the reason for the stopping of processes.

- The integration processes have to be modified so that it is possible to terminate them by sending an event to them. Add process logic to one of the previous exercise results that can receive a "terminate" event. Upon receiving the terminate event, the integration process must be terminated.
- A further improvement of this change is to restart the same process before the stopped process is terminated. The stopped process starts a separate integration process based on the same start event.

After this addition, it is possible for you to send a terminate event to a stopped process that also causes a new integration process to be started based on the same start event. This allows you to keep your business going while you can continue to figure out the cause of the problem.

4.5.5 Event Mediation

Imagine that you are working in the travel industry and have linked a travel agency with a hotel provider. Both are using different standards for defining the event types and you have successfully implemented the required event mediation. More specifically, you implemented a hotel room request event that is sent by the travel agency to reserve a room for a person at certain dates in a specific hotel.

The hotel now upgrades to that 5-Star level and, as a consequence, different types of rooms are introduced. The hotel notifies you that they will upgrade their reservation system soon to reflect the different types of rooms. This means that they will change the hotel room request event definition by adding one more field that specifies the room type. This in consequence means that you have to change the event mediation rules so that the hotel room request event contains a valid hotel room type.

- In this short time frame, it is not possible to change any reservation system the travel agency uses. Therefore, it is not possible to ask the customers for their preferred room types. A new set of mediation rules is required that populates the room type field of the hotel room request event.

- However, as it turns out, the travel agency is willing to upgrade their reservation system in order to reflect the room types. But they only allow you to use existing fields like a comment field for achieving this. What do the mediation rules look like in this case?

4.5.6   Trading Partner Extension

The travel agency from the previous exercise decides to engage with another hotel chain to improve their services for their customers by providing a bigger selection of hotel rooms. This additional hotel chain is also at the upper end of the scale and has different room types. However, the room types of this additional hotel chain are different than those of the already existing hotel provider.

- In this situation, two different solutions are possible. One solution is to best-effort map the existing room types to the new room types. A travel agent serving a customer only has one set of room types available for selection. The mediation takes care of mapping those to the appropriate ones for the different hotel chains. However, this best effort might not reflect the real situation in terms of the room types.
- The other solution would be to teach the travel agents that different hotel chains have different room types and, consequently, they have to type in the appropriate room types in the comment field. What would the mediation rules look like in this case?

REFERENCES

1. G. Alonso, F. Casati, H. Kuno, and V. Machiraju. *Web Services Concepts, Architectures and Applications,* Springer-Verlag, Berlin, 2004.
2. P. Biron, and A. Malhotra, (Eds.). XML Schema Part 2: Datatypes. W3C Recommendation, 2001.
3. C. Bussler, *B2B Integration.* Springer-Verlag, Berlin, 2003.
4. W. Caelli, D. Longley, and M. Shain. *Information Security Handbook.* Macmillan, New York.
5. Data Interchange Standards Association, Inc. *ASC X12 Workbook.* 2004. Available from https://webster.disa.org/apps/workbook (restricted access).
6. M. Dean and G. Schreiber (Eds.). OWL Web Ontology Language Reference. W3C Recommendation, 2004.
7. D. Fallside (Ed.). XML Schema Part 0: Primer. W3C Recommendation, 2001.
8. G. Hohpe, and B. Woolf. *Enterprise Integration Patterns.* Addison-Wesley, Reading, MA, 2003.
9. H. Thompson, D. Beech, M. Maloney and N. Mendelsohn (Eds.). XML Schema Part 1: Structures. W3C Recommendation, 2001.

PART II

MODELING LANGUAGES

CHAPTER 5

Process Modeling Using UML

GREGOR ENGELS, ALEXANDER FÖRSTER, REIKO HECKEL,
and SEBASTIAN THÖNE

5.1 INTRODUCTION

The Unified Modeling Language (UML)[1] is a visual, object-oriented, and multipurpose modeling language. Primarily designed for modeling software systems, it can also be used for business process modeling.

Since the early 1970s, a large variety of languages for data and software modeling like entity-relationship diagrams [2], message sequence charts [5, 10], state charts [9], and so on, have been developed, each of them focusing on a different aspect of software structure or behavior. In the early 1990s, object-oriented design approaches gained increasing attention, for instance, in the work of James Rumbaugh (Object Modeling Technique or OMT [21]), Grady Booch [1], and Ivar Jacobson [12].

The UML emerged from the intention of Rumbaugh, Booch, and Jacobson to find a common framework for their approaches and notations. Furthermore, the language was also influenced by other object-oriented approaches like that of Coad and Yourdon [3]. The first version, UML 1.0 [20], was released in 1997 and accepted as a standard by the Object Management Group (OMG)[2] the same year. The OMG, which took over the responsibility for the evolution of the UML from then on, is a consortium from both industry and academia and is also responsible for other well-known initiatives like CORBA, MDA, and XMI. OMG specifications have to undergo a sophisticated adoption process before being agreed upon as a standard by the OMG members. Since many important tool builders and influential software companies are involved in the OMG, UML has quickly been accepted by the software industry, especially since version UML 1.3 appeared in 1999. When writing this book, the current UML version was UML 2.0 [18], a major revision of the language.

UML is a conglomeration of various diagram types. Therefore, the challenge is to provide a uniform framework for all these heterogeneous diagram types and accounting for relationships between them. In UML, this is solved by a common *meta-model* that formally defines the abstract syntax of all diagram types. The

[1]www.uml.org
[2]www.omg.org

Process-Aware Information Systems. Edited by Dumas, van der Aalst, and ter Hofstede
Copyright © 2005 John Wiley & Sons, Inc.

meta-model is defined with the help of the OMG Meta-Object Facility (MOF) [16]. Such a declarative meta-model is an alternative to grammars usually used to define formal languages.

Besides the meta-model and a notation guide defining a concrete syntax for the meta-model elements, the UML specification also informally describes the meaning of the various meta-model elements. In the past, this informal semantics description has raised many issues about how to interpret certain details of the language. Even in the latest revision, UML 2.0, there are still a number of contradictions and ambiguities to be found in the specification. At some points, the UML 2.0 specification is intentionally left incomplete, providing so-called *variation points* that allow tool builders and modelers to interpret the language according to their specific purposes.

This chapter provides an introduction to UML, focusing especially on those parts relevant for process modeling. It covers five major aspects of process models, namely (1) actions and control flow, (2) data and object flow, (3) organizational structure, (4) interaction-centric views on business processes, and (5) system-specific process models used for process enactment. Although not every detail of the language can be presented, we intend to provide at least the most important concepts required for UML-based process models.

For discussing the various process modeling aspects, we use *activity diagrams* as fundamental tools for process modeling with UML. Section 5.2 explains the control flow concepts of activity diagrams, and Section 5.3 extends the process models by integrating object flow. Aspect (3), the modeling of underlying organizational structures, is covered by Section 5.4 with the help of *class* and *object diagrams*. Section 5.5 then covers aspect (4) and deals with a different view of business processes, focusing more on the interactions among involved business partners. To model such an interaction-centric view, we introduce *sequence diagrams*. To facilitate process enactment according to aspect (5), system-specific models should describe how to relate existing software components to the desired process activities. Thus, Section 5.6 introduces *structure diagrams* for describing available software systems and for specifying provided operations, which are then integrated into the considered process models. The chapter is concluded by a summary and exercises of varying degrees of difficulty.

Throughout the chapter, the different diagram types are illustrated by a running example that deals with an e-business company selling hardware products. For simplicity reasons, the company's product range is limited to monitors and computers only. It processes incoming orders by testing, assembling, and shipping the demanded products.

5.2 MODELING CONTROL FLOW WITH ACTIVITY DIAGRAMS

The basic building block of a process description in UML is the *activity*. An activity is a behavior consisting of a coordinated sequencing of actions. It is represented by an *activity diagram*. Activity diagrams visualize sequences of actions to be performed, including control flow and data flow. This section deals with the control flow aspect of process models in UML.

5.2.1   Basic Control Flow Constructs

Figure 5.1 shows a first small example of an activity. This activity describes a business process of our exemplary e-business company, which sells computer hardware products. The activity is visualized by a round-edged rectangle. If the activity has a name, it can be displayed in the upper-left corner of the rectangle. The name of the example activity in Figure 5.1 is "Sell computer hardware." Inside the activity rectangle we find a graphical notation consisting of nodes and edges that represents the activity's internal behavior. There are two kinds of nodes to model the control flow: *action nodes* and *control nodes*.

As a first step in the formulation of a business process, we need to model what tasks the process has to perform while executing. In an activity diagram, this is described by *actions*. An action stands for the fact that some transformation or processing in the modeled system has to be performed. Activities represent the coordinated execution of actions. Action nodes are notated as round-edged rectangles, much like that of an activity, but smaller. Actions have names that are displayed inside the action symbol, for instance, "check order" or "get products" in our example. Actions can manipulate, test, and transform data or can be calls to another activity. What has to be done when executing an action can be described by the name of the action such as "check order." Actions can also be specified using programming language expressions such as c:=a+b or formal expressions. The execution of actions takes place over a period of time.

Actions need to be coordinated. This coordination of actions within an activity is expressed by control flow edges and control nodes. The most fundamental control structure is the *sequence,* in which one action can start executing when another action stops executing. A simple example of a sequence of actions can be seen in Figure 5.2. The arrows between the action nodes are called *activity edges* and specify the control flow.

In UML 2.0, the semantics of activities are defined based on token flow. Tokens can be anonymous and undistinguishable; in that case, they are called *control to-*

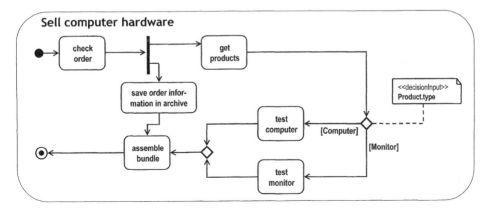

Figure 5.1   First example: computer hardware sales.

Figure 5.2 Sequence of actions.

kens. Tokens can also reference data objects. These tokens are called *object tokens.* See Section 5.3 for an introduction to the concept of object flow.

Tokens flow along the control edges, thus determining the dependencies in the execution of the actions. Actions can only begin execution when tokens are available from all preceding actions along the incoming edges (step 1 in Figure 5.3). When the execution of the action starts, all input tokens are consumed and removed from the incoming control flow edges simultaneously (steps 1 and 2 in Figure 5.3). After completion of the action, tokens are offered to all outgoing edges simultaneously (step 3).

In a control flow, actions sometimes have to be executed alternatively depending on conditions. This corresponds to the control structure often called "XOR-split" or "simple choice" (see Chapter 8), which is represented in activity diagrams by decision nodes, merge nodes, and guards. The diamond symbol in Figure 5.4 represents a *decision node* if one edge enters the node and multiple edges leave it. In the opposite case, if multiple edges enter the diamond symbol and one leaves it, it is a *merge node,* which corresponds to an "XOR-join." Diamond symbols with both multiple edges entering and multiple edges leaving them are combined decision and merge nodes.

In order do describe the conditions for the choice of the alternative control flows, the edges leaving a decision node are usually annotated by *guards.* Guards are logical expressions that can differentiate true from false. They can be formulated using natural language, programming language constructs, or formal expressions such as mathematical logic or OCL. OCL stands for Object Constraint Language [17], which was also developed by the OMG. It is a language for describing constraints whenever expressions over UML models are required. In an activity diagram, guards have to be enclosed in square brackets. An edge can only be traversed if the guard attached to that edge, if any, is true.

If a guard expression becomes very lengthy, one can also attach a ≪decisionInput≫ note box to the diamond containing the text of the guard condition. This

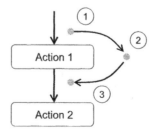

Figure 5.3 Token flow.

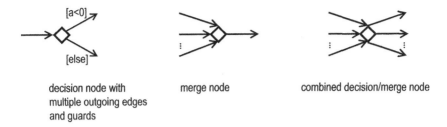

| decision node with multiple outgoing edges and guards | merge node | combined decision/merge node |

Figure 5.4   Decision node notations.

note box is connected to the decision node with a dashed line, as in Figure 5.1. In the example, a product is either a computer or a monitor. As there exist two different test facilities for monitors and computers, the control flow has to be split into two different alternatives.

A special case of a guard is [else], which is true if and only if all other guards on all other edges leaving the same node are false. The use of guards is not restricted to edges leaving decision nodes. As a general rule, control edges can only be traversed if their guard conditions are true.

In process models, one frequently has to model concurrent control flows. Concurrency in activity diagrams can be expressed by using fork and join nodes. They are equivalent to the concept of "AND-splits" and "AND-joins" described in Chapter 8. A thick-lined bar is a *fork node* if one edge enters it and multiple edges leave it, as in Figure 5.5. At a fork node, the control token becomes duplicated and the control flow is broken into multiple separate control flows that execute in parallel. In order to simplify the model, one can also draw multiple outgoing edges leaving an action node (implicit fork). In our example in Figure 5.1, the action "save order information in archive" can be executed in parallel with the action "get products" and the product tests, as indicated by the fork node.

A *join node* is used to combine the concurrent control flows. It is represented by a thick-lined bar with multiple edges entering it and one edge leaving it. It synchronizes the control flows at the incoming edges since the execution is stopped until there are tokens pending along all incoming edges. A thick-lined bar with multiple incoming and outgoing edges is a combined join and fork node, as depicted in Figure 5.5. Actions with multiple incoming edges represent implicit joins as the action

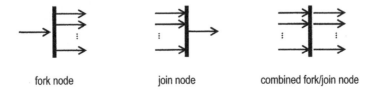

| fork node | join node | combined fork/join node |

Figure 5.5   Fork/join node notations.

Figure 5.6   Action with multiple incoming and outgoing edges and implicit fork/join.

"assemble bundle" in our example in Figure 5.1. Figure 5.6 shows an action with implicit fork and join.

In Figure 5.1, there are two more control nodes. A solid circle indicates an *initial node,* which is the starting point for an activity. A solid circle surrounded by a hollow circle is the *final node,* indicating the end of the control flow. It is possible to have more than one final node in one activity. In that case, the first final node reached stops all flows in the activity. A detailed analysis of control structures in workflow models can be found in [13].

5.2.2   Advanced Concepts

Pre- and Postconditions.   In process models, it is often required to formulate assertions and conditions that need to hold locally at certain points in the control flow, at the overall beginning of an activity, or at its end.

In order to express global conditions for an activity, the activity can be constrained with pre- and postconditions. Whenever the activity starts, the precondition is validated. Whenever the activity ends, the postcondition has to be fulfilled. Both pre- and postconditions are modeler-defined constraints. They are indicated by the keywords ≪precondition≫ and ≪postcondition≫, typically in the upper part of an activity box, as in Figure 5.7a.

Local pre- and postconditions can be attached to actions. They are displayed as note boxes containing the keywords ≪localPrecondition≫ or ≪local-Postcondition≫, as in Figure 5.7b. A token can only traverse an edge when it satisfies the postconditions of the source node, the guard condition for the edge, and the precon-

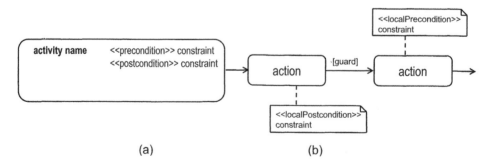

Figure 5.7   Pre- and postconditions.

ditions for the target node all at once. The constraints can be formulated in natural language, programming language expressions, or any formal language like OCL, mathematical logic, and so on.

Hierarchical Process Composition. Business processes can easily become very complex. It is advantageous for a process description language to allow hierarchical nesting in order to reduce the complexity. Thus, actions as part of a UML activity can be calls to other activities. The nesting of activities results in a call hierarchy in which activities can be found on different levels of abstraction. An action that calls another activity is symbolized by a hierarchy fork within the action symbol (see action "test computer" in Figure 5.8.)

Edge Weights. In business processes, it is sometimes necessary to describe a situation in which a defined number of objects or tokens have to accumulate at a certain point in the process before the execution can continue. In our example, one needs to collect all monitors and computers of an order before they can be bundled for shipment. With activity diagrams, it is possible to describe such situations. Edges can carry multiple tokens at the same time. They can also have weights that are displayed by writing {weight=n} next to an edge. The weight expression by which n is replaced determines the number of tokens that are consumed from the source node on each traversal. The traversal of the edge is delayed until the required number of tokens is offered by the source node.

Connectors. If edges cross large parts of a diagram, one can use connectors to split a control flow edge into two parts (see Figure 5.9). Connectors are circles containing a label. The label has to match uniquely with the label of one other connector.

Process Interaction and Signaling. If the modeled system contains multiple threads of control or different activities or instances of activities running at the

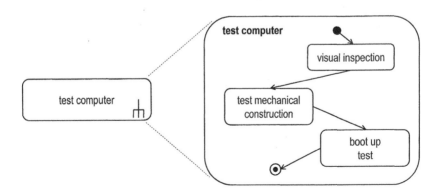

Figure 5.8   Example of an activity call.

Figure 5.9   Example of an activity edge split into two parts by a labeled connector.

same time, process interaction may be required to coordinate the execution between these control flows. Process interaction can be facilitated by sending and receiving signals. In activity diagrams, there are two special nodes representing this functionality, as shown in Figure 5.10: *send signal action* and *receive signal action*.

If a token reaches the send signal action, it triggers the emission of the signal. Signals can be received by receive signal actions. Corresponding send and receive actions can be determined by the signal name and optionally by a dashed line connecting sender and receiver. As soon as the signal is sent, the control token can pass on.

Receive signal actions may be included in the control flow, that is, they have an incoming control edge. In that case, they become activated as soon as there is a token available along their incoming edge. When the incoming signal is received, the execution can continue and the control token will be passed on. Receive signal actions without incoming edges become activated as soon as the activity starts execution. After that, activities can always receive signals.

Constructs to Model Exception Handling.   The UML provides constructions for exception handling. A common problem is that in part of a process an exceptional condition can arise that requires actions to be performed apart from the regular workflow. This situation can be reflected in activity diagrams by introducing an *interruptible activity region*. Such a region contains one or more actions. It is displayed by a round-edged dashed rectangle surrounding the actions that form the interruptible region. A lightning-bolt-shaped edge called the interrupting edge leaves the interruptible region. The semantics of this construction is that if the interrupting edge is traversed, all other actions within the region are canceled and all remaining tokens within the interruptible region become abandoned. Two alternative notation options are available for the interrupting edge, as shown in Figure 5.11.

Another exception handling situation occurs when an exceptional condition arises within one single action. For example, the action could be a mathematical divi-

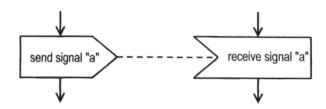

Figure 5.10   Signal send and receive actions.

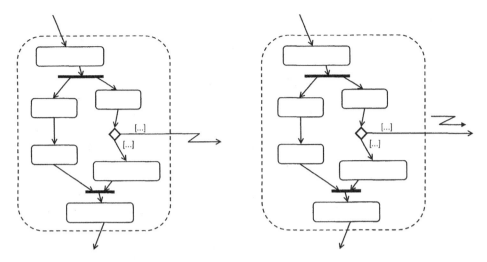

Figure 5.11   Two alternative notations for an interruptible activity region.

sion operation, possibly leading to a division by zero. In activity diagrams, an *exception handler* can be attached to single actions, as in Figure 5.12. In this case, the exception handler is a behavior that is executed whenever a predefined exception occurs while an action is being executed.

Multiple exception handlers can be attached to catch different types of exceptions. The execution of the exception handler substitutes for the execution of the action during the time it is running. After the execution of the exception handler has terminated, the control flow is continued, at the point where the execution was triggered.

The exception handler does not have own incoming and outgoing control edges since it only replaces the execution of the interrupted action. In the cases in which an exception cannot be caught, it becomes propagated to the next-higher nesting or abstraction level; that is, if the action raising the exception is part of an activity A that has been called by an activity B, then the exception is propagated to B

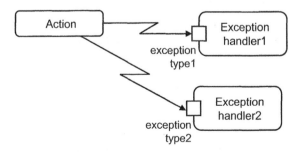

Figure 5.12   Exception handler.

if it is not caught by A. If no exception handler can be found, the system behavior is undefined.

5.3   MODELING OBJECTS AND OBJECT FLOW

All processes perform operations on physical objects. For example, goods are produced from raw materials or logical objects like information and data. With UML, it is possible to model the types, properties, and states of those objects as well as to integrate corresponding object flows into the activities.

For instance, consider the order handling process of our computer hardware company (see Figure 5.1), which comprises the packing of product bundles for incoming orders. This process involves two basic object types, namely, hardware products and order forms. From this simple scenario, we can derive the following three requirements for modeling objects and object flow:

1. We want to model data structures, objects types (in object-oriented languages called *classes*), and relationship types in order to classify objects, define common properties, restrict possible relationships, and explain internal structures. For instance, we want to describe that order forms always contain a list of order items and that each item refers to a certain product type. For this purpose, we will introduce *UML class diagrams*.

2. We want to represent individual objects with their concrete properties and relationships. For instance, we want to describe pending orders and available products at a particular point in time. For this purpose, we will introduce *UML object diagrams*.

3. We want to define the dependencies between objects and actions occurring in activities, in particular input and output relationships as well as object flow dependencies. For instance, we want to describe that our packaging process requires a new order as input and how this order is processed at the different stages of the process. For this purpose, we will explain *object flow concepts* as part of UML activity diagrams.

5.3.1   Object Types and Instances

Since UML is an object-oriented language, *objects* and their types are fundamental concepts of the language. They can be used to represent physical entities like products or persons, information like data or documents, as well as logical concepts like product types or organizations. Object types, also called classes, are defined in UML class diagrams. Objects are instances of these types, and they are represented in UML object diagrams.

Figure 5.13 summarizes the basic constructs that can be used within a class diagram. In principle, each class diagram is a graph with classes as nodes and relationships as edges. A class defines a set of common properties, also called *attributes*, that all instances of the class assign concrete values to. A property is defined in the

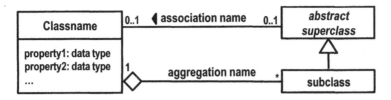

Figure 5.13 Basic class diagram constructs.

second compartment of a class symbol by a property name and a property type like string, integer, and so on.

Besides the classes as object types, a class diagram can contain three different kinds of relationship types (see Figure 5.13):

- A *generalization relationship* (depicted as a triangle-shaped arrow) is used to factorize common properties of different classes in a common superclass. The subclasses inherit all the properties and associations of their superclasses. If it is not intended or meaningful to create own instances of the superclass; it can be declared to be an *abstract class* (indicated by its name printed in italics).

- An *association* (depicted as a line between classes) is used to define possible links between objects. The usual form are binary associations between exactly two classes. Besides a name, an association has cardinality constraints at its ends, which are given as a fixed value or as a range of lower and upper bounds (the symbol * means "unbounded"). For each association end, the cardinality constraint restricts the number of objects that can be associated to an instance of the opposite association end. A small solid arrowhead next to the association name can be used to indicate a reading direction for ambiguous association names.

- An *aggregation* (depicted as an association with a diamond symbol next to the container class) is a special association indicating a containment relationship. It is used to model object types that have other objects as parts.

Coming back to our example, consider the class diagram in Figure 5.14. It states that every Order is submitted by a Customer and that it is composed of one or more OrderItems. The Producttype class and its subclasses Computertype and Monitortype are used to describe the product range of the company. Every OrderItem refers to a Producttype that the customer wants to order. The Product class and its subclasses Computer and Monitor are used to describe the physical products to be sold. The association isOfType between Product and Producttype is used to assign a type to every product. Both Product and Producttype are abstract classes so that only their subclasses can have instances. Due to the generalization, the subclasses inherit the isOfType association and the name attribute. Products can be aggregated to a Bundle.

Objects, being instances of the defined classes, have unique identifiers and concrete values for their properties. A snapshot of the objects existing at a certain point

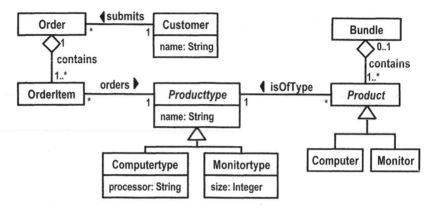

Figure 5.14   Class diagram example.

in time is modeled by a UML object diagram, as shown in Figure 5.15 for our application example. In contrast to classes, objects are depicted with underlined identifiers and type names. Objects that are parts of composite objects can be shown within the rectangle of the container object.

5.3.2   Extending Activities with Object Flows

In Section 5.2, we introduced UML activities that focus solely on the control flow aspect. Now, we can combine the control flow with object flow.

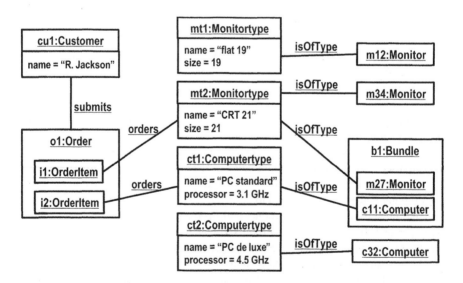

Figure 5.15   Object diagram example.

In UML activities, we use *object nodes* to model the occurrence of objects at a particular point in the process. If we expect objects of a certain type only, we can typify object nodes by one of the classes defined in the class diagram. Since business processes usually perform transformations on physical objects or data objects, it is often useful to add information about the current *state* in the object life cycle to an object node. In general, an object node is depicted as a rectangle containing the type name and, in square brackets, the state information, as shown in Figure 5.16a.

In order to also capture object flow, the token flow semantics of activity diagrams is extended by *object tokens*. An object token behaves like a control token, but, in addition, it carries a reference to a certain object. Edges between object nodes represent flows of such object tokens. If the target object node of such an edge has a type, it can only accept tokens with objects that are instances of this type. Thus, the modeler has to consider type compatibility, and an object flow edge is only allowed if the type of the target object node is the same as or a supertype of the type of the source object node.

Whenever an object token arrives at an object node, it is immediately offered along outgoing edges to downstream nodes. If the node has more than one outgoing edge, they have to compete for the object token and only one of them can retrieve it. If no guard condition is given, the winning edge is determined nondeterministically. Otherwise, if we want to allow all downstream nodes to have concurrent access to the object, we can insert an explicit fork node since this causes a duplication of the object token. Then, each downstream node receives a token referring to the same object.

However, if none of the downstream nodes is ready to accept tokens, the object node can temporarily store the tokens and pass them on in the same order (FIFO or "first in first out"). Instead of FIFO, one can also specify a different kind of queuing order like LIFO ("last in first out"), "by priority," and so on by a suitable ≪selection≫ note as shown in Figure 5.16b. Moreover, an *upper bound* can be given that restricts the number of tokens allowed to accumulate in an object node. Object tokens cannot flow into the node if that limit has already been reached.

With the help of object nodes and object flows, we can model how objects are directed through the different actions of an activity and how they are assigned to the input and output parameters of the various actions. To facilitate the latter, object nodes can also appear in the form of *input pins* and *output pins,* which are directly attached to an action node. Input pins are assigned to the input parameters of the ac-

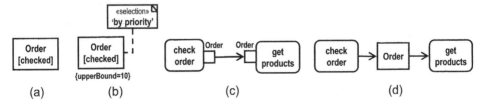

Figure 5.16   Object nodes (a and b), connected pins (c), and stand-alone notation (d).

tion, and output pins to its output parameters. As shown in Figure 5.16c, pins are depicted as small hollow squares with their types written above the square.

An action can start execution only if all its input pins hold an object token. Then, the action consumes the tokens from its input pins and, after completion, places new object tokens on all of its output pins. Figure 5.16c shows two actions whose output and input pins are connected by an object flow edge. If the connected output and input parameters have the same name and type, the standalone notation can be used instead of the two pins, as shown in Figure 5.16d.

In the following paragraphs, we show how these object flow concepts apply to our example business process of Figure 5.1, and we explain the different usages of object nodes in more detail. The resulting extended activity model is shown in Figure 5.17.

Similarly to the individual actions, the overall activity can have input and output parameters, too. Those activity parameters are modeled with object nodes playing the role of *activity parameter nodes*. In our order process, for instance, each arrival of an "Order" object places a corresponding object token at the input parameter node of the activity. From there, the token is directed to the first action, and the process is executed until the last action places a token with the "Bundle" object at the output parameter node.

The first action "check order" of the process validates an incoming "Order" object and, if successful, passes it on through its output pin. Since the downstream actions require "Order" objects in the state "checked," too, we can use the standalone notation for object nodes here.

If the check is not successful, we want the process to terminate and to reject the invalid order. We can model this as an *exception output parameter:* Both actions and activities can have such output parameters, which are used only when an excep-

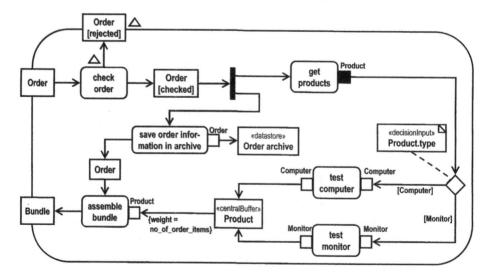

Figure 5.17   Example activity with object flow.

tion occurs. As shown for "check order" in Figure 5.17, the output pins and output parameter nodes for exceptions are indicated by a small triangle. A token is placed there only after an abnormal termination. Otherwise, if the action or activity completes successfully, it does not place any object token there.

According to the control flow model of Figure 5.1, the downstream actions are divided into two parallel paths. Since both of them need information from the "Order" object, we let the fork node duplicate the object token. One copy of the token goes to the "get products" action and another copy to the "save order information in archive" action.

The "get products" action takes the "Order" object and retrieves the ordered products from the warehouse. The resulting "Product" objects are placed on the output pin of the action. Since this is the only output pin of the action, the first "Product" token placed on that pin would cause the termination of the action. However, we want the action to continue until it has provided individual tokens for all "Product" objects to be retrieved.

To generalize the problem, we require special input and output pins that associate incoming and outgoing tokens to the same execution of an action. In UML activity diagrams, this is done by declaring pins a *stream* (depicted by a filled square). For instance, the output pin of the "get products" action in Figure 5.17 is a stream. Actions with streaming output pins can continue to place tokens there while they are executing. Similarly, actions with streaming input pins can continue to accept new input tokens while a single execution of the action is running.

Coming back to our example, we want the subsequent testing actions to treat the retrieved "Product" object tokens separately because each product has to pass its own quality test. Consequently, the pins of the testing actions are not declared as streams again. Since different tests are required for computers and monitors, a decision node is used to direct the products to the right test action. As shown in the example, we can use information about objects and their attributes in the branching conditions.

After the quality test, the products should be collected again before they are assembled into bundles. We can model this by a *central buffer* node, which is a special object node (labeled as ≪centralBuffer≫) that can be used to manage object flows of various incoming and outgoing edges. Central buffer nodes are not directly connected to actions but to other object nodes or pins. Thus, they provide additional, explicit means for queuing object tokens. In our example, the buffer type "Product" is compatible with both upstream types "Computer" and "Monitor" since they are subtypes according to the class diagram of Figure 5.14.

The "save order information in archive" action has to store statistical information about the order in an archive. If we want to model such persistent storage of data, we can use *data store nodes,* which are specialized central buffer nodes (labeled as ≪datastore≫). In contrast to central buffer nodes, a data store node keeps all tokens that enter it, copying them when they are chosen to move downstream.

Eventually, the "assemble bundle" action packages all "Product" objects into a "Bundle" object and passes it to the output parameter node of the activity. The action must not start execution unless all ordered products have finished the quality test and are available from the central buffer. This can be guaranteed by the weight expression, which delays the object flow until as many "Product" tokens are available as order items are contained in the "Order."

5.4 MODELING ORGANIZATIONAL STRUCTURE

The actions included in activities that describe business processes are executed by specific persons or automated systems within a company. Companies are complex sociotechnical organizations. It is necessary to link the underlying organizational structure of a company to the activities of its business processes in order to describe which actions have to be performed by which organizational entities. This corresponds to the resource and organizational perspectives of workflow modeling discussed in Chapter 2. This section describes how UML can be used to address the following key requirements for modeling organizations and resources:

1. Companies consist of a multitude of organizational entities such as persons, machines, and systems. Actions, for example, in an activity diagram, can be associated with any of these organizational entities. To build a coherent model of a company, all these different organizational entities should be described in one single model together with their specific properties and relationships. Examples of such relationships are leadership hierarchies, ownership and shareholder relationships, department affiliation, project group affiliation, and communication structures. We will use UML object diagrams to model concrete organizations.

2. The organizational structures in companies usually follow typical patterns, such as hierarchically organized leadership structures, functional division of labor in departments, or matrix organizations. With UML, it is possible to flexibly model the majority of these general organizational structures in such a way that concrete organizations can be treated as instances of these structures. General organizational structures can be modeled by UML class diagrams.

3. Finally, the control and object flow description contained in activity diagrams and the organizational view expressed in class and object diagrams have to be linked to each other because actions and activities need to be assigned to the organizational entities that are responsible for their execution. For this purpose, we will introduce the concepts of *activity partitions* and *swim lanes* in activity diagrams.

A general introduction to UML object and class diagrams has been presented in Section 5.3. In this section, we will focus on the usage of object and class diagrams for organizational modeling.

5.4.1 Modeling Organizational Structures with Object and Class Diagrams

Figure 5.18 shows an example of an object diagram describing the concrete organization of our exemplary computer hardware sales company. Object diagrams are al-

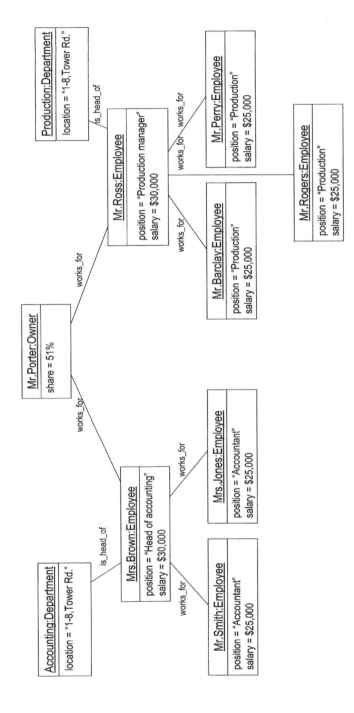

Figure 5.18 Object diagram describing a simple concrete organization.

ways instances of corresponding class diagrams. In this diagram, there are objects of three different classes:

- Objects of the class Employee for concrete persons
- Objects of the class Department for departments
- Objects of the class Owner for legal persons that own an equity stake of the company

These three classes represent three types of organizational entities in our example company. Different organizational entities can have a different set of properties. In UML, these properties are described by attributes. By associating different kinds of organizational entities with different classes, one can have different attribute sets for each kind of organizational entity. This is reflected in our example in Figure 5.18. Employees have the attributes "position" and "salary." Departments have the attribute "location" and owners have the attribute "share," describing the equity share they own of the company. These observations lead us to the corresponding class diagram as in Figure 5.19, which describes the general organizational structure of the company. The object diagram in Figure 5.18 is an instance of this class diagram.

To show the full potential of organizational modeling with class diagrams, we make some more observations about our example company:

- Departments have a number of employees that work for them. The organizational structure consists in our simplified example only of departments.
- Each department has exactly one employee or one owner as head of the department.
- The company has a board of directors that consists of the owners of the company.
- The owners of the company form the board of directors.
- Each employee can work for either another employee or an owner.

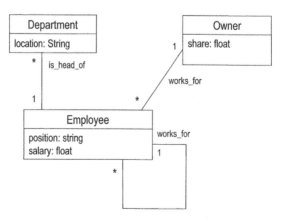

Figure 5.19   Class diagram representing a simple organizational structure.

Figure 5.20 shows a class diagram that integrates all the observations about our example organization. Now there are classes for the organizational entities: Employee, Owner, CompanyMember, BoardOfDirectors, and Department.

In the version of the class diagram in Figure 5.19, "Employee" has two distinct associations called "works_for." Employees can either work for another employee or an owner. It is possible to introduce an abstract superclass, "CompanyMember," making "Employee" and "Owner" subclasses of "CompanyMember." Then the class diagram can be optimized by having only one association called "works_for," from "Employee" to "CompanyMember."

With the abstract superclass "CompanyMember," it is also possible to model the fact that owners as well as employees can be the head of a department by changing the association "is_head_of" to be between "Department" and the new class "CompanyMember." As "CompanyMember" is an abstract class, in an object diagram describing a concrete organization, an "Employee" or an "Owner" has to take the place of the "CompanyMember." The cardinality "1" at the association "is_head_of" expresses that there has to be exactly one head of a department. The hierarchy of the company is built up by the departmental structure of the organization and by the association "works_for."

In the class diagram of Figure 5.20, we introduce a new class representing the organizational unit "BoardOfDirectors." The board of directors is built from the set of owners, which is reflected by the aggregation relationship "belongs_to" symbolized by the association line with the diamond symbol. The "Department" class has an aggregation relationship to the class "Employee" because departments consist of employees who work for the department. The cardinality "1" expresses that every employee belongs to exactly one department.

We can now describe the complete concrete organizational structure of our example. If we add the "BoardOfDirectors" and the "belongs_to" associations to the object diagram of Figure 5.18, we get the diagram in Figure 5.21.

Additional Remarks. The structure of the class diagram in Figure 5.19 indicates that, in principle, every employee can be subordinate to every other employee or owner, but every employee can only belong to one department. Therefore, the class diagram stipulates a hierarchical department structure.

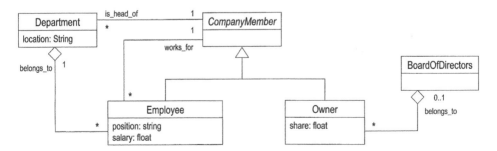

Figure 5.20   More sophisticated organizational structure.

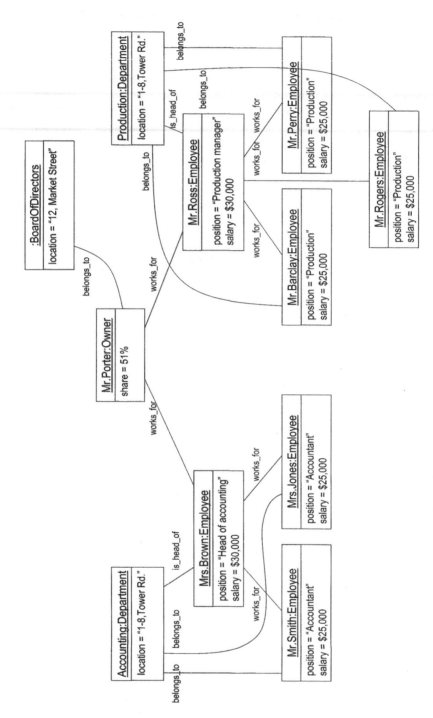

Figure 5.21 Complete object diagram for the example company.

It is also possible to describe organizational structures other than hierarchies. For example, many companies have on the one hand functional departments like "production," "accounting," and "development," and on the other hand departments for different product lines. This leads to a two-dimensional matrix organization. To model such an organizational structure, the cardinality "1" between "employee" and "department" has to be changed to "2." Sometimes, not every position in the matrix is staffed. For example, some employees fulfill the same function for different products. In that case, the cardinality between "employee" and "department" can be changed to "2..*" for a two-dimensional matrix. Figure 5.22 shows an example of an excerpt of an object diagram for a matrix organization with two product lines for monitors and computers. Some objects and associations are left out in the diagram to account for clear arrangement. In this example, we added the department "Procurement" and the employee "Mr.Taylor." Mr. Taylor is responsible for the procurement for both product lines, so the corresponding class diagram can be the same as in Figure 5.20, but the cardinality between "employee" and "department" has to be "2..*."

5.4.2 Integration of Organizational Structures in Activity Diagrams

Now that we have seen how organizational structures can be modeled using class diagrams and concrete organizations can be described using object diagrams, we have to connect these organizational models to the process models. In UML, this connection is done within an activity diagram using the notational elements *activity partition* and *swim lane.*

Activity partitions divide the set of nodes within an activity into different sections. Their use is not restricted to modeling organizational units. For example, they can also be used to constrain other resources among the nodes of an activity.

Activity diagram nodes can belong to none, one, or more partitions at the same time. Partitions can be divided into subpartitions. Partitions can be visualized in two different ways. The partition name can be written in brackets over the action name within the action symbol as is Figure 5.23a. The other possibility is the use of swim lanes as in Figure 5.23b.

Swim lanes are lines that are drawn through the activity diagram dividing it into different sections. The name of the partition is displayed on the top of the swim lane. In our case, that would be the name of the organizational unit that is responsible for execution of the actions in that partition.

With swim lanes, simple organizational structures can be reflected. In the previous section, we introduced hierarchical and matrix organizations. Simple situations of the two organizational structures can also be displayed by swim lanes. They can be hierarchically structured as shown in Figure 5.24a. Swim lanes can also intersect each other, as in Figure 5.24b, to represent example matrix organizations. Then the actions are associated with multiple partitions at the same time.

The model of the organizational structure can now be integrated into the business process models of our running example. The activity depicted in Figure 5.17 contains a number of actions that have to be executed either by the accounting de-

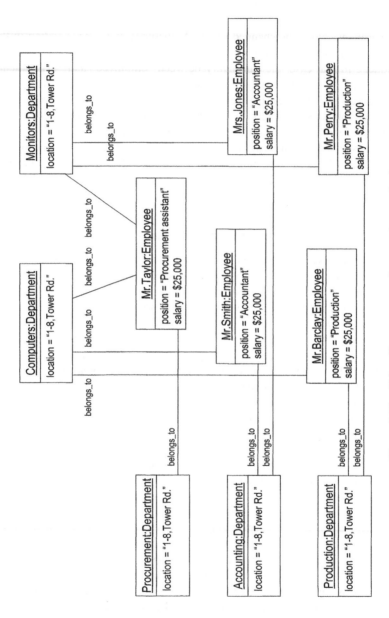

Figure 5.22 Object diagram excerpt for a matrix organization.

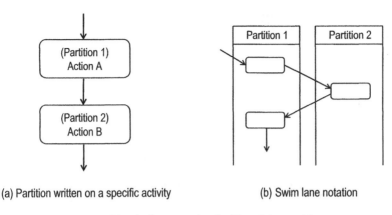

(a) Partition written on a specific activity   (b) Swim lane notation

Figure 5.23 Actions associated with activity partitions.

partment or the production department. In Figure 5.25, swim lanes are included in the activity diagram to describe that the actions "check order" and "save order information in archive" are performed by the accounting department and the other actions are performed by the production department.

5.5 MODELING BUSINESS PARTNER INTERACTIONS

So far, we have concentrated on modeling the various dependencies between the different actions of a business process. However, a complementary view of business processes is more centered around the *interactions* that take place between different participants. Such interactions occur, for example, among the employees of a certain department as well as across department and company borders. In a supply chain, for instance, the involved business partners have to interact in order to coordinate demand and supply of certain materials.

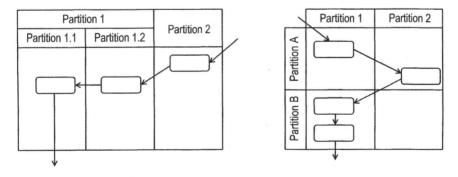

Figure 5.24 Simple organizational structures and swim lanes.

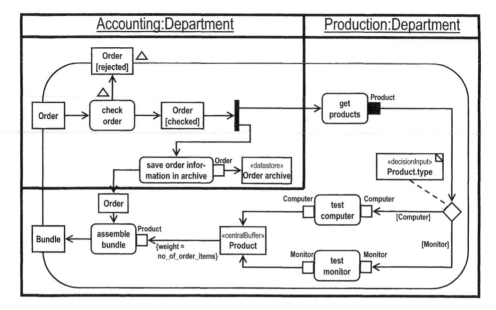

Figure 5.25   Exemplary activity with swim lane notation for the organizational entities.

In such cases, the involved participants have to agree on the way they will inter-act. An *interface process,* as defined in Chapter 4, constitutes an approach to define the interactions between partners, represented by their provided endpoints. In an in-terface process, interactions are described from the perspective of one of the in-volved endpoints. As shown in Chapter 4, an interface process can be described through an activity diagram in which the activities produce or consume interaction events. In some situations however, a more interaction-centric (rather than activity-centric) view of the relevant processes is more appropriate. This view allows mod-elers to focus on the interactions themselves, and provides a more global perspec-tive on how multiple partners interact, as the description does not focus on the events produced or consumed by a specific participant.

For this purpose, UML provides so-called *sequence diagrams.* They comprise the participants involved in an interaction. Each of them has a *lifeline* representing its progress in time (usually from top to bottom). Arrows between the lifelines indi-cate the passing of a message. The sequence of arrows along the lifelines represents the order of message exchanges.

As an example, we consider the interactions of a hardware sales company with its customers, its warehouse, and a shipping service that is in charge of delivering ordered products to customers. Figure 5.26 shows the corresponding sequence dia-gram.

This sequence diagram is named "order interactions" and comprises a "Customer," the "Company," its "Warehouse," and the "ShippingService" as partic-ipants. Every participant is depicted as a rectangle that contains the name of the par-

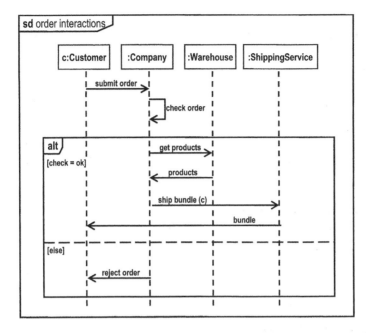

Figure 5.26   Sequence diagram for interactions related to order processing.

ticipant and its type. In contrast to the notation of objects in object diagrams, these names and types are not underlined because they represent a certain *role* rather than a concrete instance. At process enactment time, the role names have to be bound to concrete entities of the specified type. For instance, a concrete customer submits the order and a specific shipping service is selected. If the role name is not referenced later in the diagram, one can also omit it and specify just the role type.

The messages attached to the arrows represent, for example, a request for some provided service, a response to the requester, the transmission of a certain signal, the sending of a certain return value, the transportation of some objects, and so forth. As with action names in activities, there are different degrees of formalization possible, starting from simple keywords down to operation calls with formal parameters.

In UML sequence diagrams, one distinguishes between *synchronous* (filled arrowhead) and *asynchronous* (open arrowhead) message passing. The synchronous mode means that the sender stops its activity after sending the message and waits until the corresponding response message arrives. In our example, we use only asynchronous message passing, meaning that the partners remain active after having sent a message independent of the response.

In our example, the "Customer" at first submits the order, which is then checked by the "Company." Although one should usually abstract from internal actions like "check order" and concentrate on external interactions in sequence diagrams, we can still model such internal actions as self-related messages if they have an impact on the remaining part of the interaction. In our example, this is the case because the downstream interactions are divided into two alternative *interaction fragments* (indicated by the keyword "alt" and the subdivided rectangle) which are chosen according to the outcome of the check order action. Either the order is valid and the products can be retrieved from the "Warehouse" and delivered by the "ShippingService," or the order is not valid and rejected by the "Customer."

Besides the "alt" operator for alternatives, sequence diagrams also provide other *interaction operators* that can be used in combination with interaction fragments, for example, the "loop" operator that indicates that a certain fragment is repeated as long as a certain condition holds, or the "par" operator that indicates that several fragments are executed in parallel. Different fragments can also be nested to model more complex interactions.

Such interaction models provide a complementary view on the business processes modeled before. In contrast to the activity diagrams, they usually hide internal actions that do not affect other participants (e.g., the testing actions of Figures 5.1 and 5.17). Nevertheless, the two different views of the business process must be consistent with each other, which means that they have to preserve the order of overlapping actions and events. For example, in both Figure 5.1 and Figure 5.26, the "check order" action comes before the "get products" action.

5.6 SYSTEM-SPECIFIC PROCESS MODELS

The business process models presented so far can be used for design, analysis, or documentation purposes. However, another purpose of process models is to support *process enactment*. In this case, they have to be refined into activities with atomic actions that are not further subdivided. These actions can then either be performed by humans or executed by machines and computers.

At this point, we want to focus on the latter case, in which processes mainly transform information and can, therefore, be enacted with the help of computer systems (i.e., application-to-application processes). We model the available software components of an enterprise and relate their services to the actions of our process model. Thus, we receive a refined, system-specific model that can be used for process enactment. In the terminology introduced in Chapter 4, this type of model corresponds to an *integration process*.

In principle, such system-specific process descriptions can be used in two ways. The first option is to feed them into a central process engine that has access to all available software components and invokes their services according to the process description (see Figure 5.27a). Thus, the process engine is responsible for managing the various process instances, the control flow, and the object flow.

The second option is to take the more local point of view of a single component that realizes a new service by using a set of services provided by other components. Then, the process model can be used to describe how the invocations of the required services are coordinated in order to realize the desired service (see Figure 5.27b).

For instance, *service-oriented architectures* consist of distributed software components that make use of existing third-party services in order to provide new services. Since this usually involves components of different business partners,

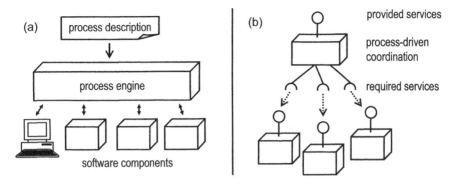

Figure 5.27 Central process engine (a) and process-driven service coordination (b).

process descriptions are needed to adjust the invocation behavior among the different partners. The Business Process Execution Language for Web Services (see Chapter 12) is a textual language for implementing such architectures in which process-driven coordination of services takes place.

As an example of system-specific models, we refine the "check order" action used in the order processing activity of Figure 5.17 and specify how existing services are combined to realize this action. For this purpose, we have to break the action down into atomic subtasks like evaluating the customer's credit rating and checking the available product supplies. We assume that there are software components such as warehouse and customer management systems that provide services for these tasks. This leads to the following requirements:

1. We require a model of available systems and components that abstracts from their internal computations but specifies their provided and required services. For instance, we want to describe that there is an order management system that provides the service to check incoming orders, and, in order to do so, it requires certain warehouse and rating services. For this purpose, we will introduce *UML structure diagrams and interface descriptions.*

2. Having specified the provided and required services, we want to integrate them into our process models in order to coordinate their invocation. For instance, we want to describe in which way the services required by the order management system are invoked in order to realize the provided order checking service. Since inputs required by one service might be provided as outputs by other services, we have to consider both control and object flow dependencies. The resulting system-specific process models should serve as a basis for computer-based process execution.

UML structure diagrams provide a high-level view of existing information systems, as shown in Figure 5.28. Components are depicted as boxes, omitting details about their internal computations. Provided and required services, in UML called *operations,* are summarized as *interfaces* of the components. Provided interfaces

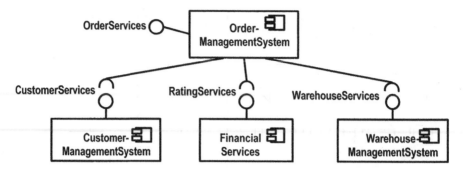

Figure 5.28 Structure diagram example.

are depicted as a circle connected to the providing component, and required interfaces as a half-circle connected to the requiring component.

For each required interface, another component is needed that can provide a matching interface. In our case, the CustomerManagementSystem provides CustomerServices to the OrderManagementSystem, the FinancialServices component provides the RatingServices interface, and the WarehouseManagementSystem provides the WarehouseServices interface.

Interfaces are specified in a simple form of class diagram, as shown in Figure 5.29. In contrast to classes used for modeling object structures, the focus is not on structural properties and relationships but on operations. An operation signature is defined in the second compartment of the interface symbol by a name and a set of input and output parameters. If there is no more than one output parameter, we can list the input parameters in parentheses and append the output parameter as the return type of the operation at the end. Otherwise, we have to distinguish input and output parameters by the keywords "in," "out," or "inout" (see, e.g., the "checkOrder" operation of the "OrderServices" interface).

In contrast to ordinary classes, interfaces cannot be instantiated but can only be used to indicate that a class or component either provides or requires the set of operations defined in the interface. In order to integrate the invocation of these operations in our process models, we introduce *call actions* for activity diagrams.

In general, call actions represent the invocation of certain behaviors defined in accompanying diagrams. In our case, we use them to call operations of component

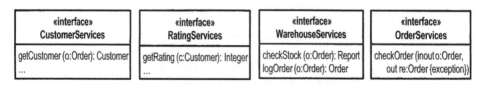

Figure 5.29 Interface specifications.

interfaces, as shown in the system-specific "checkOrder" activity (Figure 5.30). In contrast to ordinary action nodes, the node symbol contains the exact name of the operation to be called. Below, the operation name, the name of the interface or component type providing the operation, is added in brackets. All input and output parameters defined in the operation signature are transformed into input and output pins of the action node. Thus, when defining the control flow between the call action nodes, one has to consider object flow dependencies that arise from the operation's input and output behavior.

Since, according to the interface description in Figure 5.29, the "checkOrder" operation has an inout parameter of type "Order," the activity gets a corresponding input parameter node, too. From there, incoming "Order" objects are passed on to the action nodes of the activity until they are eventually placed on the output parameter node shown at the bottom of Figure 5.30.

If any of the involved checks returns a negative result, then the "Order" is rejected and placed at the second output parameter node (shown at the top of Figure 5.30), which is an exception, as indicated by the small triangle. Note that exactly this arrangement of parameter nodes is required if we want to use the activity as an refinement of the "checkOrder" action of Figure 5.17.

The activity involves two checks that can be performed in parallel: First, the customer's credit rating should have a positive value, and second, the available product supplies of the warehouse should be sufficient to satisfy the demand. Since the "ge-

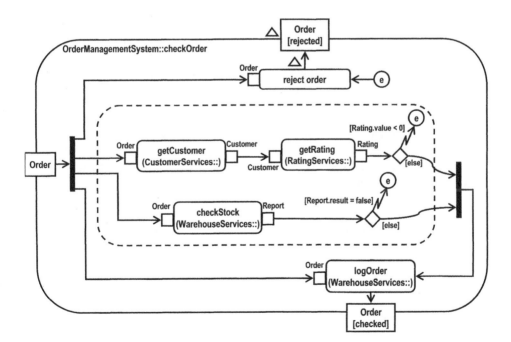

Figure 5.30 System-specific activity diagram for the checkOrder service.

tRating" operation of the "RatingServices" interface requires a "Customer" object as input parameter (see Figure 5.29), we have to insert an action calling the "get-Customer" operation first. This "CustomerManagementSystem" operation retrieves the corresponding "Customer" object from an associated database, which is then passed on to the "getRating" operation.

The two parallel action flows for order checking are enclosed in an interruptible region so that any negative result prevents further effort and directly leads to the "reject order" action causing an exception. However, if both parallel checks are successful, the interruptible region is left, and the "logOrder" operation of the "WarehouseServices" interface is invoked to update the product information stored in the "WarehouseManagementSystem." Eventually, the checked "Order" is returned as output to the superior process.

As revealed by this example, system-specific process models refine actions and activities of more abstract, business-level process models. Given a mapping of the interfaces to real components with physical addresses (also called *deployment description*), such system-specific process models can be used for process enactment and coordination of the involved software components. For related work about using activity diagrams in order to integrate applications and software components, the interested reader is referred to [6] and [22].

5.7 SUMMARY

Modeling processes require the description of a number of different perspectives of the process [11, 4]. We have covered five major perspectives of process modeling with UML diagrams: the description of actions and control flow, data and object flow, organizational structure, interaction-centric views, and application integration through system-specific, refined process models for process enactment. Table 5.1 summarizes which UML diagrams we have employed to describe these process modeling perspectives.

For further studies of the UML, the interested reader can find detailed insights into the language concepts in the book by Pender [19]. How to apply the UML for developing information systems from requirements analysis to system design is described, in the work by Maciaszek [14].

Table 5.1 Overview of the different UML diagrams

	Activity diagram	Class diagram	Object diagram	Sequence diagram	Structure diagram
Actions and control flow	X				
Data and object flow	X	X	X		
Organizational structure	(X)	X	X		
Interaction-centric view				X	
System-specific models	X	(X)			X

There are strong efforts underway to further increase the usability of UML for process modeling. The recent revision, UML 2.0, has already improved, among other things, the suitability of activity diagrams. In order to further extend the language according to business process modeling requirements, one can also use the built-in *extension mechanisms* of UML. These extensibility features allow designers to adapt certain parts of the language to their domain-specific needs while still remaining within the framework of the UML meta-model. For this purpose, so-called *stereotypes* can be defined that describe semantic extensions as well as syntactical modifications of dedicated meta-model elements. A set of related stereotype definitions forms a UML *profile.*

Work in progress includes the development of a specialized business process definition profile by the OMG [15]. The objective is to allow groups using a variety of process models, including UML activity diagrams and other process modeling notations, to map to a common meta-model and thus facilitate communication among themselves.

Among others, there are efforts underway to increase the support for collaborating business processes, business process patterns, runtime implications of process definitions, resource assignments, access control, and so on. The extensibility feature of UML will facilitate the efforts to further develop extensions of the UML for business process modeling in order to make it even more powerful and user-friendly.

5.8 EXERCISES

1. Consider the "test computer" and "test monitor" actions in Figure 5.17 and model the case when such a product test fails. For this purpose, you could, e.g., add output pins returning a test report. If the report reveals a negative test result, a substitute product has to be retrieved from the warehouse and the test has to be redone.

2. As preparation for modeling the internals of the testing actions, extend the class diagram of Figure 5.14 as follows. A checklist is associated to each product type. Every such list contains a set of items that describe the properties to be checked for the associated product type. Each item has a property name and a reference value as attributes.

3. Now, refine the "test computer" action of Figure 5.17 into an activity, showing the internals of the action. Model the input and output parameter nodes of the activity according to the pins of the corresponding action node. The activity should contain an archive for all the checklists for the various product types. Whenever a new computer object arrives, the right checklist has to be selected from the archive. You can then freely design your own control and object flow to realize the testing activity.

4. Extend the interaction model of Figure 5.26 with the company's bank as additional business partner. After ordered products have been delivered to the

customer, the company sends a bill to the customer containing a reference to the bank. Then, the customer can transmit the payment to the company's bank account. In a second step, try to model that the delivery of the products and the payment can also happen in parallel.

5. Consider the object diagram for the example company in Figure 5.21 and the matrix organization excerpt in Figure 5.22. What would a complete object diagram of the company look like if you combined the two existing diagrams?

6. In the matrix organization in Figure 5.22, we use the organizational entity "Department" both for the functional entities of the company like procurement and accounting, and for the product-oriented entities like monitors and computers. Devise an organizational structure that contains departments and product lines as two distinct organizational entities. Extend the organizational model developed in Section 5.4 with the necessary additional classes. What additional associations have to be defined? How would the object diagram in Figure 5.22 be affected?

7. In Figure 5.21, Mr. Ross is an employee. Now assume that Mr. Ross is not only an employee but also an owner of the company at the same time. How could this be modeled in the class diagram? (Hint: consider multiple inheritance.) How would the object diagram in Figure 5.21 be affected?

REFERENCES

1. G. Booch. *Object-Oriented Analysis and Design with Applications,* 2nd ed. Addison-Wesley, 1994.

2. P. Chen. The entity-relationship model—Toward a Unified View of Data. *ACM Transactions on Database Systems, 1,* 1:9–36, 1976.

3. P. Coad and E. Yourdon. *Object-Oriented Analysis,* 2nd ed. Yourdon Press, 1991.

4. B. Curtis, M. I. Kellner, and J. Over. Process Modeling. *Communications of the ACM, 35*(9), 1992.

5. W. Damm and D. Harel. LSCs: Breathing Life into Message Sequence Charts. *Formal Methods in System Design, 19*(1):45–80, 2001.

6. R. Depke, G. Engels, M. Langham, B. Lütkemeier, and S. Thöne. Processoriented, consistent integration of software components. In *IEEE Proceedings of the 26th International Computer Software and Applications Conference (COMPSAC),* pp. 13–18, 2002.

7. G. Engels, R. Heckel, and J. M. Küster. The Consistency Workbench: A Tool for Consistency Management in UML-Based Development. In *Proceedings UML 2003—The Unified Modeling Language,* Springer LNCS 2863:356–359, 2003.

8. A. Förster. Quality Ensuring Development of Software Processes. In *European Workshop on Software Process Technology (EWSPT),* Springer LNCS 2786:62–73, 2003.

9. D. Harel. Statecharts: A Visual Formalism For Complex Systems. *Science of Computer Programming, 8*(3):231–274, June 1987.

10. ITU-TS, Geneva. *ITU-TS Recommendation Z.120: Message Sequence Chart (MSC),* 1996.

11. S. Jablonski and C. Bussler. *Workflow Management: Modeling Concepts, Architecture and Implementation.* International Thomson Computer Press, London, 1996.

12. I. Jacobson, M. Christerson, P. Jonsson, and G. Overgaard. *Object-Oriented Software Engineering—A Use Case Driven Approach.* Addison-Wesley, 1992.

13. B. Kiepuszewski, A.H.M. ter Hofstede, and W.M.P. van der Aalst. Fundamentals of Control Flow in Workflows. *Acta Informatica, 39*(3):143–209, 2003.

14. L. A. Maciaszek. *Requirements Analysis and System Design: Developing Information Systems with UML.* Addison-Wesley, 2001.

15. Object Management Group. *Business Process Definition Metamodel RFP.* http://www.omg.org/docs/bei/03-01-06.pdf.

16. Object Management Group. *Meta-Object Facility (MOF) Specification, Version 1.4.* http://www.omg.org/cgi-bin/doc?formal/2002-04-03.

17. Object Management Group. *UML 2.0 OCL 2nd revised submission.* http: //www.omg.org/cgi-bin/doc?ad/2003-01-07.

18. Object Management Group. *UML 2.0 Superstructure Final Adopted specification.* http://www.omg.org/cgi-bin/doc?ptc/2003-08-02.

19. T. Pender. *UML Bible.* Wiley, 2003.

20. J. Rumbaugh, G. Booch, and I. Jacobson. *Unified Modeling Language, Notation Guide, Version 1.0.* Rational Software Corporation, Santa Clara, 1997.

21. J. E. Rumbaugh, M. Blaha, W. J. Premerlani, F. Eddy, and W. Lorensen. *Object-Oriented Modeling and Design.* Prentice-Hall, 1990.

22. S. Thöne, R. Depke, and G. Engels. Process-oriented, Flexible Composition of Web Services with UML. In *Proceedings of the International Workshop on Conceptual Modeling Approaches for e-Business (eCOMO 2002),* Springer LNCS 2784:390–401, 2002.

CHAPTER 6

Process Modeling Using Event-Driven Process Chains

AUGUST-WILHELM SCHEER, OLIVER THOMAS, and OTMAR ADAM

6.1 INTRODUCTION

The event-driven process chain (EPC) was developed in 1992 in an R&D project involving SAP AG[1] [9, 7] at the Institute for Information Systems (IWi) of the University of Saarland, Germany. It is the key component of SAP R/3's modeling concepts for business engineering and customization [10, 8, 11] and has also been integrated in SAP's NetWeaver System. It is also the modeling notation supported by the ARIS Process Platform, which provides an integrated toolset for designing, implementing, and controlling business processes.[2]

In the 1990s, and following the evolution of the ARIS toolset, the basic EPC notation has been extended with a number of symbols corresponding to various aspects of business modeling. This has led to what is known as the extended EPC (or eEPC) notation, which is the subject of this chapter. In the rest of the chapter, we will not make a distinction between the basic and the extended EPC notation, as the extended version encompasses the basic one. Accordingly, the term EPC will be used to refer to the original EPC notation and its extensions.

EPC is based on the concepts of stochastic networks and Petri nets (see Chapter 7). However, using the EPC notation does not require a strong formal framework, among other things, because the notation does not rigidly distinguish between output flows and control flows or between places and transitions, as these often appear in a consolidated manner. Perhaps it was precisely this kind of simplification that led to the successful adoption of EPCs in practical applications.

The chapter is structured as follows. Section 6.2 introduces EPC through a working example, showing how a business process model can be obtained by combining several views. Section 6.3 introduces a meta-model of EPC that forms the basis of

[1]http://www.sap.com
[2]http://www.ids-scheer.com

Process-Aware Information Systems. Edited by Dumas, van der Aalst, and ter Hofstede
Copyright © 2005 John Wiley & Sons, Inc.

the ARIS toolset. Section 6.4 discusses methodological issues, providing guidance on how to design processes using EPC. Section 6.5 then discusses the ARIS architecture for business process design and implementation. Finally, the chapter concludes with an exposition of relevant trends.

6.2 OVERVIEW OF EPC

Generally speaking, a business process is a continuous series of enterprise activities, undertaken for the purpose of creating output. The starting point and final product of the business process is the output requested and utilized by corporate or external "customers." Business processes often enable the value chain of the enterprise as well as help to focus on the customer when creating output. Business process models are the core objects when improving the outputs (and their creation) of a company and can be seen as a starting point for IT development.

In the following, we explain the key issues in modeling business processes using event-driven process chains with a simple example from customer order processing [14]. First, let us outline the scenario.

A customer wants to order several items that need to be manufactured. Based on customer and item information, the feasibility of manufacturing these items is studied. Once the order has arrived, the necessary materials are obtained from a supplier. After arrival of the material and subsequent order planning, the items are manufactured according to a work schedule and shipped to the customer, along with the appropriate documentation.

This scenario is now discussed from various points of view. As we have already seen in system theory [2, 16, 12], we can distinguish between system structures and system behavior. The starting point is the description of the responsible entities and relationships involved in the business process; then, by means of function flows, we will describe the dynamic behavior. Output flows describe the results of executing the process; information flows illustrate the interchange of documents involved in the process.

Functions, output producers (organizational units), output, and information objects are illustrated by various symbols. Flows are depicted by arrows.

6.2.1 Responsible Entities and their Relationships

Figure 6.1 depicts the responsible entities (organizational units) involved in the business process, along with their output and communication relationships, illustrated as context or interaction diagrams. The sequence in which processes are carried out is not apparent. Nevertheless, this provides an initial view of the business process structure. In complex processes, the myriad interchanges among the various business partners can become somewhat confusing. In addition to the various interactions, it is also possible to enter the activities of the responsible entities. This has been done in only a few places.

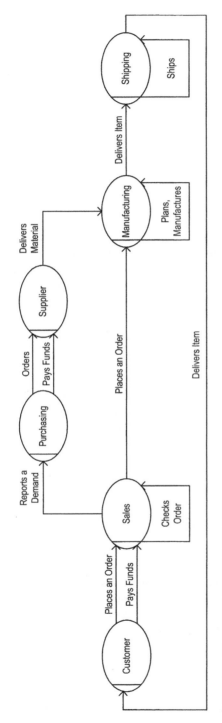

Figure 6.1 Interaction diagram of the business process "order processing."

121

6.2.2   Function Flow

Figure 6.2 describes the same business process by depicting the activities (functions) to be executed, as well as their sequence. In contrast to the interaction diagram, the focus is not the responsible entities but the dynamic sequence of activities. For illustration purposes, the organizational units are also added in Figure 6.2. Being function sequences for creating output, function flows characterize the business process. The output flows themselves will be displayed individually.

6.2.3   Output Flow

The purpose of a business process is the creation of an output in order to receive an equivalent. In our example, the output of the enterprise process is to execute the customer order and the equivalent is the receipt of funds. In addition, within the enterprise itself intermediate output is created because of executing functions.

The designation "output" is very heterogeneous. Business output is the result of a production process, in the most general sense of the word. Output can be physical (material output) or nonphysical (services). Whereas material output is easily defined, for example by the delivery of material, manufactured parts or even the finished product, the term "services" is more difficult to define because it comprises heterogeneous services, for example, insurance services, financial services, and information brokering services. Figure 6.3 illustrates this simplified classification of "output," which can also be "input," as a hierarchical diagram.

Concerning our business process example, the result of the function "manufacture item" in Figure 6.4 is the material output, defined by the manufactured item.

Quality checks are carried out and documented during the manufacturing process. All data pertinent to the customer is captured in "order documents." This collection of data is a service by virtue of the information it provides. After every intercompany function, an output is defined that indicates the specific deliverable entering the next process as an input. To avoid cluttering the diagram, the organizational units involved are not displayed. It is not possible to uniquely derive the function sequence from the illustration of the output flow.

6.2.4   Information Flow

In addition to information services, also other kinds of information, used as environment descriptions during the business processes, are components of the ordering process. Figure 6.5 illustrates the information objects of our business process example, along with the data interchanged among them.

Objects listed as information services have double borders. Information objects describing the environment of the business process are shown as well; for example, data regarding suppliers, items, or work schedules. This data is necessary to create information services. For example, when checking orders, the customer's credit is checked and inventory is checked for availability.

Due to the fact that data flow is triggered by the functions that are linked to the information objects, it is more or less possible to read the function flow in Figure

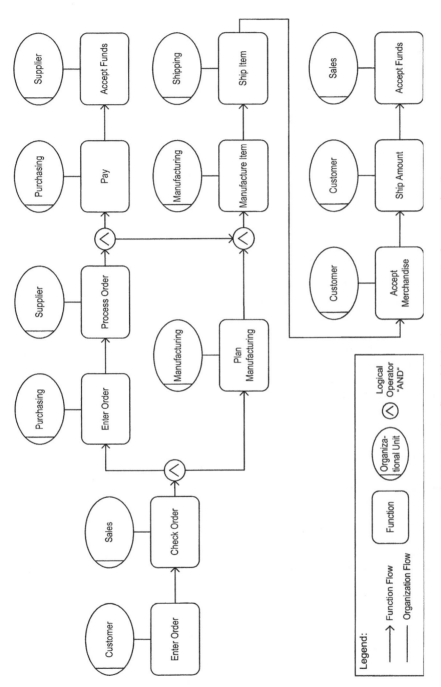

Figure 6.2 Function flow of the business process "order processing."

123

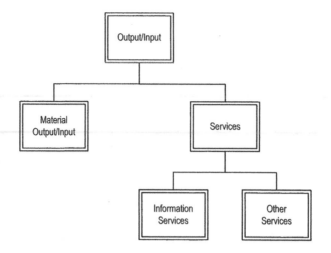

Figure 6.3   Types of inputs and outputs.

6.5. However, if multiple functions are applied to an information object or if multiple data flows are requested by a function, the function process cannot be uniquely deduced.

Besides information flow modeling, the (static) description of data structures is a very important modeling task. Static enterprise data models are used to develop proper data structures in order to implement a logically integrated database. Chen's Entity Relationship Model (ERM) [5] is a widespread method for the conceptual modeling of data structures.

6.2.5   Consolidated Business Process Model

Building various views serves the purpose of structuring and streamlining business process modeling. Splitting up views has the advantage of avoiding/controlling redundancies that can occur when objects in a process model are used more than once. For example, the same environmental data, events, or organizational units might be applied to several functions. View-specific modeling methods, which have proven to be successful, can also be used. Particularly in this context, view-based approaches are easier to handle than more system-theoretical modeling concepts, in which systems are divided into subsystems for reducing complexity. In principle, however, every subsystem is depicted in the same way as the original system. This is why it is not possible to use various modeling methods in the same system.

It is important to note that none of the flows (organization, function, output, and information flow, respectively) illustrated above is capable of completely modeling the entire business process. We must, therefore, combine all these perspectives. One of the views should be selected as a basis and then integrated into the others. The

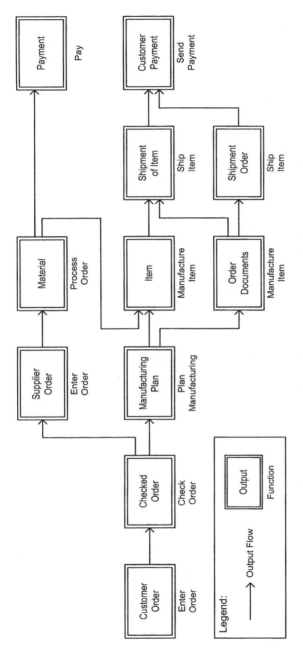

Figure 6.4 Output flow of the business process "order processing."

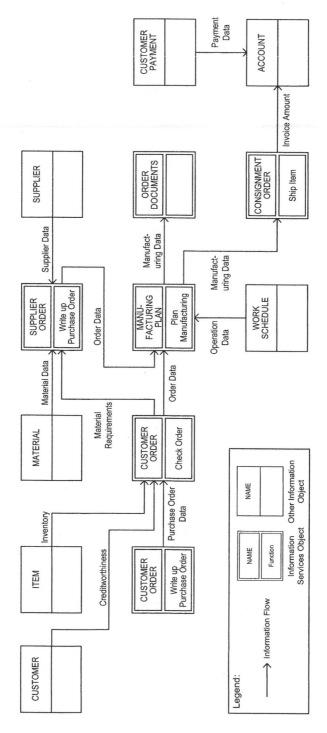

Figure 6.5 Information flow of the business process "order processing."

function view is closest to the definition of a business process and is, therefore, typically used as a starting point. However, in the context of object-oriented enterprise modeling, information flows (see Chapter 5) can serve as a starting point as well.

Figure 6.6 provides a detailed excerpt of our business process example, focusing on the function "manufacture item" with all flows described above.

Figure 6.6 shows the consolidated process model in the EPC notation. In this notation, function flows are enhanced by event and message controls. This makes it possible to better describe the process sequence. Events describe condition changes. They can characterize the result of an activity, in turn triggering the next function. In addition to simple events, there are also compound events. For example, for the function "manufacture item," planning needs be concluded and the necessary parts need to be available. This is expressed by the logical "AND" operator between the events.

Control flows regulate how events are triggered in accordance with sensible process logic. Sequential, parallel, alternative, and merged methods, along with logical links, may be used. Control flows are executed by events and messages that they trigger, after which information regarding the beginning of the event is transferred to the next entity. In the illustrations, messages are depicted by letter symbols. They determine how functions react to events. Messages can also contain additional attributes besides information regarding the beginning of the event.

After the events "manufacturing plan completed" and "(supplier) order processed" have triggered the function "manufacture item" by means of messages, the event "item completed" is created. Then the process is concluded. This event activates successive events by means of messages.

Only events pertinent for the continuation of the business process are illustrated here. These events are known as relevant events.

6.3 THE ARIS BUSINESS PROCESS META-MODEL

Business process models can be designed at various abstraction levels. The previously discussed business process models were based on an order processing application. This example did not describe the procedure of an order for a certain customer but, rather, the generic order processing process, which is an abstraction of several actually realized processes. This kind of description is known as a business process type.

Figure 6.7 illustrates an excerpt of the manufacturing process of an individual order processing process. Here, every object involved in the business process is instantiated by the affixed name or names. Individual business process models are used for controlling individual processes. In manufacturing, this is customarily carried out by creating work schedules as the manufacturing process descriptions for individual parts or manufacturing orders.

In office management, individual business process models are executed through workflow control systems (see Chapter 3). Workflow systems control document flows and work flows electronically. Therefore, they must have access to information regarding the respective control structure and responsible entities or devices for

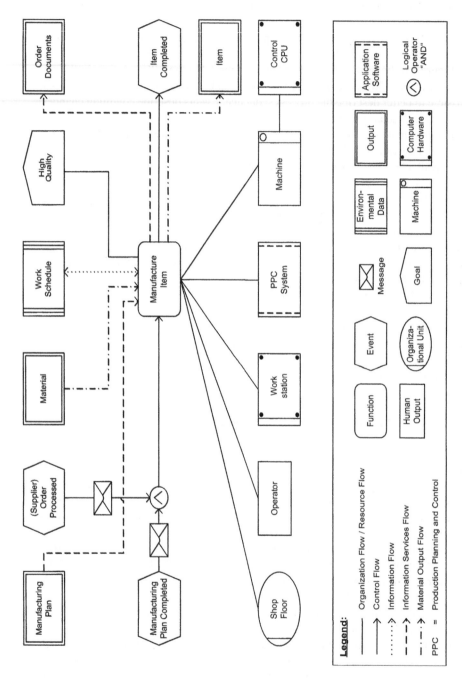

Figure 6.6 Detailed excerpt of the business process "order processing."

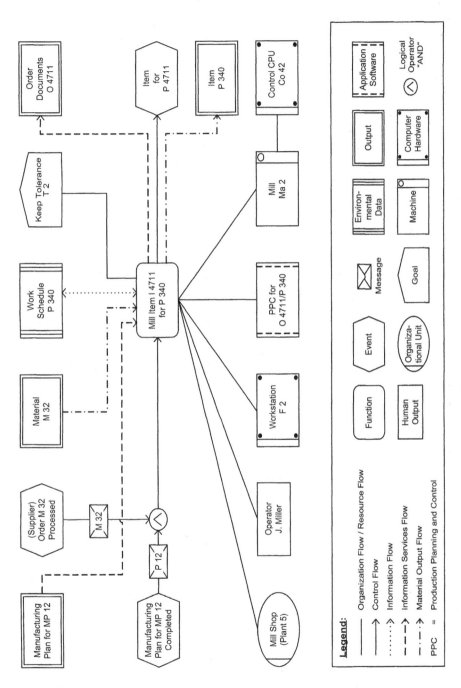

Figure 6.7 Business process model of an order instance.

129

every single business event. Individual business processes are known as instances. There is a class–instance relationship between the business process type in Figure 6.6 and the process instance in Figure 6.7.

All individual order processes make up the class or type "order processing business process." The individual processes are instances (elements) of this class. Classes take on the characteristics of the elements, although the individual instances are abstracted.

Type levels are the most important levels in business process modeling. In order to support (re)organization measures, not only is know-how regarding each business process necessary, but also know-how regarding the entire general process structure. After all, the organizational changes are carried out in order to improve the process as a whole. Instances, thus, proceed according to the new schema. Due to exception handling regarding the process structure, individual variances of the instances can be taken into account.

The illustration of instances is known as description level 1, whereas type levels are known as description level 2. Levels 1 and 2, thus, have the same relationship as instances and classes. Every class is characterized by its name and the enumeration of its attributes, by which the instance is described. For example, the class "customer" is characterized by the attributes "customer number," "customer name," and "payment period." The instances of these characteristics are the focus of the descriptions at level 1. Figure 6.8 depicts a few examples of levels 1 and 2.

Grouping classes is always a complex task. Therefore, when defining order designations, we will only abstract specific properties of cases 4711 or 4723 (see Figure 6.8), respectively, leading to the classes "completed order" or "finished order." At level 2, we will abstract the "completed" and "finished" properties and create the parent class "order" from the subset. This operation is known as generalization and is illustrated by a triangular symbol.

When quantities are generalized, they are grouped as parent quantities. This makes order instances of level 1 instances of the class "order" as well. The class "order" is characterized by the property "order status," making it possible to allocate the process state "completed" or "finished" to every instance. Materials and items are also generalized, making them "parts" and "resources."

Thus, level 2 contains application-related classes of business process descriptions. On the other hand, with new classes created from similar classes of level 2 by abstracting their application relationships, these are allocated to level 3, the meta-level, as illustrated in Figure 6.8. Level 2 classes then become instances of these meta-classes. For example, the class "material output" contains the instances "material" and "item" as well as the generalized designation "part." The class "information services" contains the class "order," along with its two child designations and the class "certificate." The creation of this class is also determined by its purpose. Thus, either the generalized classes of level 2 or their subclasses can be included as elements of the meta-classes.

When creating classes, overlapping does not have to be avoided at all costs. For example, from an output flow point of view, it is possible to create the class "information services" from the classes "order" and "certificate." Conversely, from the

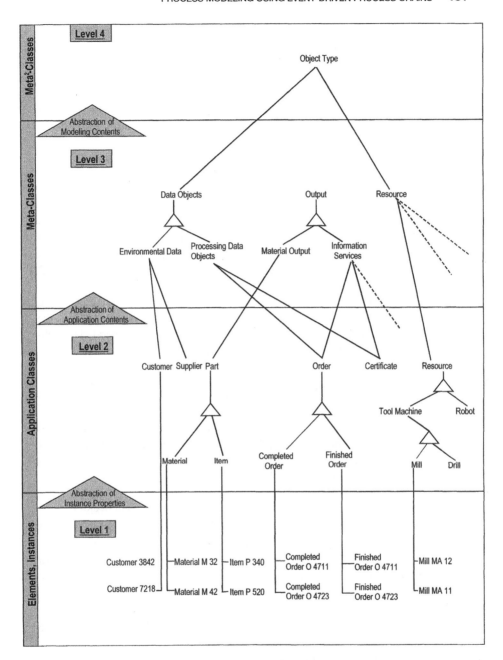

Figure 6.8   Abstraction levels in modeling.

data point of view, these are also data objects, making them instances of the class "data objects" as well.

This leads to the general ARIS business process meta-model at level 3, depicted in Figure 6.9. This figure contains the general description classes of business processes, along with their relationships. The relationships depicted by arrows could also be expressed as classes (relationship classes). For simplification, this has been avoided. When we subsequently speak of meta-classes, we mean every representation object (classes and relationships).

In addition to the relationships displayed here, other relevant relationships between the classes are feasible. It is also possible to create subclasses from the classes of the meta-level. The model in Figure 6.9 shows the essential objects necessary for illustrating business processes, although this figure is not necessarily complete.

Thus, the classes at modeling level 3 define every object necessary for describing the facts at level 2. These objects make up the building blocks for describing the applications at level 2. Objects at level 3 are also the framework for describing the individual business processes, as the classes at level 2 comprise the terminology at level 1.

This abstraction process can be continued by once again grouping the classes at level 3 into classes, which are then allocated to the second meta-level. Next, the content-related modeling views are abstracted. In Figure 6.8, the general class "object type" is created, containing all the meta-classes as instances.

6.4 HOW TO CORRECTLY MODEL EPCs

Each EPC must follow some simple design rules to avoid or reduce undesirable behavior like deadlocks right from the beginning. Therefore, no rigorous and complex system of rules and design patterns is proposed, as avoiding all possible conflicts would limit the user too much. The rules are:

1. The three core nodes of an EPC are activities, events, and connectors.
2. The name of an event should reflect its characteristic as a point in time, for example, "item completed." It is represented by a hexagon.
3. The name of an activity should consider the time-consuming perspective of the accomplished task, for example, "manufacture item." An activity is represented by a rectangle with rounded edges.
4. Connectors are represented by a circle. Within the circle, the type of connector is defined through the corresponding symbol. The connector can be split into an upper and lower part, reflecting differences between incoming and outgoing connection rules.
5. To clearly define when a business process is intended to begin and what is the final result, each EPC starts and ends with one or more events.
6. An EPC contains at least one activity.
7. An EPC can be composed of several EPCs.

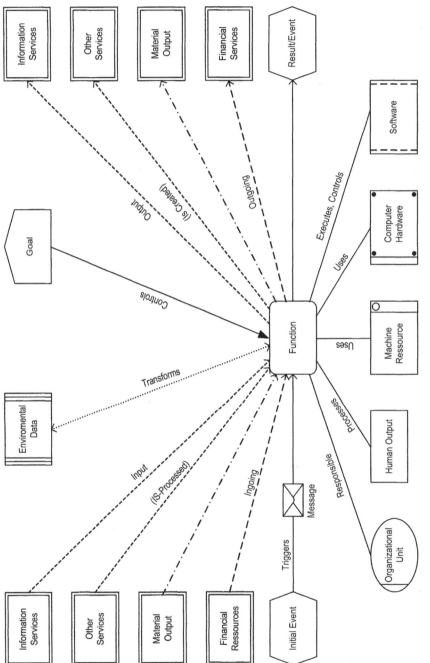

Figure 6.9 The general ARIS business process meta-model.

133

8. Edges are directed and always connect two elements corresponding to the sequence of activation.

9. An event cannot be the predecessor or the successor of another event.

10. An activity cannot be the predecessor or the successor of another activity.

11. Each event and each activity have only one incoming and/or one outgoing edge.

As mentioned above, connectors describe the links between events. A distinction is made between the combination of inputs into an operator and the combinations of outputs. Inputs as well as outputs can be linked by logical "AND" ($\wedge$), "OR" ($\vee$), and "exclusive OR" ($\times$) relationships. The node for representing the linking operators for input and output is divided. The upper area contains the logical symbols for the input links; the lower level, those for the output links. If only one input and/or output occurs, a logical symbol is eliminated. If only one input and one output exist, the node is eliminated.

Figure 6.10 shows several cases of conceivable process structures. In Figure 6.10(d), which contains an "OR" link to the successors, the rule governing selection can be represented as an independent (decision) function, as shown in Figure 6.10(e). Function F contains the rules for making decisions in an "OR" scenario, which leads to the intermediate events E3 and E4, which are followed by functions F1 or F2.

When more complex relationships exist between the completed and the starting activities, for example, different logical relationships between groups of functions, a connector can be backed by input and output decision tables. Linking different operators is also permissible, as the example in Figure 6.10(f) shows. The logical relationships between functions and events can be explained by analogy to the links between events and functions, as in the example shown in Figure 6.10(d).

In computer science, control flows can be described by ECA (event–condition–action) rules. This technique can be transferred to business processes. An event denotes a fact regarding what and when something happens. For events that represent a point in time, both aspects are combined into one. The condition defines circumstances under which an event is relevant for further steps. If such a setting of circumstances occurs, the action part determines the next activities.

In business process models, events are the output of activities or are externally produced. At the time of modeling an EPC, the relevant events for certain actions can be defined and linked to subsequent activities. Thus, only those events that are relevant for the further steps are used in the model. The condition is combined with the event so that the ECA rule can be reduced to the EA rule. Thus, the representation shown in Figure 6.10(e) is the preferred form.

Instead of defining an event "Order value known" and subsequently checking the condition "Order value $>=$ 5.000 EURO" in Figure 6.11(a), the relevant events are identified and modeled as two different outputs of the preceding activity [Figure 6.11(b)].

By following these basic rules, most of the modeling problems can be avoided. Furthermore, using a standardized procedure to analyze natural-language descriptions

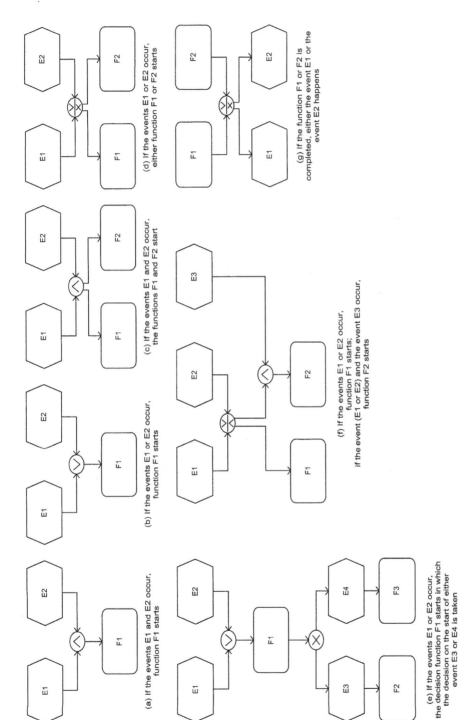

Figure 6.10 Basic control-flow structures.

(a) If the events E1 and E2 occur, function F1 starts

(b) If the events E1 or E2 occur, function F1 starts

(c) If the events E1 and E2 occur, the functions F1 and F2 start

(d) If the events E1 or E2 occur, either function F1 or F2 starts

(e) If the events E1 or E2 occur, the decision function F1 starts in which the decision on the start of either event E3 or E4 is taken

(f) If the events E1 or E2 occur, function F1 starts; if the event (E1 or E2) and the event E3 occur, function F2 starts

(g) If the function F1 or F2 is completed, either the event E1 or the event E2 happens

135

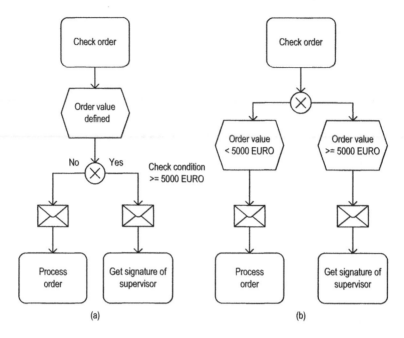

Figure 6.11   Example for the reduction of the ECA rule to the EA rule in EPCs.

of business processes helps to manage this task in a structured way. As mentioned before, the complexity of business processes can be reduced by splitting the real-world descriptions into views. To model EPCs, you can use the following guidelines:

- Determine the exact name of the business process to be modeled. This is a simple but very important step, as it must be clear which process is meant. Thus, choose a title everybody involved in this modeling task agrees on.
- From this definition, derive the initial event(s) (when or under which circumstances does this business process start?), as well as the final event(s) (when does it stop?).
- Now, fill the space between the initial and final events with the basic control flow. In this context, use the function view models, especially the function flow. To do so, look for all relevant verbs in the process description (e.g., by underlining them). Then, transfer these verbs into activities (rectangles with rounded edges) and order them according to their execution procedure. Where appropriate, you should use adequate connectors (AND, OR, XOR).
- Determine one or more events for each transition from one activity to the next one. Ensure that each event is produced by the preceding activity and that the same event triggers the execution of the next activity. If this is not fulfilled, think about whether the event is not correctly named or whether you need an additional activity in between.

- Go through the whole control-flow model again and test its compatibility with the eleven abovementioned structural rules so that you have modeled a correct control flow. In this stage, it is recommended to get the approval of people who are involved in executing the modeled process, so that you have also modeled the correct process.

- Now, add all relevant entities from the other perspectives. The best way is to start with the organizational view by adding the responsible entities like departments, roles, or employees to the activities. Afterwards, the data view as well as the output view should be added.

6.5   THE ARIS ARCHITECTURE

In practice, business processes are highly complex. To cope with this complexity, the process is divided into different views, as shown in the starting paragraphs of this chapter. This makes it possible to describe individual views by using specialized methods without having to incorporate the corresponding relationships into the other views. Ultimately, however, the relationships between the views are reintroduced. At this point, the EPC is used as the core modeling method, integrating different perspectives in the control view. To get a better understanding of this framework, its two basic principles—the view concept and the life-cycle phases—are presented in the following. They build the ARIS architecture and, thus, can serve to position the EPC method.

Figure 6.12 shows the ARIS house, in which the views are visualized [13, 14]. In addition to this division, the second basic thrust of ARIS involves the concept of different descriptive levels. Information systems can be described with respect to their proximity to information technology. Development of a phase concept ensures a consistent description from the business problem all the way to implementation.

The implementation of business processes with the aid of information technology is generally described by differentiated life-cycle models in the form of levels or phases. In ARIS, however, the life cycle does not have the meaning of a procedural model for developing an information system; rather, it defines the different levels based on their proximity to information technology. This follows a three-tiered model (see Figure 6.13).

The description of the operational business problem is the starting point in systems development. This step incorporates the information technology options for supporting business processes and decisions through ICT-oriented business administration. The description encompasses rough business processes that are oriented very closely to user objectives and user language. Therefore, only semiformal descriptive methods are used to represent the description of the business problem. Because of their lack of detail and their highly technical vocabulary, they cannot serve as a starting point for a formalized translation into implementation.

The requirements definition has to describe the business application to be supported in such formalized language that it can be used as the starting point for a consistent translation into information technology. The requirements definition is

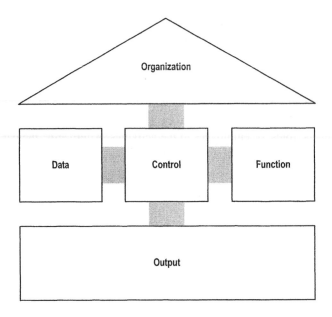

Figure 6.12   ARIS house view of a business process.

very closely associated with the problem description, as expressed by the width of the double-headed arrow in Figure 6.13.

At the design specification level, the conceptual environment of the requirements definition is transferred to the categories of the data processing conversion. The modules or user transactions that execute the functions are defined, instead of the functions themselves. This level can also be thought of as an adaptation of the requirements description to general information technology interfaces. The requirements definition and the design specification should be loosely linked. This means that a design specification can be changed without modifying the requirements definition. This loose link should not mean, however, that the requirements definition and the design specification can be developed in isolation from one another. After completing the requirements definition, it is much more important that the business content be determined in such a way that ICT-oriented considerations such as system output do not have an influence on the requirements content.

In the third step, the implementation description, the design specification is transferred to concrete hardware and software components, thus establishing the physical link to information technology.

The levels are characterized by different update cycles (see Figure 6.13). The updating frequency is highest at the information technology level and lowest at the requirements definition level.

The implementation description is very closely linked to the development of information technology and is, thus, subject to ongoing revision as a result of techno-

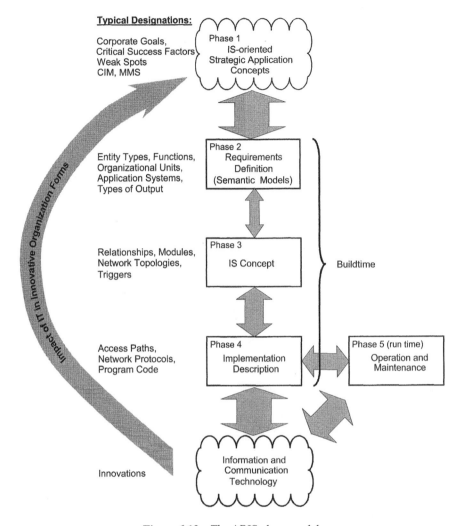

Figure 6.13   The ARIS phase model.

logical changes such as the development of new database systems, networks, and hardware.

The requirements definition level is particularly significant because it is both a long-term repository of collective business knowledge and, at the same time, a point of departure for further steps in generating the implementation description. For this reason, emphasis is placed on the view of developing requirements definitions or semantic models.

The focus is, therefore, on creating requirements definitions, since they possess the longest life cycle within the information system and, through their close affinity to the description of the business problem, they also document the heaviest use of the information system. The requirements definition is the link between users and the initial implementation of their problem description into a data processing language.

The ARIS architecture is developed using the division process (views) and the descriptive levels (phases), including the initial business problem (see Figure 6.14).

6.6 FUTURE EXTENSIONS

EPCs are widely used as a modeling method to analyze and design business processes. They are basis for the optimization of manual and automated procedures, serve as the foundation for implementing ERP systems, and enable customer-cen-

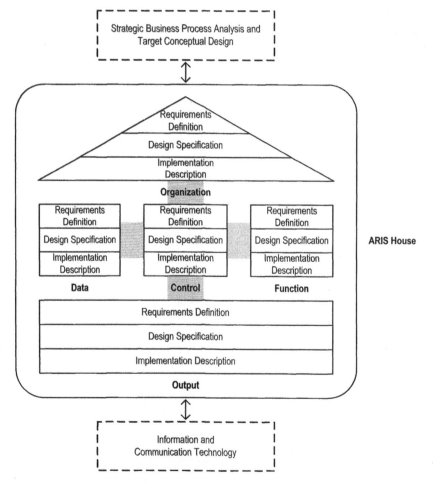

Figure 6.14   The ARIS architecture.

tered service offers in global enterprises and public authorities. Two trends are having a major influence on the further development of the EPC.

First, business process modeling will enable seamless integration between enterprises, leading to collaborative business scenarios [17]. The emergence of dynamic business networks and virtual enterprises as well as extended supply chain scenarios has strengthened the research field of business integration. New business application systems like collaboration-ready ERP software, called ERP II, have been developed [4]. The necessity for interoperability between modeling methods has led to representations of business process models in open standards like BPMI's Business Process Modeling Language (BPML) [15]. Thus, business processes can be designed in distributed, decentralized environments and can be exchanged between process participants. The first approaches incorporate an ontology into the modeling techniques to define common or compatible semantics between cooperation partners. In these scenarios, the automatic configuration of interfaces between IT systems by using Web service technology is targeted. EPCs can be enriched with technology-oriented parameters or can be extended with elements explicitly considering process interfaces to cooperating partners on a business level. This implies that EPCs are mediators between the business perspective and the system perspective.

Second, business process modeling is becoming more intuitive, adaptive, and intelligent. The original idea of introducing a semantic business process modeling technique was aimed at bridging the gap between employees who have the knowledge to fulfill their business task, method experts who can optimize procedures, for example, through simulation or activity-based costing, or IT experts who can build IT systems to support the business task. In order to fully incorporate human knowledge about procedures and to make models more flexible, the rigorous character of decision rules, the claim for static validity of a model over a longer period of time, and monolithic approaches must be dissolved. Intuitive, adaptive, and intelligent modeling can be reached through the integration of artificial intelligence in process models. Traditional concepts like fuzzy set theory to handle vagueness, learning with neural networks, and managing complexity with multiagent systems are being applied to business process modeling [1, 2, 6].

6.7 EXERCISES

Exercise 1. Modeling in General, ARIS Concept

Debacle Plc. is an internationally operating large-scale enterprise headquartered in Saarbruecken. According to a management decision of 1 January 2004, the most important business processes are to be reorganized and supported by IT. The chairman of the board charges you, an external consultant, with the development of a frame concept for the implementation of a respective information system. This information system should refer to the presentation of the basic business information as well as point out possibilities for the operational management of business processes. Today's board meeting expects you to provide models of this system.

- At the beginning of your presentation, state why you think that modeling of business information is such an essential part of the project and the aims which it pursues.
- Explain by means of different viewpoints and layers of the ARIS architecture how it provides the basis for a complete description of information systems.

Exercise 2. Function Structure

The sales figures of myLibro Plc. are not as profitable as desired. After a detailed analysis concerning the sales module of myLibro Plc., you come to the conclusion that the system used up to now does not meet the present requirements at all. A number of weak points exist that considerably prolong the processing times for orders. The order entry, surprisingly, takes four times as long as the order processing. To get an overview over the functions incurred in the scale of sales logistics, you ask the head of the sales department, Sabrina Roofer, for an appointment. Because she is on a business trip in Huerth, near Cologne, at the moment, you conduct the conversation with the sales assistant, John Carpenter. During the conversation, you were able to take some notes, as follows:

- The area sales has these basic functions: issue of an offer, book an order, pursuance of an order, dispatch, invoice, and receipt of payment.
- In context of booking an order, the order registration takes place in a first step. In a second step, the order processing, which is composed of costing and reservation of the ordered items, takes place.
- The pursuance of an order includes the processing of an inquiry and a reminder.
- In the dispatch department, the quality check of the delivery note, product picking, route planning, and booking of products are carried out.
- Invoicing consists of registration, issuing an invoice, and booking following the dispatch of the invoice.

Illustrate the facts described above using a function tree. Identify, thereby, elementary functions, subfunctions, functions, as well as function bundles.

Exercise 3. Target Structure

On the occasion of the semiannual strategy meeting of the management and the heads of departments you, as consultant, are asked to depict the aims mentioned in the meeting in a well-structured diagram, which also elucidates the relations between the aims.

This is an excerpt from the minutes of the strategy meeting:

Everybody unanimously stated that for myLibro Plc. the main goal is the optimization of its enterprise profits. This is to be accomplished in the following ways: increase of domestic market share, opening up new markets, and cost reduction are considered essential. Concerning the increase of domestic market share, necessary measures should be taken that particularly aim at the increase in customer satisfaction by improvement of product quality and reduction of throughput time.

What would the diagram look like?

Exercise 4. Organization View

Dr. Percy, manager of myLibro Plc., would like to have a depiction of the current organizational structure because, due to rapid growth, function processing is all haywire at the moment. You have to depict the hierarchical organization using an organizational chart. Your interview with Dr. Percy produces the following information.

> Management is in charge of the company. All strategic decisions are made by management. The sales, marketing, IT development, and finance departments are subordinate to management. Sabrina Roofer is in charge of the sales department. This department has the subdepartments customer service and logistics subordinate to it. A further breakdown did not seem necessary because of the size of the enterprise. The sales assistant, John Carpenter, is in charge of the logistics department. Mr. Carpenter frequently stands in for Ms. Roofer because she also works for a start-up in San Francisco. Customer service thereby takes care of booking customer orders, processing, as well as customer support. The person in charge of the customer service department is Alexander Barkeeper. Marketing requires creative heads. Experience has shown that for this area it is necessary to have a combination of Internet marketing specialists as well as classic product managers. We are lucky to have Manuela Hangover for this demanding job. Zlatko Shakesbier is in charge of IT development. He is in charge of three programmers and one web designer. Because he, an IT freak, does not have a clue about the requirements that the business demands of its systems, he has to rely on the specification booklets of Ms. Hangover. John Endemol, the meticulous bookkeeper, is in charge of the finance department. In addition to these standard activities we also carry out projects. Now we have the project "Virtual Book Community" going, which is setting up a virtual magazine subscription club on the Internet. This project is headed by Mr. Shakesbier and Mrs. Hangover.

What will your diagram of the organization look like?

Exercise 5. Control View—EPC

As a consultant to myLibro Plc., you were engaged to reorganize the process concerning business trips. For a basis, you first want to ascertain the actual process. Therefore, you question not only Ms. Roofer, who does a lot of traveling because of her job in the sales department, but also the young secretary, Ms. Kermit, who receives the requests of employees for approval by the manager. Model the process as an event-driven process chain (EPC). Use the modeling rules from Chapter 3.

The interview with Mrs. Roofer follows.

> When a business trip is necessary, I make an application for a business trip, which I hand in at the secretary's office for approval. The application is normally approved by

the manager after a while, and approval comes back to my post box. Then I ask around to see whether our company car is available. If it is available, I book it, otherwise I order a rental car. Then I make the business trip, and after it is over the accounting takes place. If a business trip is not approved, I check whether I have to abandon it or integrate it into another business trip at some other time. Otherwise, I go over my report and hand it in at the secretary's office.

The interview with Ms. Kermit follows.

If an employee places a request for a business trip, I check at first to see whether the business trip is in accordance with the requirements for business trips at myLibro Plc. Then I hand in the request with a note at the manager's office. Only if the business trip is accepted do I make a note of the employee and date of the business trip in the file. I put the request, approved or rejected, into the mailbox of the employee.

What will your diagram of the business trip process look like?

REFERENCES

1. O. Adam, O. Thomas, and G. Martin. Fuzzy Workflows—Enhancing Workflow Management with Vagueness. In *EURO/INFORMS Istanbul 2003 Joint International Meeting, 2003,* July 6–10. Istanbul, 2003. URL: http://www.tk.uni-linz.ac.at/EUROIN-FORMS2003_Workflow/rc30_1.pdf [14.07.2003].

2. W. R. Ashby. General Systems Theory as a New Discipline. *General Systems, 3* pp. 1–6, 1958.

3. M. B. Blake. Forming Agents for Workflow-Oriented Process Orchestration. URL: http://www.cs.georgetown.edu/~blakeb/pubs/blake_ICEC2003.pdf [05.03.2004].

4. B. Bond, Burdick, D. Miklovic, K. Pond, and C. Eschinger. *C-Commerce: The New Arena for Business Applications.* Stamford, CT: Gartner Research, 1999.

5. P. P.-S. Chen. The Entity-Relationship Model—Toward a Unified View of Data. *TODS, 1* 1, 9–36, 1976.

6. E. Denti, A. Ricci, and R. Rubino. Integrating and Orchestrating Services upon an Agent Coordination Infrastructure. URL: http://www.ai.univie.ac.at/~paolo/conf/ESAW03/ pre-proc/E0011.pdf [05.03.2004].

7. W. Hoffmann, J. Kirsch, and A.-W. Scheer. Modellierung mit Ereignisgesteuerten Prozeßketten: Methodenhandbuch; Stand, Dezember 1992. In Scheer, A.-W. (Ed.): *Veröffentlichungen des Instituts für Wirtschaftsinformatik,* no. 101, Saarbruecken: Universität des Saarlandes, 1992 (in German).

8. G. Keller, A. Lietschulte, and T. A. Curran. Business Engineering mit den R/3-Referenzmodellen. In A.-W. Scheer and M. Nüttgens (Eds.): *Electronic Business Engineering.* Heidelberg: Physica, 1999 (*4. Internationale Tagung Wirtschaftsinformatik 1999*), pp. 397–423 (in German).

9. G. Keller, M. Nüttgens, and A.-W. Scheer. Semantische Prozeßmodellierung auf der Grundlage "Ereignisgesteuerter Prozeßketten (EPK)." In A.-W. Scheer (Ed.): *Veröffentlichungen des Instituts für Wirtschaftsinformatik,* no. 89, Saarbruecken: Universität

des Saarlandes, 1992. URL http://www.iwi.uni-sb.de/Download/iwihefte/heft89.pdf [20.02.2003] (in German).

10. G. Keller and T. Teufel. *SAP R/3 prozeßorientiert anwenden: Iteratives Prozeß-Proto-typing zur Bildung von Wertschöpfungsketten,* 2nd ed. Bonn: Addison-Wesley, 1998 (in German).

11. G. Keller and T. Teufel. *SAP R/3 prozessorientiert anwenden: Iteratives Prozess-Proto-typing mit Ereignisgesteuerten Prozessketten und Knowledge Maps,* 3rd ed. Bonn: Addi-son-Wesley, 1999 (in German).

12. A. Rapoport. *General System Theory: Essential Concepts and Applications.* Tunbridge Wells: Abacus Press, 1986.

13. A.-W. Scheer. *Architektur Integrierter Informationssysteme: Grundlagen der Un-ternehmensmodellierung.* Berlin: Springer-Verlag, 1991 (in German).

14. A.-W. Scheer. *ARIS—Business Process Modeling,* 2nd ed. Berlin: Springer, 1999.

15. H. Smith and P. Fingar. *Business Process Management: The Third Wave.* Tampa, FL: Meghan Kiffer Press, 2003.

16. L. von Bertalanffy. *General System Theory: Foundations, Development, Applications.* New York: Braziller, 1968.

17. T. Whitney. Collaboration Meets Process Integration. *Transform Magazine, 10,* 9, 32–37, 2001. URL: http://www.transformmagazine.com/db_area/archs/2001/09/tfm0109f1.shtml?enterprise_468 [01.01.2004].

CHAPTER 7

Process Modeling Using Petri Nets

JÖRG DESEL

7.1 INTRODUCTION

Petri nets can be seen as a modeling language and formalism, as a methodology supported by toolsets developed by commercial vendors and by academic institutions, as a theory with a long tradition and hundreds of theorems, as a scientific discipline, and, sometimes, even as a philosophy. This chapter is about process modeling using Petri nets, so not all the aspects mentioned above will be addressed, but only those relevant in the context of processes and information systems. However, instead of just providing the necessary definitions and notions, I also try to present the underlying ideas—the "spirit of Petri nets"—and provide some illustrative examples.

The title of this chapter includes the word "using" rather than "in," "by," or "by means of." This is due to the fact that Petri nets are employed in process modeling in different ways. Process models can be directly expressed as Petri nets, and there are numerous examples of successful applications of Petri net models in practice. Since Petri nets do not only have a nice and intuitive graphical notation but also a sound mathematical foundation, including deep results regarding the relations between process models and their behavior, they are a frequent choice as a reference formalism for other languages that share some common principles with them. This applies in particular to numerous Petri net variants, which are sometimes tailored for specific application domains. It also applies to notations that are not termed "Petri nets," but share some similarities with them, such as UML activity diagrams[1] or event-driven process chains presented in previous chapters of this volume. Furthermore, Petri nets share some fundamental similarities with process algebras.

Section 7.2 gives a very short introduction to Petri nets. Since Petri nets will be considered as a semantical foundation and as a modeling language in their own right, we distinguish the mathematical, graphical, and programming views.

---

[1]As discussed in Chapter 5, activity diagrams in UML 2.0 have a semantics informally defined in terms of Petri net concepts.

Process-Aware Information Systems. Edited by Dumas, van der Aalst, and ter Hofstede   **147**
Copyright © 2005 John Wiley & Sons, Inc.

Section 7.3 shows how single processes can be modeled by elementary Petri nets. In this section, phenomena like choice, concurrency, and deadlocks are discussed. Resources play no important role for single processes, but do for sets of processes that compete for resources, as will be shown in Section 7.4. In Section 7.5, we introduce high-level Petri nets for single models of multiple processes and for the integration of data. Section 7.6 is devoted to behavioral aspects and refinement. A short introduction to analysis issues is given in Section 7.7. Finally, we introduce structurally restricted net classes in Section 7.8. Workflow nets, defined in this latter section, will be employed in the three subsequent chapters, namely Chapter 8 (on process patterns), Chapter 9 (on process design), and Chapter 10 (on process mining). At the end of this chapter, five exercises are provided.

7.2 PETRI NETS

Petri nets are models of distributed and concurrent discrete dynamic systems for which local consequences of operations and local influences of object states play the most important roles. In particular, Petri nets are useful for modeling systems for which behavior is dominated by the flow of information, objects, control, and so on, that is, by *give* and *take* operations. In contrast, *read* and *write* operations are not representable in a canonical way, though they can be simulated by Petri nets.[2]

Petri nets allow the illustrative and precise representation, simulation, and analysis of information and control flow of concurrent components and of synchronization phenomena. Applications include systems and processes in the area of automatic control, distributed algorithms solving synchronization problems in communication systems, and information systems, together with their business processes. In this chapter, emphasis is on the last topic.

Carl-Adam Petri's Doctoral Thesis, published in 1962, was the foundation of what was later called Petri net theory. Since then, numerous variants of Petri net classes have been established. The general popularity of Petri nets in science and in practice is due to the simple and mathematically sound language, the uniform graphical representation of Petri nets, and the formal semantics of Petri net models, which is a prerequisite for simulation and analysis methods. Extensions of the original Petri net model allow one to express specific phenomena of the various application areas.

For a general introduction to Petri net theory, see [19, 20, 15, 16, 8]. Application of Petri nets in the area of business processes can be found in [18, 3, 5] and in books on business process management [2, 4, 14]. Performance analysis is relevant to business process modeling when throughput times, average load, and so on are considered. To this end, Petri nets equipped with time or stochastic time distributions are employed. See [6] for an introduction to stochastic (timed) Petri nets and [17] for a survey on results of the class of generalized stochastic Petri nets. Reference [3] provides some examples for the application of stochastic nets in the area of business

---

[2]Other formalisms like abstract state machines [7] are primarily based on read and write operations on variables and can express give and take operations only indirectly.

process models. For links to bibliographies on Petri nets, see the virtual world of Petri nets[3] on the Web.

The distinction between Petri nets as a visual language, as a mathematical theory, and as a formal language is important when Petri nets are compared or combined with other approaches. Taking Petri nets as a modeling language, they are viewed as a visual language. When Petri nets are used as an underlying semantics of another language, then the mathematical theory of Petri nets is considered. Applications need a computer-supported representation of Petri nets to apply simulation and analysis techniques. Therefore, Petri nets have to be given as a formal language in this area. No matter how a Petri net appears, there are some dominant underlying principles that make a net a *Petri* net. These aspects will be explained in the remainder of this section, after presenting an introductory example. For a deeper discussion, see [12].

7.2.1 Introductory Example

Let us consider a very simple business process of a company that deals in cars. The process is always initiated by the request of a customer who wants to buy a car with some specification (task *init*). The company has a department U for available used cars and another department N for new cars that have to be ordered. For simplicity, we assume that, for each specification, either a used car exists or a new car can be ordered that matches the specification. So we have the task *find used,* performed by an employee of department U, which, after successful termination, is followed by the task *sell used.* Analogously, an employee of department N can execute the task *find new,* possibly followed by *order new.*

To reduce waiting time for the customer, the tasks *find used* and *find new* are enabled concurrently after the initial task *init.* As soon as *find used* terminates successfully, *sell used* is executed. Execution of the latter task also requires and removes the request from department N, that is, it disables *find new.* Similarly, the task *order new* can be executed after successful termination of *find new,* and it removes the request from department U.

The normal and successful execution sequences of this process are *init, find used, sell used,* and *init, find new, order new.* The reader might have noticed that, due to the concurrent *find* tasks, two more execution sequences might happen, namely, *init, find used, find new* and *init, find new, find used.* After these sequences, the task *sell used* is not enabled because it requires the request from department N, which is no longer available. Similarly, *order new* is also not enabled because it requires the request from department U. Therefore, the process got stuck in a *deadlock.*

It is very possible that the reader does not agree with the above observation because he or she might have a different understanding of the process description. Hence, it is necessary to formulate this description in a more precise way. The main aim of business process modeling is to provide formal means for a precise description of business processes, using languages with formal syntax and semantics.

[3]http://www.informatik.uni-hamburg.de/TGI/PetriNets/

Moreover, the models defined in such a language should be both easy to generate and to understand, that is, they should support the communication of business process models. Another aim is the identification of possible flaws in process descriptions, such as the possible deadlock in the example above. Furthermore, formal process models can be used for performance analysis, for resource planning, and as inputs for workflow management systems.

There are several ways to describe the above process description in a formal way. Concentrating only on the tasks and their respective order would lead to the picture shown in Figure 7.1. The facts that *sell used* can only occur if *find new* did not occur and that *order new* can only occur if *find used* did not occur is not represented in this figure because the representation of the requests is missing.

If we emphasize states instead of tasks, then we might come up with the picture shown in Figure 7.2. In this representation, the deadlock is clearly depicted. However, the information about the concurrency of *find used* and *find new* is lost. It rather looks like there was a decision made right after the initial task to start either with *find used* or with *find new* (in both cases, it would be easy to disallow the subsequent occurrence of the other task).

Petri nets represent tasks by transitions (rectangles) and distributed states by sets of places (circles). A Petri net picture of our example is given in Figure 7.3. An arc leads from a place to a transition representing a task if this place represents a precondition for the execution of the task. Similarly, arcs from transitions to places indicate the set of postconditions. Places can carry a token, as is the case for the place *customer request* initially. The occurrence rule states that a transition can occur if (and only if) all its preconditions carry a token. The occurrence removes a token from each precondition and adds a token to each postcondition. In our example, initially only the transition representing the task *init* is enabled, and its occurrence leads to the state in which the places *request for U* and *request for N* carry one token each (and all other places are unmarked). At this state, both transitions representing the *find* tasks are enabled. They are enabled concurrently because they do not share a common precondition, that is, they do no not compete for a token. If both transitions occur, then afterward no transition is enabled and we reach a deadlock. Otherwise, execution of *find used* and *sell used* leads to the state *successful termination,* and so does execution of *find new* and *order new*.

7.2.2   Petri Nets as a Visual Language

There is no need to argue that the two dimensions provided by a graphical language have significant advantages for specifying, communicating, and understanding concepts and ideas in general, compared to one-dimensional textual languages. Specification and communication of dynamic systems is one of the main applications of Petri nets in practice.

A typical definition of a Petri net based on its graphical appearance is:

> A Petri net is a directed graph with two different types of nodes: *places,* represented by circles (or ellipses), and *transitions,* represented by rec-

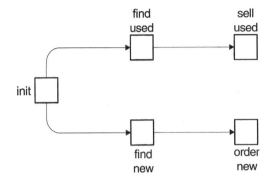

Figure 7.1   Representing the tasks and their order.

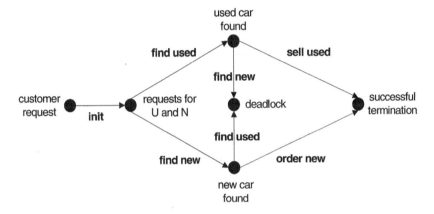

Figure 7.2   Representing the states and state transitions.

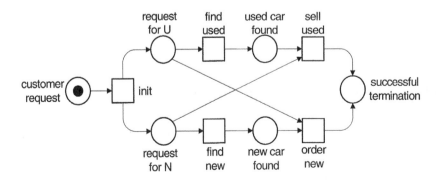

Figure 7.3   Representing the business process by a Petri net.

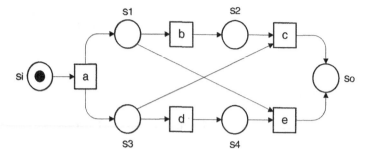

Figure 7.4   A Petri net.

tangles (or bold bars). Petri nets are bipartite, that is, no arc connects two places or two transitions. Nodes and arcs can have various annotations.

Figure 7.4 shows the same Petri net as Figure 7.3. Here we consider the Petri net more formally, without referring to the previous interpretation as a business process of a car vendor company. Places and transitions of the Petri net are annotated by their names, s_i, s_1, s_2, s_3, s_4, s_o and a, b, c, d, e. Since all names are different, that is, since the annotation mapping is injective, we can identify the places and transitions with their names. In other words, the net has places s_i, ..., s_o and transitions a, ..., e. The places also have annotations representing a *marking*. The place s_i carries one token and all other places carry no token. Arcs have no annotations in this example.

Petri nets are not only supported by graphics but each Petri net *is* a special directed (sometimes annotated) graph. In general, annotations play an important role, but they can often be replaced by graphical means. Formal annotations have to be distinguished from informal decorations that might help understanding the components of a Petri net (as is the case for the net shown in Figure 7.3). In contrast to semiformal languages, in Petri net theory there is always a very clear distinction between elements belonging to the language (and hence having influence on behavior) and formally irrelevant comments.

In some sense, we just confused mathematical graphs with graphical notations. Mathematical graphs do not indicate the relative position of nodes, the style of arcs, and the like. This is the case for Petri nets, too.[4] However, the topology of a drawn Petri net is important from a pragmatic perspective.

7.2.3   Petri Nets as Mathematical Structures

Mathematically, a Petri net graph is given by two sets of nodes (places and transitions) and a binary relation representing the arcs. So a typical mathematical definition of a Petri net is:

---

[4]In contrast, graphical languages such as SADT [21] distinguish arcs touching a node at its right, left, upper, or lower sides.

A Petri net consists of two disjoint sets S (*places*) and T (*transitions*) and a binary relation $F \subseteq (S \times T) \cup (T \times S)$ (*flow relation*).

The letter S for places is from the original German word "Stelle" for "place." Often, P is used instead.

In the above example, we have

$$S = \{s_i, s_1, s_2, s_3, s_4, s_o\}$$

$$T = \{a, b, c, d, e\}$$

$$F = \{(s_i, a), (s_1, b), (s_1, e), (s_2, c), (s_3, c), (s_3, d), (s_4, e), (a, s_1), (a, s_3),$$
$$(b, s_2), (c, s_o), (d, s_4), (e, s_o)\}$$

For technical convenience, it is often required that the sets S and T are finite and not empty. Moreover, it is useful to consider *connected* Petri nets, that is, Petri nets satisfying $(F \cup F^{-1})^* = (S \cup T) \times (S \cup T)$. Graphically speaking, a connected Petri net is a Petri net that cannot be drawn as two nets without connection, both having at least one node.

When are two nets identical? Taking nets as mathematical structures, two nets are identical if their sets of places, transitions, and arcs coincide. However, the graph layout of two identical nets might be completely different. Conversely, two drawn Petri nets that look identical might differ because the nodes of the nets represent different sets of places and transitions. If a drawn net has no annotations representing names of nodes, then it refers to a class of isomorphic nets, where *isomorphism* is defined by suitable bijections between places and between transitions such that the flow relation is preserved.

Usually, a Petri net can be equipped with *markings,* where a marking is a mapping from the set of places to some domain. The definition of a Petri net frequently includes one *initial marking.* This marking is often denoted by m_0, and it is depicted in the graphical representation of the net.

7.2.4   Petri Nets as a Formal Language

Formal languages and mathematical structures are not the same. A formal language *syntactically* describes a mathematical structure, which has its identity on a *semantic* level. To understand the difference, consider the two character strings $\{1, 2, 3\}$ and $\{3, 2, 1\}$. They are syntactically different but describe the same set of three numbers. Since the only way to talk or write about mathematical structures is to use some kind of syntactically defined language, the distinction between the semantical and the syntactical level is often neglected. In fact, the previous section also used syntactical means to describe a mathematical structure.

A usual way to formalize Petri nets, which is relevant only on the syntactical level, is to present their components in triplets: $N = (S, T, F)$. The initial marking is included in the definition if we use a 4-tuple: $N = (S, T, F, m_0)$.

In general, there are many ways to describe Petri net components syntactically such that this representation can be in- and output to Petri net tools. For example, the Petri net markup language PNML [22] is based on XML. Sometimes, the information about the graphical layout of a Petri net is included in the syntactical representation, and sometimes it is not.

Petri nets and similar languages are often used as a process description language in *workflow systems* [3]. In this application area, a Petri net is input to a *workflow management system* that executes the automatized business process, based on the process description (see Chapter 2).

7.2.5 Principles of Petri Net Theory

Different Petri net classes allow modeling on different abstraction levels. Independent from the concrete net class and from representation issues, the Petri net components have a clear interpretation:

- A place represents one or many *objects.* Each object is always in some state.
- A transition represents one or many *operations,* which are only possible at specific states of objects and which change the state of specific objects.
- The *occurrence rule* determines under which object states a transition is enabled to *fire (enabling condition)*, that is, the respective operation can occur, and which state changes are caused by the firing of the transition.
- By the *principle of locality,* the enabling condition of a transition and the state changes caused by its firing concerns only places that are directly connected to the transition by an arc (in either direction). Conversely, a state of a place only influences transitions in the immediate vicinity of the place, and it can only be changed by firings of these transitions.

The last item is of particular importance. In large and complex systems with increasing indirect dependencies, it seems that everything is somehow related to everything and no detail can be understood. This holds in particular for graphical models based on automata, where each node represents a global state. For Petri nets, the local vicinity of each node corresponds to the logical or physical vicinity of the represented object or task, which is usually bounded even for complex systems. So the local vicinity of net elements does not increase in general. Hence, even large Petri net models, composed from many components, remain comprehensible because, for each detail, emphasis is on the local relations between elements.

7.3 PETRI NET CLASSES AND BEHAVIOR

Before discussing Petri net models of business processes in the next sections, we provide some general definitions for the behavior of elementary Petri nets and high-level Petri nets.

The following notations will be useful. We will always consider a net with S being the set of places, T the set of transitions, and F the set of arcs.

For a transition t, the *preset* of t is the set of places s from which a directed arc leads to t. It is denoted by ${}^{\bullet}t$, that is, ${}^{\bullet}t = \{s \mid (s, t) \in F\}$. Similarly, $t^{\bullet} = \{s \mid (t, s) \in F\}$ is the *postset* of t. The elements in the preset of a transition are called *preconditions*, the elements in the postset *postconditions*. Analogously, ${}^{\bullet}s$ and $s^{\bullet}$ denote the pre- and postset of a place s.

A *path* of a net is a sequence of net elements (places and transitions) such that each element except the first one is in the postset of its predecessor in the sequence. A path is a *cycle* if its last element is identical to its first element.

7.3.1 Elementary Petri Nets

Elementary Petri nets are the original variant of Petri nets.[5] Tokens on a place of an elementary Petri net cannot be distinguished. Therefore, in an elementary Petri net the state of a place is given by a nonnegative integer, representing the number of tokens on the place. Consequently a *marking*, representing a global state, is a mapping $m: S \rightarrow \{0, 1, 2, \ldots\}$.

Graphically, a marking m is depicted by $m(s)$ black dots in the place s, for each place s. Mathematically, if some order is defined on the set of places, a marking can also be represented by a vector. For example, $(1, 0, 0, 0, 0, 0)$ is the marking of a net with six places $s_i, s_1 \ldots, s_4, s_o$, which assigns one token to the place s_i and no token to the other places (compare Figure 7.4).

The *enabling condition* requires that a transition t is only enabled at (or by) a marking m if each place in the preset of t carries at least one token, that is, $m(s) > 0$ holds for each place s in ${}^{\bullet}t$. By the *occurrence rule*, the *firing* of t at m yields the marking m' (notation: $m \xrightarrow{t} m'$), defined by

$$
m'(s) = \begin{cases} m(s) - 1 & \text{if } s \in {}^{\bullet}t \text{ and } s \notin t^{\bullet} \\ m(s) + 1 & \text{if } s \notin {}^{\bullet}t \text{ and } s \in t^{\bullet} \\ m(s) & \text{otherwise} \end{cases}
$$

In the example, the marking $(1, 0, 0, 0, 0, 0)$ enables transition a. Firing a leads to the marking $(0, 1, 0, 1, 0, 0)$, which enables transitions b and d. Firing b at this marking yields $(0, 0, 1, 1, 0, 0)$.

An *occurrence sequence*, enabled at a marking m, is a sequence $\tau = t_1 t_2 t_3 \ldots t_n$, $n \geq 0$, of (not necessarily distinct) transitions such that

$$
m \xrightarrow{t_1} m_1 \xrightarrow{t_2} m_2 \xrightarrow{t_3} \cdots \xrightarrow{t_n} m_n
$$

holds for suitable markings $m_1, m_2, \ldots, m_n$ (notation: $m \xrightarrow{\tau} m_n$). For $n = 0$ the sequence τ is the empty sequence, enabled at any marking.

A marking m' is *reachable* from a marking m if there exists an occurrence sequence τ satisfying $m \xrightarrow{\tau} m'$. Given two markings m and m' of a Petri net, it is de-

---

[5]The name *elementary Petri net* should not be confused with *elementary net systems*. We consider elementary net systems as well as place/transition Petri nets to be elementary.

cidable whether m' is reachable from m (this is far from trivial and it was an open problem for many years).

Now we are ready to give the necessary formal definitions:

> An *elementary Petri net* is a Petri net equipped with an *initial marking* that assigns to each place a nonnegative integer.

> The set of *reachable markings* of an elementary Petri net is the set of markings reachable from the initial marking.

An elementary Petri net is called *bounded* if for each place s there is a natural number k such that $m(s) \leq k$ for each reachable marking m. If the set of places of the net is finite, boundedness coincides with finiteness of the set of reachable markings.

7.3.2 High-Level Petri Nets

At this point, the reader might wish to jump to Section 7.4 if he or she is not yet interested in the more involved definitions for high-level Petri nets, and come back here later.

In a high-level Petri net, a place can carry different tokens, which can be distinguished. Graphically, instead of drawing black dots, tokens are represented by suitable symbols for elements of the respective domains. A place can also carry the same element more than once. Since colors can be used to distinguish tokens, variants of high-level Petri nets are also called *colored Petri nets* [16].

Each place s has an associated *domain* dom(s), which is the underlying set for the possible tokens on that place. Formally, the current state of a place s is given by a finite bag of tokens, that is, a mapping m: dom(s) $\rightarrow \{0, 1, 2, \ldots\}$ satisfying $m(x) \neq 0$ only for finitely many x in dom(s) (which ensures that no place carries infinitely many tokens). Elementary Petri nets occur as a special case. The domain of each place of an elementary Petri net is the singleton set $\{\bullet\}$.

A *marking,* representing a global state, is given by the states of all places. Taking dom(S) as the union of all domains of places, a marking is formally defined by a mapping m: $S \rightarrow$ (dom(S) $\rightarrow \{0, 1, 2, \ldots\}$) satisfying, for each place s: $m(s)(x) \neq 0$ only for x in dom(s), and only for finitely many x in dom(s). As for elementary nets, a high-level net is equipped with an initial marking.

A transition of a high-level net can fire in different *firing modes.* In each firing mode, specific tokens are moved from and to the places in the pre- and postset of the transition. The enabling condition requires for a firing mode that the corresponding tokens exist in the preconditions of the transition. Formally, a mapping assigns for each firing mode of a transition, and for each pre- and postcondition, a finite bag over the domain of the place (see the above definition of the place's state). When the transition occurs in a given mode, then tokens are removed or added to the respective place, according to this mapping.

Whereas high-level Petri nets have been defined mathematically, *predicate/transition nets* are syntactical variants that combine formal language with a graphical

Petri net representation. Each arc (from or to a place s) is annotated by a term that might include variables. By assigning suitable values to these variables, the terms are evaluated to bags over the domain of the place s. A firing mode of a transition is given by a complete and consistent assignment of values to all variables appearing at arcs in the vicinity of the transition, such that the domains of the respective places in the pre- and postset are respected. The mappings mentioned in the previous paragraph on firing modes correspond to the interpretation of the terms. Additionally, a logical expression using the same variables can restrict legal assignments of variables. This so-called *guard* of a transition must evaluate to *true* for a variable assignment to enable the transition. An example for a predicate/transition net will be given in Section 7.5.2.

7.4   MODELING SINGLE PROCESSES WITHOUT RESOURCES

Now we turn to modeling of processes. Since we aim at providing a suitable modeling language, we have to discuss what we mean by a process first.

> A *process* consists of *tasks* that have to be executed. These tasks can be in some order (*sequentially*), stating that one task can only be executed after the execution of another task is finished. If two tasks are not ordered, then they can be executed *concurrently*. Tasks can also be *alternative,* that is, if one task is executed, then the other task is not executed and vice versa. Tasks can be executed more than once in general.

> A process can be in different states. A process starts with an initial state (which is not necessarily unique) and might end with a final state (which is also not necessarily unique). Usually, it passes through several intermediate states.

In this definition, we did not consider usual ingredients of process definitions that have to do with business issues because they will not be represented in our first model. However, one should bear in mind that each process run (or process execution) should eventually reach a final state because the final state represents successful termination of the process. Other issues, partly addressed in later sections, concern the usage of resources and performance aspects.

According to [1], we call the execution of a task an *activity*. Therefore, a process consists of tasks, whereas each possible run of the process consists of activities referring to the tasks. The order between tasks mentioned in the above definition carries over to activities in runs; activities can be ordered, concurrent, or exclude each other in case of alternative tasks.

7.4.1   Elementary Building Blocks

The order between tasks is constituted by conditions. If one task can only be executed after another task, then a postcondition of the former task is a precondition of the latter. We will employ particular elementary Petri nets to model single process-es. Tasks are represented by transitions and activities by transition occurrences. Pre- and postconditions of tasks are modeled by places that are in the post- and pre-sets of the respective transitions. The idea is that for ordered tasks, the first transi-tion occurrence produces a token on the place, whereas the second transition is only enabled after this token is produced, and it consumes the token (see Figure 7.5).

Since places play the role of conditions now, they can only be in two distinct states: *true* and *false*. Therefore, a place should have only two possible states, *marked* (with one token) and *unmarked*. In other words, we require that the elemen-tary Petri net is not only bounded but has the bound 1 for all places. Petri nets en-joying this property are called *1-safe* (or just *safe*).

Unfortunately, it is not obvious to see (though decidable) whether a given ele-mentary Petri net is 1-safe or not because, by definition, all reachable markings have to be considered. Therefore it makes sense to demand 1-safety for each place. We just do not allow a transition firing that would add a second token to a place (this is called a *contact* situation). Alternatively, we can modify each elementary Petri net in such a way that the resulting net is 1-safe, which is discussed next.

For each place s which might violate 1-safety, we add a complement place $\bar{s}$ with ${}^{\bullet}\bar{s} = s^{\bullet} \backslash {}^{\bullet}s$ and $\bar{s}^{\bullet} = {}^{\bullet}s \backslash s^{\bullet}$ (see Figure 7.6). The initial marking m_0 is extended to the additional places by $m_0(\bar{s}) = 1 - m_0(s)$. It is not difficult to observe that, in this en-larged net, every reachable marking m satisfies $m(s) + m(\bar{s}) = 1$. Hence s carries at most one token. Moreover, a transition is enabled at a marking in the original net (respecting contacts) if and only if the transition is enabled at the corresponding marking in the extended, *contact-free,* net. The behavior is not changed by the mod-ification.

Concurrency and alternatives are also easily represented by transitions and places. Alternative tasks are modeled by transitions that share a common place in their presets.

The transitions can only be enabled when the place is marked. Since it carries at most one token, the occurrence of one transition disables the other transition. So there is a *choice* (or *conflict*) between the transitions. Sometimes, the forward-branching place is called an OR-split. Corresponding OR-joins are modeled by backward-branching places (see Figure 7.7). Nothing is said about the actual choice taken in a particular run. This might happen alternatingly, with some given proba-bilities, always in favor of one transition, or following any other strategy.

Concurrent tasks are modeled by transitions with disjoint pre- and postsets, which are both enabled at some marking. Figure 7.8 shows a forward-branching transition (an AND-split) distributing tokens to two places such that both subse-

Figure 7.5   Sequence of tasks.

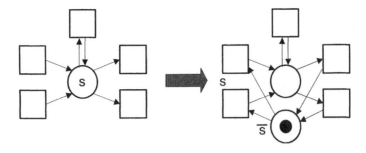

Figure 7.6   Adding a complement place.

quent transitions are enabled concurrently. Concurrency comes with *synchronization;* the backward-branching transition can only fire after both the upper and the lower lines of tasks have finished (an AND-join).

7.4.2   More-Involved Building Blocks

As a process should have an initial state, so should its model. We could choose any initial marking as a representation of the initial state. However, it is convenient to have one special *input place* s_i with an empty preset. This place (and no other place) carries a token initially. Therefore, only transitions in $s_i^\bullet$ are enabled initially. Likewise, the final states are represented by markings assigning a token (only) to a special *output place* s_o which has an empty postset.

The example shown in Figure 7.4 represents a process in its initial state (only place s_i marked). After firing transition a, representing an initial task, transitions b and c can fire sequentially and, alternatively, d and e. Both sequences lead to a marking in which only s_o is marked, representing the legal final state. As discussed in the introduction, there are more runs of this process: After firing a, transitions b and d are enabled concurrently. After firing both, a marking is reached that enables no transition and does not mark the place s_o. Hence, this run never reaches a final

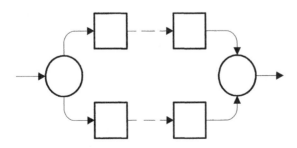

Figure 7.7   Choice.

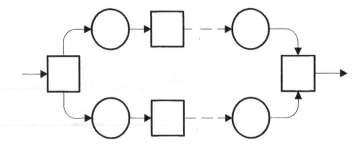

Figure 7.8 Concurrency.

state but, rather, a deadlock. The process is considered ill-designed because a deadlock can be reached.

In the following, we suggest four possible solutions to repair the process, at the same time explaining some modeling building blocks of Petri nets.

The first solution, shown in Figure 7.9, adds arcs (s_1, d), (d, s_1), (s_3, b), and (b, s_3). For the sake of readability, these two small *loops* are depicted by arrows with two arrowheads. Due to the loops, transition d is no longer enabled after b has fired, and vice versa. So the bad run is excluded.

In the second solution, shown in Figure 7.10, the additional place s_5 ensures that, after firing transition a, transitions b and d are in a conflict situation. So only one of both can occur, and we are done.

The third solution adds the transition d', the complement to transition d (Figure 7.11). This transition ensures that after reaching the deadlocked state, the effect of transition d can be reversed, and c can become enabled again, leading to the final state. In this solution, the number of possible runs is infinite, because transitions d and d' can always continue to fire alternatingly. It is even possible that d always fires after d', and vice versa, so that the final state is never reached. Since either c or e is also enabled infinitely often during the sequence, *fairness assumptions* ensure that eventually one of these transitions fire. Another solution to this problem is to

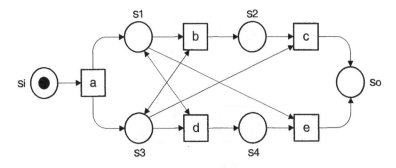

Figure 7.9 Adding loops.

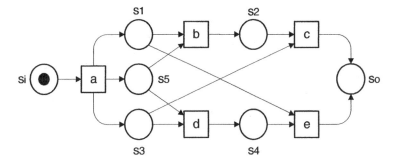

Figure 7.10   Adding a conflict.

assume *priority* of transition *c*. If both transitions *c* and *d* are enabled, then only *c* should fire.

Figure 7.12 shows the last solution. Here, the possible behavior is restricted. Transitions *b* and *e* occur alternatingly, and so do transitions *c* and *d*. The depicted initial marking will lead to consequences of *d* and *e*, whereas for the next run of the process transitions *b* and *c* will occur. In this example, the initial marking not only marks s_i but also either of the two additional places (and similarly for the final markings and s_o).

7.4.3 Modeling Repetitive Processes

In the previous example, we talked about a subsequent run after the first run. In other words, we added a new token to the initial place s_i after the first token was consumed for the first run. Hence, it makes sense to add an input transition t_i to s_i that generates input tokens, and, similarly, an output transition t_o to s_o (see Figure 7.13).

In this Petri net, transition t_i can fire arbitrarily, spoiling 1-safety. Moreover, deadlocks cannot be identified anymore. After firing the occurrence sequence t_i *a b d* (which previously led to a deadlock), now transitions t_i *a c* t_o *e* t_o can occur,

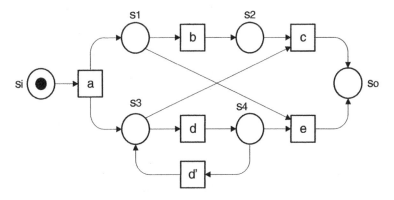

Figure 7.11   Adding a complement transition to *d*.

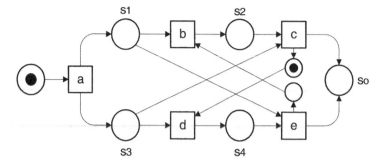

Figure 7.12 Adding places for alternation.

leading back to the empty marking. So the input transition should only occur again after the first run has come to an end. Therefore, we glue input and output transitions, and this way close the circuit between s_o and s_i. This is shown in Figure 7.14 for the Petri net of Figure 7.12.

The obtained Petri net has a property that is important in Petri net theory: it is *live*. Liveness means that not only every reachable marking enables some transition, but also enables a sequence containing all transitions. In other words, every transition can always become enabled again. Liveness together with safety is called *soundness* for a slightly restricted class of Petri nets in [1]; see Subsection 7.8.4.

7.5 MODELING PROCESSES WITH RESOURCES

Tasks of processes need *resources* to be executed. Resources can be humans, machines, computing time, and so on. Resources can be grouped in *resource classes*. The execution of a task needs specific numbers of resources from respective resource classes. Typically, a single resource cannot be used by more than one task simultaneously.

Several processes might compete for resources. That is, during the execution of several processes, resources are assigned to tasks. Some tasks might not be executable yet because necessary resources are missing. This appears only for scarce

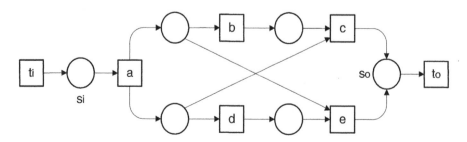

Figure 7.13 Adding input and output transitions.

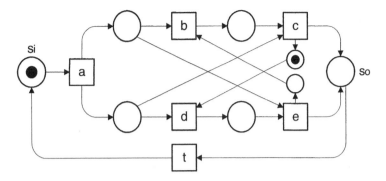

Figure 7.14   Adding a transition closing the circuit.

resources, that is, resources of some resource class that are not available in a sufficient number to handle all concurrent tasks of all processes. Conversely, resources
that are always sufficiently available are not relevant in this sense (they are important when *cost* aspects are considered; see [11, 13]).

Resource competition might also occur for single processes if two concurrent
tasks need the same resource. More severe problems arise when many processes run
concurrently, using the same restricted set of resources.

7.5.1   Modeling Resources with Elementary Petri Nets

The most obvious way to model resources with elementary Petri nets is to use additional places, one place for each resource class. The initial marking of such a *re-*
source place is the number of resources available for the class. Then, a complete
Petri net model consists of a number of Petri nets representing the concurrent
processes, partially connected by the resource places. These processes can be identical (if several instances of one process run concurrently) or different.

Figure 7.15 shows an example. We have two instances of the same process and
one instance of a different process. Transitions a_1, a_2, and d cannot occur concurrently because they access a resource which is available only twice. The resource
used by b_1, b_2, and f is available only once. It is easy to observe that the resource
places are always bounded but, in general, not 1-safe. In fact, they never change
their number of tokens.

In this example, we have two disjoint resource classes. In real processes, however, it often appears that resource classes are not disjoint. Consider, for example, a
task that needs to be performed by a male employee and a concurrent task that
needs an employee who is not older than 55 years. Then, the resource class of all
employees of the company involved has to be divided in four disjoint subclasses:
young males, old males, young females, and old females. The former task accesses
a resource from young males or from old males, whereas the latter accesses a resource from young males or from young females. See Figure 7.16, where *ym* stands
for young males, *om* for old males, *yf* for young females and *of* for old females.

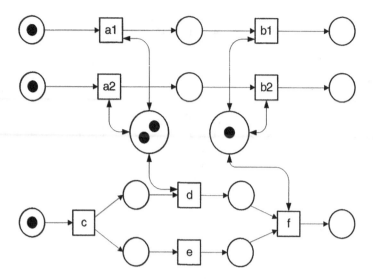

Figure 7.15   Adding resource places.

Until now, we modeled access to resources by a transition representing a task and by places representing the corresponding resource classes. The semantics of this approach is that either all necessary resources are taken simultaneously or none of them are allocated. The matter becomes more complicated if the resources are allocated in some order, which is a more realistic setting. Then, a refined view of a task requiring two resources could look like that in Figure 7.17.

Combining two processes that compete for the same two resources may lead to the Petri net shown in Figure 7.18.

After firing transitions *a* and *e*, no transition is subsequently enabled. The Petri net reached a deadlock, due to the resources. This phenomenon appears for all systems of processes that access resources concurrently. It is called the *resource allo-*

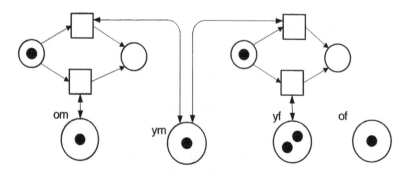

Figure 7.16   Four resource classes.

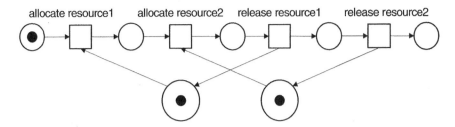

allocate resource1   allocate resource2   release resource1   release resource2

Figure 7.17   Separate allocation and release of resources.

cation problem. Many solutions to avoid deadlocks or to identify and solve a deadlock by resetting a resource allocation have been suggested in the literature.

The same problem occurs when resources are not allocated to single tasks but to sets of tasks in a single process. For example, it might make sense to allocate a human resource for three subsequent tasks of a process, without releasing the resource after each task execution. The danger of deadlocks is always present when a process can demand an additional resource while it is already utilizing a resource.

7.5.2   Modeling Processes and Resources with High-Level Petri Nets

Modeling several instances of the same process by separate elementary Petri nets, as shown in Figure 7.15, is clumsy and leads to huge nets that are not easy to understand in general. The same holds for resources with multiple classification, as in Figure 7.16. A very elegant way to model the same behavior in a more compact way is to use high-level Petri nets. Figure 7.19 provides a predicate/transition net model for four process instances $p_1, \ldots, p_4$, where p_1 and p_2 refer to the same process and p_3 and p_4 refer to the same process. The resource place contains all employees in terms of pairs, where the first component is m for males and f for fe-

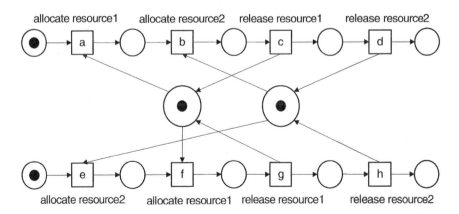

allocate resource1   allocate resource2   release resource1   release resource2

allocate resource2   allocate resource1   release resource1   release resource2

Figure 7.18   Deadlock caused by resource allocation.

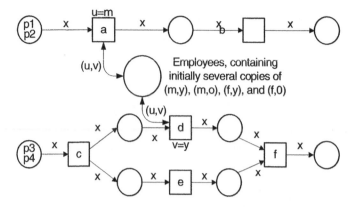

Figure 7.19 Modeling with high-level Petri nets.

males, whereas the second component is y for young employees and o for old ones. Then, transition a can only fire for males and transition d only for young employees.

An elementary Petri net for the process representation with two (black) tokens in the initial place does not serve the same purpose. In the above model, after firing transition d for p_3 and transition e for p_4, transition f is not enabled, because the two tokens on its input places do not match—none of the two processes have completed the tasks represented by d and e. Taking an elementary Petri net instead, the tokens in the preset of transition f could not be identified as mismatched, and so the transition f would be enabled, which is not desired.

Another advantage of high-level Petri nets for modeling a collection of processes is the ability to create new process instances during run time. Adding an input transition to the initial place of a process, new and different instances are created with every occurrence of the transition (see Figure 7.20). To ensure that new process instances are represented by different high-level tokens, we chose increasing natural numbers as names. This input transition could also be triggered by transitions of other processes.

Since high-level Petri nets can also be used to assign data to tokens, they allow an integrated view of processes and data. In [18], each place is considered a relation scheme and each marking of a place a relation with tokens being tuples. In contrast,

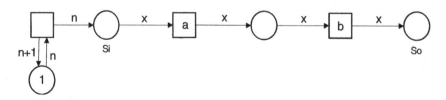

Figure 7.20 Creation of process instances.

[1] assigns only the part of data to tokens that is necessary to make choices (*routing information*). For example, if an alternative in a process depends on some number relevant to the process instance, then this number—but not all data of the process instance—is represented by the token. Guards at the conflicting transitions make sure that, according to the value of the number, always only one of the conflicting transitions can occur.

7.6 BEHAVIOR AND REFINEMENT

We mentioned runs of processes in Section 7.4, and we defined sequences of transition occurrences called occurrence sequences in Section 7.3.1. We claim that these two notions do not really agree, although, in most contributions in this area, occurrence sequences are used for formalizing behavior of Petri nets. Since an occurrence sequence is just a sequential representation of behavior, causality and concurrency are not expressed explicitly. In other words, if some transition appears after another transition in an occurrence sequence, then it might be causally dependent on the previous transition occurrence (that is, it uses a token produced by the first transition), or both transitions occur concurrently and appear in the sequence in an arbitrary order. In this section, we argue that runs are more appropriately modeled by particular Petri nets called *causal nets*.[6]

7.6.1 Causal Nets

As an example, let us consider the net shown in Figure 7.15 again. In each possible run of all three processes, two of the three transitions, a_1, a_2, and d, occur concurrently. Now, consider the case in which a_1 and a_2 occur concurrently, whereas d uses the resource token after a_2. Moreover, b_1, b_2, and e share one resource and have, thus, to be ordered in a run. Let us assume that this resource is first used by b_2, then by b_1, and finally by f. The net shown in Figure 7.21 represents this run.

Each transition of a causal net represents a transition occurrence of the Petri net modeling the process. Each place represents a token that either exists initially (these places of the causal net are initially marked) or is produced by a transition occurrence. Formally, a causal net is a 1-safe elementary net that has no cycles. Its flow relation represents immediate causality. The partial order given by the transitive closure of the flow relation is the causality relation. In other words, a transition occurrence depends on another transition occurrence if both are connected by a sequence of directed arcs. Since transitions of a Petri net representing a process can occur more than once, there might be several copies of such transitions in a causal net. Since each token is produced by at most one transition occurrence and is consumed by at most one transition occurrence, places of causal nets are neither forward nor backward branched. In particular, if the Petri net (the process model) contains a choice represented by a forward-branching place, then at most one of its

---

[6]We avoid the usual term *process net* here because it is easily confused with a Petri net modeling a process.

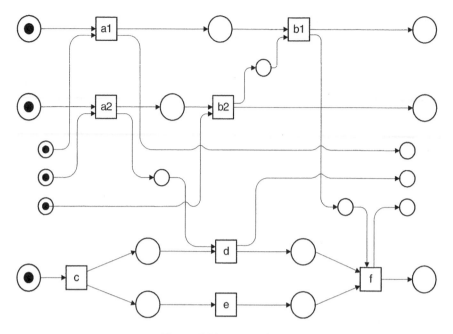

Figure 7.21   A causal net.

output transitions appears in the postset of a respective place of a causal net, that is, conflicts caused by choices are solved in runs. Finally, a transition occurrence precisely follows the specification given by the preset and postset of a transition. The preset of a transition in a causal net corresponds to the preset of the corresponding transition in the Petri net representing the process, and similarly for postsets.

The token game can be played for causal nets as for the original nets. Every occurrence sequence of a causal net is also an occurrence sequence of the original net. Conversely, for each occurrence sequence of the original net there exists a causal net possessing the same occurrence sequence. So this partial-order semantics agrees with the previously defined sequential semantics.

One main advantage of causal nets is the efficient representation of runs, compared to occurrence sequences. If there are many concurrent transitions, then their concurrent firing is expressed by only one causal net but corresponds to many different occurrence sequences, where concurrent transition occurrences are arbitrarily interleaved. Moreover, causal nets distinguish causality and concurrency very clearly.

7.6.2   Refinement

As single tasks can be divided into subtasks, transitions can be refined to subnets. Each subnet contains transitions, but it might also contain places constituting the order of these transitions [11]. It is useful to consider only transition refinements with

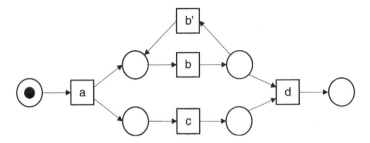

Figure 7.22   A net representing a process.

distinguished input and output transitions such that the preset of the input transition equals the preset of the refined transition, and the postset of the output transition equals the postset of the refined transition.

Figure 7.22 shows an elementary net representing a process. Figure 7.23 shows the refinement of transitions *b* and *c*. Figure 7.24 shows one causal net of the original net, and Figure 7.25 a causal net of the refined net. The interesting point is that this last net can also be obtained by refining the transition occurrences of the causal net shown in Figure 7.24 by causal nets of the refinement nets.

7.7 ANALYSIS

This chapter on Petri nets does not focus on analysis issues, although most contributions in Petri net theory concern analysis. Instead of discussing analysis methods in detail, we only give a rough classification of analysis methods and some examples.

7.7.1 Simulation

Simulation means creation of runs. Hence, a simulation tool can either construct occurrence sequences or causal nets. The runs are either visualized or undergo further investigation. If one is only interested in properties of runs, then the created runs can automatically be analyzed with respect to these properties and the results of this

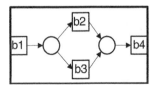 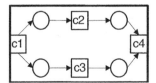

Figure 7.23   Refinement nets.

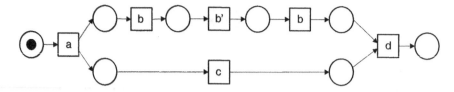

Figure 7.24   A causal net.

analysis are shown to the user. Most approaches are based on occurrence sequences. Simulation by construction of causal nets is done in the VIP tool [10].[7]

Simulation can be used to find (and repair) design errors. Conversely, simulation
cannot prove that a Petri net is free of errors unless all possible runs are created.

Simulation is particularly useful when the performance of processes is investigated based on their models. To this end, the models must have time annotations or
timing distribution functions. There is a large research field on timed Petri nets and
stochastic Petri nets; see [6, 17]. Also, probabilistic assumptions on conflict resolu-
tions belong to this area.

A combination of causal net simulation and performance analysis is given in
[13].

7.7.2   Model Checking

Model checking means entering a model and a specification into some program that
outputs positively if the model has the properties formulated in the specification, or
otherwise presents a counterexample to the specification in terms of a violating run.
Specifications can either be expressed in terms of some logical formulae or can be
standard properties such as liveness or 1-safety. Usually, model checking is a very
complex task, especially for high-level nets.

Most algorithms are based on the construction of all reachable markings and the
reachability graph of a Petri net. The nodes of this graph are the reachable mark-

[7]See http://www.informatik.ku-eichstaett.de/projekte/vip.

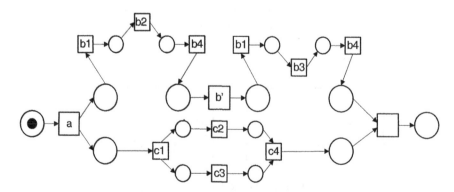

Figure 7.25   A causal net of the refined net.

ings of the net, with distinguished initial marking. For each reachable marking m and each transition occurrence $m \xrightarrow{t} m'$, the reachability graph has a directed arc leading from the node m to the node m', annotated by t. For elementary Petri nets, the reachability graph is finite if and only if the Petri net is bounded. It is acyclic if and only if the Petri net has no infinite run. It has a node without an exiting arc (which does not represent an intended final state) if and only if it is not deadlock - free. It is strongly connected if and only if the initial marking can always be reached again.

The reachability graph of the elementary Petri net shown in Figure 7.4 is given in Figure 7.26.

7.7.3 Proofs

Often, an automatic proof is hard to find for a model checker, but the process designer can provide a short proof of desired properties. Proofs are hard to find but easy to verify once they are found.

The most prominent example for this kind of proof method is given by *place invariants*. A place invariant assigns a weight to each place such that the weighted sum of all tokens is not changed by any transition occurrence. A place invariant assigning a positive number to each place proves that an elementary net is bounded.

In the example net of Figure 7.4, the assignment of the number 2 to the places s_i and s_o and the number 1 to the places s_1 to s_4 is a place invariant. Since the weighted sum of tokens is initially 2, it will always be 2. This place invariant proves that the net is bounded. It also proves that, whenever s_0 is marked, all other places are unmarked (because one token on s_0 multiplied by the weight 2 yields the sum 2).

Transition invariants are assignments of numbers to transitions such that whenever each transition occurs as often as this number indicates, starting with a given marking, this occurrence sequence leads back to the same marking. Hence, transition invariants correspond to occurrence sequences reproducing a marking. In the example given in Figure 7.14, the only way to reproduce the initial marking is to fire transitions a and t twice and transitions b, c, d, and e once. The corresponding

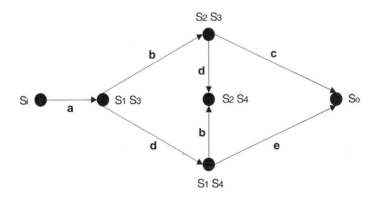

Figure 7.26 The reachability graph of the net of Figure 7.4.

transition invariant assigns the number 2 to a and to t, and the number 1 to the other transitions. If some transition does not have a positive assignment for each nonnegative transition invariant, then this transition does not occur in any occurrence sequence from the initial marking back to the initial marking. If we consider process models with a transition closing the circuit from the output place s_o to the input place s_i, such a transition indicates that there is a modeling error. For live and bounded nets, no such transition can exist.

Place and transition invariants can be interpreted as solutions to homogenous equation systems based on the so-called *incidence matrix* of a Petri net. For these and more linear algebraic approaches to Petri net analysis, see [9].

A *trap* is a set of places that remains marked (i.e., contains a marked place) once it is marked. This is ensured by the defining condition of a trap, which demands that each transition removing a token from a place of the trap also adds a token to a place of the trap. Similarly, a *co-trap* is a set of places that remains unmarked once has it lost its last token, because each transition having a place of the co-trap in its postset also has a place of the co-trap in its preset. More on traps and co-traps can be found in [8].

In the example of Figure 7.4, the set of places $\{s_i, s_1, s_3, s_o\}$ constitutes a co-trap. After firing transitions a, b, and d, this co-trap is unmarked, and will always remain unmarked. At this marking, a deadlock is reached. For this reason, co-traps are sometimes called deadlocks. In the net of Figure 7.9, the same set of places is also a trap. So here, these places cannot loose all their tokens. Therefore, the marking with only one token on s_2 and one token on s_4 is not reachable in this net.

7.8   NET CLASSES

We conclude this chapter with some classes of Petri nets. Each class is defined by a structural restriction that has some semantical interpretation. For more information about state machines, marked graphs, and free-choice nets see [8]. More on workflow nets can be found in [3].

7.8.1   State Machines

A *state machine* is a Petri net in which each transition has exactly one precondition and exactly one postcondition. Moreover, state machines have initially one marked place (which carries only one token). In a state machine, the single token can be passed from place to place in the direction of the arcs. Sequence and choice can be modeled, but there is no concurrency and no synchronization. Each state machine is 1-safe, that is, bounded with the bound 1.

A place of a state machine can be marked if and only if there is a path from the initially marked place to this place. State machines are very simple to analyze. Removing the places that can never be marked and all transitions in their postset yields a state machine with the very same behavior. Its marking graph is obtained by replacing each transition together with its ingoing arc and outgoing arc by a single

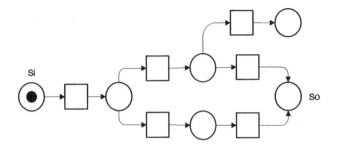

Figure 7.27   A state machine.

arc, annotated by the name of the transition. Hence, a state machine is live if and only if it is strongly connected, and in this case the initial marking can always be reached again.

The example shown in Figure 7.27 is a state machine modeling a simple process. Each place can become marked but there is more than one terminating state.

7.8.2 Marked Graphs

In a *marked graph,* each place has exactly one transition in its preset and exactly one transition in its postset. Therefore, transitions cannot compete for tokens and choice cannot be modeled. Marked graphs can model sequence and concurrency. A marked graph is live if and only if each cycle has at least one marked place initially. Since the number of tokens on cycles is not changed by transition occurrences in marked graphs, this condition remains true for all reachable markings.

A live marked graph is bounded if and only if it is strongly connected. It is, moreover, 1-safe if and only if each place is contained in a cycle with exactly one token. Figure 7.28 shows a live and safe marked graph. In this example, it is not

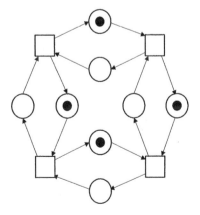

Figure 7.28   A marked graph.

possible to mark, for every cycle, only one place of the cycle without destroying liveness. However, the marked graph is covered by 1-token cycles.

7.8.3 Free-Choice Nets

Free-choice nets generalize state machines and marked graphs. They allow modeling of choice by branching places and modeling of concurrency by branching transitions. However, they are structurally restricted in the sense that a (marked) forward-branching place always models choice between all the transitions in its postset—hence the name "free choice." The formal definition of a free-choice net requires that, if a place is in the preset of more than one transition, then it is the only place in all these presets. In other words, an "N" pattern as shown in Figure 7.29 (left-hand side) does not occur. The definition of free-choice nets given in [8] is slightly more general. It requires that if such an "N" occurs, then there are arcs from all places to all transitions of the "N" (see Figure 7.29, right-hand side). This class of nets is often called *extended free choice*.

For free-choice nets, there is a rich body of theory including efficient analysis methods for behavioral properties. The most prominent result is that a free-choice net is live if and only if each nonempty co-trap includes a trap with an initially marked place (remember that co-traps and traps are sets of places). Its proof is not trivial. For the combination of liveness and boundedness, there exist characterizations based on the rank of the incidence matrix that can be checked very efficiently (in polynomial time with respect to the size of the net).

The restriction given in the definition of free-choice nets might be considered quite artificial. For a better motivation, consider the examples given in Figures 7.4, 7.9, 7.10, 7.11, and 7.12. All these nets are not free choice. For the net of Figure 7.4, the forward-branching place s_1 does not model a free choice because transition e is only enabled after d has occurred, whereas b is enabled immediately. In Figure 7.9 we have additional loops such that s_1 branches even to three transitions. Transition d can only occur when s_1 is marked. Such a conditional split demands forward-branching places with transitions controlled by other places via loops. So conditional splits can not be modeled with free-choice nets.

Another example is given by Figure 7.12. Here transitions b, c, d, and e are controlled by an additional cycle carrying only one token. Also, such *regulation cycles* destroy the free-choice property. Modeling shared resources by additional places

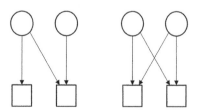

Figure 7.29 In free-choice nets, the left-hand structure is not allowed.

with free-choice nets is also not possible because the resource place itself is branching but the choice depends on the requests of the considered processes.

However, free-choice nets do properly model sequence, choice, and concurrency. Many Petri net models of business processes without resources are free choice.

In a free-choice net, one can imagine that every choice is made within branching places (that is, process elements modeled by these places). In Figure 7.30, a modified version of our running example of Figure 7.4 is shown. The choice of firing transition b or transition e' is made in s_1; the choice of firing d or c' is made independently in s_3. Hence, there is one more possible deadlock, reached after the occurrence of transitions a, e', and c'.

7.8.4   Workflow Nets

Workflow nets are nets modeling business processes without resources and without any history. In particular, a workflow net is intended to describe the behavior of a single *workflow case* in isolation. Any case handled by the procedure represented by the workflow net is created when it enters a workflow management system and is deleted once it has completed (see Chapter 2). When a workflow net is executed twice, for two *cases,* then the second case will run through exactly the same process specification as the first.

A workflow net has a distinguished input place s_i and a distinguished output place s_o, such that s_i has no ingoing arc and s_o has no outgoing arc. This condition is met by our examples shown in Figures 7.4, 7.9, 7.10, 7.11, 7.12, and 7.30.

The execution of a workflow net should lead from the marking assigning only a token to s_i to the marking assigning only a token to s_o. Our net from Figure 7.12 is different because it has initially one more token; the other examples satisfy this condition as well. One more condition for workflow nets is that each place and each transition lies on a path from s_i to s_o. This condition holds true for all examples mentioned above. Figure 7.27 shows a counterexample.

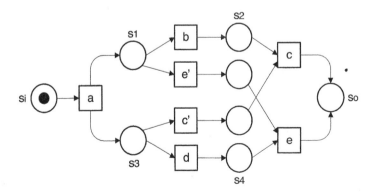

Figure 7.30   A free-choice variant of the running example.

A workflow net is considered correct if it is *sound,* a property formally defined as follows: the net is 1-safe and from each reachable marking the final marking assigning only a token to the output place s_o should be reachable. As a (nontrivial) consequence, there is no other marking assigning a token to s_o. Considering the net with an additional transition connecting s_o to s_i (see Figure 7.14), this net is live and safe if and only if the workflow net is sound.

EXERCISES

Exercise 1. Consider the following process description: After an initial task a, task b and either task c or d are performed concurrently. After b, the sequence e b of tasks is performed an arbitrary number of times. Finally, the process can reach a final state by performing task f.
  a. Do you recognize any ambiguities in the text?
  b. Draw an elementary Petri net modeling the described process.
  c. Draw the reachability graph of this elementary Petri net.

Exercise 2. Add a resource to the Petri net of Exercise 1 that is available only once and is used by tasks e and d.

Exercise 3. Create a predicate/transition net by adding annotations and a suitable initial marking to the net of Exercise 1 such that the process runs for initial tokens p and q, task c is only performed for p, and task d is only performed for q.

Exercise 4. Why did we not define the complement $\bar{s}$ of a place s simply by $^{\bullet}\bar{s} = s^{\bullet}$ and $\bar{s}^{\bullet} = {}^{\bullet}s$?

Exercise 5. Prove that, in the net of Figure 7.10, a marking assigning tokens only to the places s_2 and to s_4 is not reachable, by employing a suitable place invariant.

REFERENCES

1. W. M. P. van der Aalst. The application of Petri nets to workflow management. *Journal of Circuits, Systems and Computers* 8(1), 21–66 (1998).
2. W. M. P. van der Aalst, J. Desel, and A. Oberweis (Eds.). *Business Process Management: Models, Techniques and Empirical Studies.* LNCS 1806, Springer-Verlag (2000).
3. W. M. P. van der Aalst and K. van Hee. *Workflow Management—Models, Methods and Systems.* MIT Press (2000).
4. W. M. P. van der Aalst, A. ter Hofstede, and M. Weske (Eds.). *Proceedings of Business Process Management (BPM 2003).* LNCS 2678, Springer-Verlag (2003).
5. W. M. P. van der Aalst. Business process management demystified: A tutorial on models, systems and standards for workflow management. In J. Desel, W. Reisig, and G. Rozenberg (Eds.). *Lectures on Concurrency and Petri Nets,* LNCS 3098, Springer-Verlag (2004), pp. 1–65.

6. F. Bause and P. S. Kritzinger. *Stochastic Petri Nets,* 2nd ed., Vieweg (2002).

7. E. Börger and R. Stärk. *Abstract State Machines—A Method for High-Level System Design.* Springer-Verlag (2003).

8. J. Desel and J. Esparza. *Free Choice Petri Nets.* Cambridge University Press (1995).

9. J. Desel. Basic linear algebraic techniques for place/transition nets. In W. Reisig and G. Rozenberg (Eds.), *Lectures on Petri Nets I: Basic Models,* LNCS 1491, Springer-Verlag (1998), pp. 257–308.

10. J. Desel. Validation of process models by construction of process nets. In W.M.P. van der Aalst, J. Desel, and A. Oberweis (Eds.). *Business Process Management: Models, Techniques and Empirical Studies.* LNCS 1806, Springer-Verlag (2000), pp. 108–126.

11. J. Desel and T. Erwin. Modelling, simulation and analysis of business processes. In W. M. P. van der Aalst, J. Desel, and A. Oberweis (Eds.). *Business Process Management: Models, Techniques and Empirical Studies.* LNCS 1806, Springer-Verlag (2000), pp. 127–139.

12. J. Desel and G. Juhás. What is a Petri net? Informal answers for the informed reader. In H. Ehrig, G. Juhás, J. Padberg, and G. Rozenberg (Eds.), *Unifying Petri Nets,* LNCS 2128, Springer-Verlag (2001), pp. 1–25.

13. J. Desel and T. Erwin. Quantitative engineering of business processes with VIPbusiness. In H. Ehrig, W. Reisig, G. Rozenberg, and H. Weber (Eds.), *Petri Net Technology for Communication-Based Systems,* LNCS 2472, Springer-Verlag (2003), pp. 219–242.

14. J. Desel, B. Pernici, and M. Weske (Eds.). *Proceedings of Business Process Management (BPM 2004).* LNCS 3080, Springer-Verlag (2004).

15. J. Desel, W. Reisig, and G. Rozenberg (Eds.). *Lectures on Concurrency and Petri Nets.* LNCS 3098, Springer-Verlag (2004).

16. K. Jensen. *Coloured Petri Nets.* Volumes I–III. Springer-Verlag (1992–1997).

17. M. Ajmone Marsan, G. Balbo, S. Donatelli, and G. Franceschinis. *Modelling with Generalized Stochastic Petri Nets.* Wiley (1995).

18. A. Oberweis and P. Sander. Information system behavior specification by high-level Petri nets. *ACM Transactions on Information Systems 14*(4), 380–420 (1996).

19. W. Reisig. *A Primer in Petri Net Design.* Springer-Verlag (1992).

20. W. Reisig and G. Rozenberg (Eds.). *Lectures on Petri Nets. I: Basic Models. Lectures on Petri Nets.* II: Applications. LNCS 1491/1492, Springer-Verlag (1998).

21. D. T. Ross. Structured Analysis (SA): A language for communicating ideas. *IEEE Transactions on Software Engineering, SE-3,* 1, 16–34 (1977).

22. M. Weber and E. Kindler. The Petri net markup language. In H. Ehrig, W. Reisig, G. Rozenberg, and H. Weber (Eds.), *Petri Net Technology for Communication-Based Systems,* LNCS 2472, Springer-Verlag (2003), pp. 124–144.

CHAPTER 8

Patterns of Process Modeling

WIL M. P. van der AALST, ARTHUR H. M. ter HOFSTEDE,
and MARLON DUMAS

8.1 INTRODUCTION

The previous chapters have presented different languages for and approaches to process modeling. In this chapter, we review some issues in process modeling from a more language-independent perspective. To this end, we rely on the concept of pattern: an "abstraction from a concrete form which keeps recurring in specific non-arbitrary contexts" [18]. The use of patterns is a proven practice in the context of object-oriented design, as evidenced by the impact made by the design patterns of Gamma et al. [10].

Process-aware information systems (PAISs) address a number of perspectives. Jablonski and Bussler [11] identify several such perspectives in the context of workflow management. These include the process perspective (describing the control flow), organization perspective (structuring of resources), data/information perspective (to structure data elements), operation perspective (to describe the atomic process elements), and integration perspective (to "glue" things together).[1] In a typical workflow management system, the process perspective is described in terms of some graphical model, for example, a variant of Petri nets (see Chapter 7), the organization perspective is described by specifying and populating roles and organizational units, the data/information perspective is described by associating data elements to workflow instances (these may be typed and have a scope), the operation perspective is described by some scripting language used to launch external applications, and the integration perspective is described by some hierarchy of processes and activities. In principle, it is possible to define design patterns for each of these perspectives. However, the focus of this chapter is on patterns restricted to the process (i.e., control-flow) perspective. This perspective is the best understood as well as the dominant perspective of workflow. Patterns for the data perspective have been reported in [20], whereas patterns for other perspectives (e.g., resource

---

[1]Note that in [11], different terms are used.

Process-Aware Information Systems. Edited by Dumas, van der Aalst, and ter Hofstede
Copyright © 2005 John Wiley & Sons, Inc.

perspective) are the subject of ongoing efforts. In addition, it is worthwhile mentioning that the control-flow patterns do not pay too much attention to issues related to exception handling.

The control-flow perspective is concerned with enforcing control-flow dependencies between tasks[2]; for example, sometimes tasks need to be performed in order, sometimes they can be performed in parallel, sometimes a choice needs to be made as to which task to perform, etc. There is an abundance of approaches to the specification of control flow in PAISs in general, and workflow management systems in particular. Many commercial workflow management systems and academic prototypes use languages with fundamental differences. Concepts with similar names may have significantly different behavior, different languages may impose different restrictions (e.g., with respect to cycles), and some concepts are supported by a select number of languages only. The reader is referred to [12, 13, 14] for a fundamental discussion of some of these issues. This work identifies a number of classes of workflow languages, which are abstractions of approaches used in practice, and examines their relative expressive power. The workflow patterns initiative took a more pragmatic approach, focusing on suitability.

The workflow patterns initiative, which this chapter reviews, started in 1999. It aimed at providing a systematic and practical approach to dealing with the diversity of languages for control-flow specification. The initiative took the state of the art in workflow management systems as a starting point and documented a collection of 20 patterns, predominantly derived from constructs supported by these systems. The patterns provided abstractions of these constructs as they were presented in a language-independent format. The patterns consist of a description of the essence of the control-flow dependency to be captured, possible synonyms, examples of concrete business scenarios requiring the control-flow dependency, and, for the more complex ones, typical realization problems and (partial) solutions to these problems.

There are a number of applications of the workflow patterns. The patterns can be used for the selection of a workflow management system. In that case, one would analyze the problem domain in the context of which future workflow management system is to be used; that is, analyze the needs in terms of required support for various workflow patterns and subsequently match the requirements with the capabilities of various workflow management systems (this could be termed a suitability analysis). Additionally, the patterns can be used for benchmarking purposes, examining relative strengths and weaknesses of workflow products. Such examinations may be the basis for language development and adaptations of workflow management systems. Another use of the patterns can be found in the context of prescribing a particular workflow tool and certain patterns that need to be captured. Here, the workflow pattern collection acts as a resource for descriptions of typical workarounds and realization approaches for patterns in different workflow systems. It should also be remarked that although the patterns were developed in the context of P2A processes (e.g., workflow and case handling systems; see Chapter 1), they can also be applied to P2P and A2A processes. At least two languages aimed at de-

---

[2]In this chapter, the terms "task" and "activity" are used interchangeably.

scribing A2A processes (BPEL and BML, see references below) have been evaluated in terms of the patterns. This having been said, some of the patterns may be less relevant for P2P and A2A processes than they are for P2A processes, and an exact characterization of which patterns are more relevant for which types of processes remains an open question.

The first paper related to the workflow patterns initiative[3] appeared in the CoopIS conference in 2000 (see [2]). The main paper appeared in 2003 in the *Distributed and Parallel Databases Journal* (see [5]). Apart from a description of the complete set of patterns, this latter paper contains an analysis of 13 commercial workflow management systems and two academic prototypes in terms of their support for the patterns. The patterns have been used for analyses of UML Activity Diagrams version 1.4 (see [8]); BML (Business Modeling Language) an approach used in the context of Enterprise Application Integration (see [24]); and various approaches and proposed standards in the area of Web service composition, such as BPEL4WS (see [23] for this evaluation and Chapter 12 for an introduction to BPEL4WS). The workflow patterns formed the starting point for the development of YAWL[4] (Yet Another Workflow Language). This language extends Petri nets with constructs for dealing with some of the patterns in a more straightforward manner. Though based on Petri nets, its formal semantics is described as a transition system (see [4]) and YAWL should not be seen as a collection of macros defined on top of Petri nets. A first description of the design and implementation of the YAWL environment can be found in [1]. In this chapter, the YAWL notation will be used to explain various patterns.

The goal of this chapter is to take an in-depth look at a selection of the patterns as presented in [5] from a more didactic perspective. The organization of this chapter is as follows. A classification of the patterns is discussed, followed by a detailed discussion of a selection of patterns organized according to this classification. The chapter concludes with a brief summary and outlook. Note that whenever products are referred to in this chapter, their version corresponds to the version which was used for the evaluation in [5] unless stated otherwise.

8.2 CLASSIFICATION OF PATTERNS

As mentioned earlier, the patterns initiative has led to a set of 20 control-flow patterns. These patterns range from very simple patterns such as sequential routing (Pattern 1) to complex patterns involving complex synchronizations such as the discriminator pattern (Pattern 9). These patterns can be classified into six categories:

1. *Basic control-flow patterns.* These are the basic constructs present in most workflow languages to model sequential, parallel, and conditional routing.
2. *Advanced branching and synchronization patterns.* These patterns transcend the basic patterns to allow for more advanced types of splitting and joining be-

---

[3]www.workflowpatterns.com
[4]www.yawl-system.com

havior. An example is the synchronizing merge (Pattern 7), which behaves like an AND-join, XOR-join, or combination thereof, depending on the context.

3. *Structural patterns.* In programming languages, a block structure that clearly identifies entry and exit points is quite natural. In graphical languages allowing for parallelism, such a requirement is often considered to be too restrictive. Therefore, we have identified patterns that allow for a less rigid structure.

4. *Patterns involving multiple instances.* Within the context of a single case (i.e., workflow instance), sometimes parts of the process need to be instantiated multiple times; for example, within the context of an insurance claim, multiple witness statements may need to be processed.

5. *State-based patterns.* Typical workflow systems focus only on activities and events, not on states. This limits the expressiveness of the workflow language because it is not possible to have state-dependent patterns such as the milestone pattern (Pattern 18).

6. *Cancelation patterns.* The occurrence of an event (e.g., a customer canceling an order) may lead to the cancelation of activities. In some scenarios, such events can even cause the withdrawal of the whole case.

Figure 8.1 shows an overview of the 20 patterns grouped into the six categories. This classification is used in the next section to highlight some of the patterns.

Basic Control Flow Patterns
- Pattern 1 (Sequence)
- Pattern 2 (Parallel Split)
- Pattern 3 (Synchronization)
- Pattern 4 (Exclusive Choice)
- Pattern 5 (Simple Merge)

Advanced Branching/Synchronization Patterns
- Pattern 6 (Multi-choice)
- Pattern 7 (Synchronizing Merge)
- Pattern 8 (Multi-merge)
- Pattern 9 (Discriminator)

Structural Patterns
- Pattern 10 (Arbitrary Cycles)
- Pattern 11 (Implicit Termination)

Cancellation Patterns
- Pattern 19 (Cancel Activity)
- Pattern 20 (Cancel Case)

State-based Patterns
- Pattern 16 (Deferred Choice)
- Pattern 17 (Interleaved Parallel Routing)
- Pattern 18 (Milestone)

Patterns involving Multiple Instances
- Pattern 12 (Multiple Instances Without Synchronization)
- Pattern 13 (Multiple Instances With a Priori Design Time Knowledge)
- Pattern 14 (Multiple Instances With a Priori Runtime Knowledge)
- Pattern 15 (Multiple Instances Without a Priori Runtime Knowledge)

Figure 8.1 Overview of the 20 workflow patterns described in [5].

8.3 EXAMPLES OF CONTROL-FLOW PATTERNS

This section presents a selection of the various control-flow patterns using the classification shown in Figure 8.1.

8.3.1 Basic Control-Flow Patterns

The basic control-flow patterns essentially correspond to control-flow constructs as described by the Workflow Management Coalition[5] (see, e.g., [9] or [22]). These patterns are typically supported by workflow management systems and as such do not cause any specific realization difficulties. It should be pointed out, though, that the behavior of the corresponding constructs in these systems can be fundamentally different. In this section, the sequence pattern is discussed in terms of the format used for capturing patterns, whereas the other four patterns are discussed in a less structured manner.

Pattern 1 (Sequence)
Description: An activity should await the completion of another activity within the same case before it can be scheduled.
Synonyms: Sequential routing, serial routing.
Examples:
    After the expenditure is approved, the order can be placed.
    The activity *select_winner* is followed by the activity *notify_outcome*.
Implementation: This pattern captures direct causal connections between activities and is supported by all workflow management systems. Graphically, this pattern is typically represented through a directed arc without an associated predicate. □

There are four other basic control-flow patterns, two of which correspond to splits and two of which correspond to joins.

The *XOR-split* corresponds to the notion of an *exclusive choice* (Pattern 4). Out of two or more outgoing branches, one branch is chosen. Such a choice is typically determined by workflow control data or input provided by users of the system. In some systems (e.g., Staffware) there is explicit support for XOR-splits, whereas in some other systems (e.g., MQSeries/Workflow) the designer has to guarantee that only one outgoing branch will be chosen at run time by providing mutually exclusive predicates.

In the YAWL environment, predicates specified for outgoing arcs of an XOR-split may overlap. If multiple predicates evaluate to true, the arc with the highest preference (which is specified at design time) is selected. If all predicates evaluate to false, the default arc is chosen. This solution is similar to Eastman's solution (see [21], pp. 144–145); in which a rule list can be specified for an activity, and these rules are processed after completion of that activity. Among others, this list may contain rules for passing control to subsequent activities. Control is passed to the

---

[5]www.wfmc.org

first activity occurring in such a rule whose associated predicate evaluates to true. A default rule can be specified that does not have an associated predicate and, therefore, should be last in such a list ([21], p. 145). As an example of an XOR-split consider the case in which purchase requests exceeding $10,000 are to be approved by head office, while purchase requests not exceeding this amount of money can be approved by the regional offices.

The converse of the XOR-split is the *XOR-join* or *simple merge* (Pattern 5). An XOR-join is enabled when one of its preceding branches completes. The definition by the Workflow Management Coalition (WfMC) requires that the XOR-join (called OR-join by the WfMC) is not preceded by parallelism, that is, no two or more preceding branches of the XOR-join run in parallel at any point in time. The pattern incorporates this requirement, which can be seen as a context assumption. Without this assumption substantial differences between various workflow products would become apparent, but we will treat this as a different pattern, the multi-merge, to be discussed in the next section. In some cases, XOR-joins should have corresponding XOR-splits (e.g., Visual WorkFlo), which, combined with other similar restrictions, may guarantee that no parallelism occurs in branches preceding XOR-joins. Such structured workflows are discussed in [12, 14]. As an example of an XOR-join, consider the activity *report_outcome,* which is to be executed after activity *finalize_rejection* or activity *finalize_approval* completes. It is assumed that these two latter activities never run in parallel.

The *AND-split* can be used to initiate parallel execution of two or more branches (Pattern 2). It should be remarked that the description in [5], which was adapted from the original formulation by the WfMC, left open the possibility that no true parallelism takes place, so that these branches could only be executed in an interleaved manner. This interpretation is convenient, as it can be used in contexts in which other constraints (e.g., on resources) do not permit activities of different branches to be executed at the same time, but where the specification should be flexible enough to allow the execution environment to decide for itself which activity of which branch should be scheduled next (hence, one should not be forced to make an arbitrary decision at design time). An example of an AND-split could be in the context of an application process in which, after short-listing of candidates, referee reports need to be obtained and interviews need to be held. For a particular candidate, these activities could be done in parallel or in any order. In terms of implementation, sometimes the AND-split is supported in an implicit manner (e.g., MQSeries/Workflow), in which predicates are required to be specified for all outgoing branches. In those cases, the AND-split can be realized by associating the predicate *true* to all outgoing branches.

The *AND-join* is the converse of the AND-split and synchronizes its incoming branches (Pattern 3). All incoming branches of an AND-join need to be completed before the AND-join can proceed the flow. For example, a decision for particular candidates can only be made once their referee reports have been received and they have been interviewed. Again, there is a context assumption for this pattern. It should not be possible for a branch to signal its completion more than once before all other branches have completed, so the AND-join only needs to remember whether a particular branch has completed or not. Note that it is possible that a com-

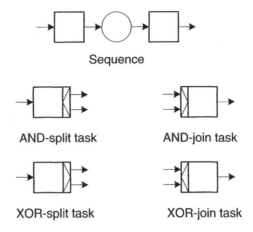

Figure 8.2 Some basic symbols used in YAWL.

pletion signal is never received for a particular branch, in which case the AND-join causes a deadlock.

Figure 8.2 provides an overview of the graphical constructs in YAWL that can be used to capture the basic control-flow patterns. In this figure, boxes denote "tasks," whereas circles denote "conditions." These correspond to the notions of transition and place in Petri nets, respectively (see Chapter 7).[6] These YAWL constructs do not require the context assumptions of some of the patterns presented in this section and, as such, have a broader interpretation than these patterns (e.g., the simple merge can be realized by the YAWL XOR-join, but this XOR-join also realizes the multimerge pattern discussed in the next section).

8.3.2 Advanced Branching and Synchronization Patterns

The patterns presented in this section deal with less straightforward split and join behavior and, though not uncommon in practice, pose more difficulties in terms of their support by contemporary workflow management systems. The multichoice (Pattern 6) and the synchronizing merge (Pattern 7) will be looked at in depth, the other patterns will be discussed more briefly.

Although only one outgoing branch is chosen in case of an XOR-split, the OR-split or multichoice allows a choice for an arbitrary number of branches.

Pattern 6 (Multichoice)
Description: Out of several branches, a number of branches are chosen based on user input or data accessible by the workflow management system.
Synonyms: Conditional routing, selection, OR-split.

---

[6]Note that in the YAWL representation of the sequence (which appears in the figure), the condition does not need to be explicitly shown.

Examples: After the execution of activity *determine_teaching_evaluation,* execution of activity *organize_student_evaluation* may commence as well as execution of activity *organize_peer_review.* At least one of these two activities is executed, and possibly both.

Problem: Workflow management systems that allow for the specification of predicates on transitions support this pattern directly. Sometimes, however, the multimerge needs to be realized in terms of the basic patterns (e.g., Staffware supports AND-splits and XOR-splits, but nothing "in between").

Implementation: As mentioned above, this pattern is directly supported by systems that allow for predicates to be specified for transitions (as is the case for Verve, Forté Conductor, and MQSeries/Workflow). As stated in the previous section, such systems provide implicit support for XOR-splits and AND-splits. However, their approach also allows for splits that are neither (selection of more than one branch, but not all). The multimerge may be considered a generalization of the XOR-split and the AND-split.

An OR-split in YAWL is shown in Figure 8.3. It should be noted that in the YAWL environment, at least one outgoing transition needs to be chosen, which makes its OR-split slightly less general than the pattern. In YAWL, the selection of at least one branch is guaranteed by the specification of a default branch, which is chosen if none of the predicates evaluate to true (including the predicate associated with the default branch!).

For those languages that only support the basic XOR-splits and AND-splits, there are two solutions:

- Transform the *n*-way multichoice into an AND-split followed by *n* binary XOR-splits, each of which checks whether the predicate of the corresponding branch in the multichoice is true or not. If a predicate evaluates to true, the corresponding activity needs to be executed; otherwise, no action is required.

- Transform the *n*-way multichoice into an XOR-split with 2^n outgoing branches. Each of these outgoing branches corresponds to a particular subset of outgoing branches that may be chosen as part of the multichoice (AND-splits would be used for those subsets that consist of at least two outgoing branches). The associated predicate should capture

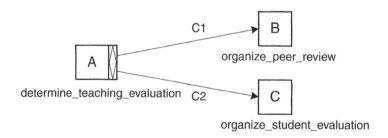

Figure 8.3   Multichoice in YAWL.

the fact that the corresponding predicates of these branches are true, but that this does not hold for any of the predicates associated with the other branches (note that this solution indeed guarantees mutual exclusion of the outgoing branches of the XOR-split). This solution is exponential in terms of the number of outgoing transitions as opposed to the previous solution.

These solutions are illustrated in YAWL in Figure 8.4, where it should be noted that, contrary to YAWL's semantics, we assume that the multichoice of Figure 8.3 also allows none of the branches to be executed (to be in line with the description of the pattern). □

Although the multichoice does not cause too many problems in contemporary workflow management systems, the same cannot be said for the so-called *synchronizing merge* (Pattern 7), which, in a structured context, can be seen as its converse. Consider the YAWL schema of Figure 8.5. After activity *A* (*determine_teaching_evaluation*), either *B* (*organize_peer_review*) or *C* (*organize_student_evaluation*) or both *B* and *C* will be executed (note that as mentioned before, in YAWL the multichoice has to choose at least one of the outgoing branches). The synchronization

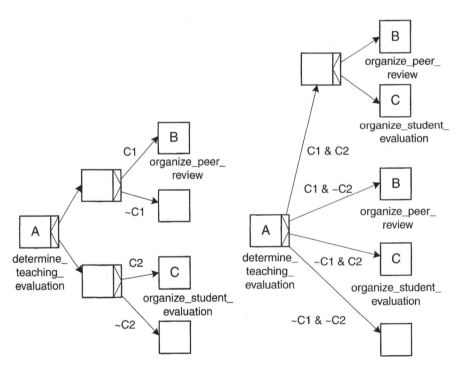

Figure 8.4 Expanding a multichoice in terms of simple choice and parallel split (illustrated in YAWL). "~C" stands for "not C."

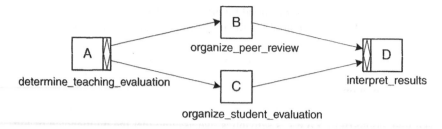

Figure 8.5 Illustration of the synchronizing merge in YAWL.

prior to the execution of activity D (*interpret_results*) should now be such that only active threads should be waited upon. In particular, if activity B is not triggered after the OR-split, then the OR-join should not wait for it and the same goes for activity C. This is achieved through the use of an *OR-join* in YAWL. The use of an AND-join here may lead to a deadlock in case either B or C was chosen (but not both), whereas the use of a synchronization construct that executes D upon completion of any of the incoming branches (multimerge) may lead to executing this activity twice if both B and C were chosen.

In the description of the synchronizing merge, a more general approach will be taken, in line with the formalization in YAWL, than the one described in [5].

Pattern 7 (Synchronizing Merge)

Description: A form of synchronization in which execution can proceed if and only if one of the incoming branches has completed and, from the current state of the workflow, it is not possible to reach a state in which any of the other branches has completed.

Synonyms: Synchronizing join, OR-join.

Examples: Consider again the example presented in Pattern 6 (multichoice). After activities *organize student evaluation* and *organize_peer_review* have finished, activity *interpret results* could be scheduled. This activity should only await completion of those activities that were actually executed and itself be performed once.

Problem: The main challenge of achieving this form of synchronization is to be able to determine when more completions of incoming branches are to be expected. In the general case, this may require an expensive state analysis.

Implementation: In workflow systems such as MQSeries/Workflow and InConcert the synchronizing merge is supported directly because of the evaluation strategy used. In MQSeries/Workflow, activities have to await signals from all incoming branches. Such a signal may indicate that a certain branch completed and that the associated predicate is true, or it indicates that a certain branch was bypassed or that the associated predicate is false. Depending on the particular combination of signals received and the evaluation of the join predicate, the activity is or is not executed and a corresponding signal is propagated. In InConcert, activities just await signals from all incoming branches and the evaluation of a precondition (not the value of these signals) will determine whether they themselves will be executed or not. In either case, no deadlock will occur as neither MQSeries/ Workflow nor InConcert allows cycles.

The interpretation of the OR-join in YAWL (as formalized in [4]) is such that it is enabled if and only if an incoming branch has signaled completion and from the current state it is not possible to reach a state (without executing any OR-join) in which another incoming branch signals completion. Although this can handle workflows of a structured nature it can also handle workflows such as the one displayed in Figure 8.6. As a possible scenario, consider the situation in which after completion of activity A both activities B and C are scheduled. If activity C completes and activity B has not completed, then activity D can not be executed as it is possible that activity F will be chosen after completion of B. In this case, if after completion of activity B, activity E is chosen, activity D can be scheduled for execution as it is not possible to reach a state in which activity F will be scheduled. So the OR-join guarantees that activity D has to await completion of activity C if it was scheduled, and if activity B was scheduled, activity D has to at least await the outcome of the decision after completion of activity B. If activity E was subsequently chosen, it does not need to wait for completion of activity F, but if activity F was chosen it will have to await completion of that activity. Current work with respect to the OR-join in YAWL is reconsidering the treatment of other OR-joins in the reachability analysis required for determining whether a certain OR-join is enabled, and is examining algorithmic solutions for this analysis.

As stated in [21] (p. 109), "non-parallel work items routed to Join worksteps bypass Join Processing" and an XOR-split followed by an AND-join in Eastman's solution does not necessarily lead to a deadlock. Hence, joins have information about the number of active threads to be expected. The situation captured by the YAWL workflow in Figure 8.5 would not cause any problems in Eastman's solution as the OR-join would know whether to expect parallel execution of B and C or not. Although this solution works fine in a structured context in which information about active threads initiated after a split can be passed on to a corresponding join, this does not work so well in a context in which workflows are not fully structured. Consider again the YAWL workflow depicted in Figure 8.6. In Eastman's solution, this specification leads to a deadlock if B was chosen and after completion of B a choice for activity E was made. The OR-join would then keep waiting for the completion of activity F.

In the context of EPCs (see Chapter 6), there exists a body of research examining possible interpretations of the OR-join [3, 7, 15, 19].

If there is no direct support for the OR-join, it may require some work to capture its behavior (and without the use of the data perspective and under a certain notion of equivalence, it may not always be possible; see [13, 12]). In a structured context, a multichoice could be replaced as indicated in the two solutions in its pattern description and synchronization could then be achieved in a straightforward manner. □

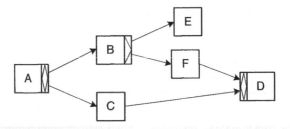

Figure 8.6   Another illustration of the synchronizing merge in YAWL.

The *multimerge* (Pattern 8) does not make the context assumption specified for the XOR-join in the previous section. It will execute the activity involved as many times as its incoming branches signal completion. This interpretation allows these incoming branches to execute in parallel. The YAWL XOR-join corresponds to the multimerge.

The *discriminator*[7] (Pattern 9) provides a form of synchronization for an activity in which, out of a number of incoming branches executing in parallel, the first branch to complete initiates the activity. When the other branches complete, they do not cause another invocation of the activity. After all branches have completed, the activity is ready to be triggered again (in order for it to be usable in the context of loops). In YAWL, one of the ways to capture the discriminator involves the use of cancelation regions (cf. [4]). The discriminator is specified with a multimerge and a cancelation region encompassing the incoming branches of the activity. In this realization, the first branch to complete starts the activity involved, which then cancels the other executing incoming branches. This is not in exact conformance with the original definition of the pattern (as it actually cancels the other branches), but this choice is motivated by the fact that it is clear in this approach what the region is that is in the sphere of the discriminator, giving it a clearer semantics. The discriminator is a special case of the *n*-out-of-*m* join (sometimes referred to as a partial join [6]), as it corresponds to a 1-out-of-*m* join.

8.3.3   Structural Patterns

This section briefly examines two so-called structural patterns: arbitrary cycles and implicit termination. Structural patterns deal with syntactic restrictions that some languages impose on workflow specifications.

Some workflow systems only allow the specification of loops with unique entry and exit points. *Arbitrary cycles,* in which there are multiple ways of exiting from the loop or multiple ways of entering the loop, are not allowed. Sometimes, this is enforced in an explicit manner, for example, in languages that are struc-

---

[7]The term discriminator originates from Verve Workflow.

tured (see, e.g., [14]), whereas sometimes the restriction comes about through the fact that iterative behavior can only be specified through postconditions on decompositions (as, e.g., in MQSeries/Workflow). The process specified in the decomposition is to be repeated until the postcondition evaluates to true. In the case of MQSeries/Workflow, this more implicit way of specifying loops is a direct consequence of the evaluation strategy used—incoming signals are expected from all incoming branches of an activity. Obviously, cycles in a specification would then cause a deadlock. YAWL allows for the specification of arbitrary cycles. In [12, 14], expressiveness issues in relation to structured workflows are investigated. It is shown that not all arbitrary cycles can be converted to structured cycles (in the context of a given equivalence notion and without considering the data perspective).

At least two different termination strategies can be distinguished for workflows. In one approach, a workflow execution is considered completed when no activity can be scheduled anymore (and the workflow is not in a deadlock). This is referred to as *implicit termination* (e.g., supported by MQSeries/Workflow and Staffware). In the other approach, the workflow is considered completed if a designated endpoint is reached. Though other activities may still be executing, they are terminated when this happens. Although the two approaches are different, in some cases workflows following one approach may be converted to workflows conforming to the other approach. In [13], it is shown how so-called standard workflows, which do not contain a deadlock and do not have multiple concurrent instances of the same activity at any stage, can be transformed into equivalent standard workflows with a unique endpoint so that when this endpoint is reached, no other part of the workflow is still active. YAWL does not support implicit termination, so workflow designers are forced to carefully think about workflow termination.

8.3.4 Patterns Involving Multiple Instances

The patterns in this section involve a phenomenon that we will refer to as *multiple instances*. As an example, consider the reviewing process of a paper for a conference. Typically, there are multiple reviews for one paper, and some activities are at the level of the whole paper (e.g., accept/reject), whereas others are at the level of a single review (e.g., send paper to reviewer). This means that inside a case (i.e., the workflow instance, in this example a paper) there are sub-instances (i.e., the reviews) that need to be dealt with in parallel (i.e., in parallel, multiple reviewers may be reviewing the same paper). Multiple-instance patterns are concerned with the embedding of subinstances in cases (i.e., workflow instances). From a theoretical point of view, the concept is relatively simple and corresponds to multiple threads of execution referring to a shared definition. From a practical point of view, it means that an activity in a workflow graph can have more than one running, active instance at the same time. As we will see, such behavior may be required in certain situations. The fundamental problem with the implementation of these patterns is that, due to design constraints and lack of anticipation for this requirement, most of the workflow engines do not allow for more than one instance of the same activity to be active at the same time (in the context of a single case).

When considering multiple instances, there are two types of requirements. The first requirements has to do with the ability to launch multiple instances of an activity or a subprocess. The second requirement has to do with the ability to synchronize these instances and continue after all instances have been handled. Each of the patterns needs to satisfy the first requirement. However, the second requirement may be dropped by assuming that no synchronization of the instances launched is needed.

If the instances need to be synchronized, the number of instances is highly relevant. If this number is fixed and known at design time, then synchronization is rather straightforward. If, however, the number of instances is determined at run time or may even change while handling the instances, synchronization becomes very difficult. Therefore, Figure 8.1 names three patterns with synchronization. If no synchronization is needed, the number of instances is less relevant: Any facility to create instances within the context of a case will do. Therefore, Figure 8.1 names only one pattern for multiple instances without synchronization.

In this section, we highlight only one of the four multiple instance patterns. This is the most complex of these patterns since it requires synchronization and the number of instances can even vary at run time.

Pattern 15 (Multiple Instances Without a Priori Runtime Knowledge)
Description: For one case, an activity is enabled multiple times. The number of instances of a given activity for a given case is not known during design time, nor is it known at any stage during run time, before the instances of that activity have to be created. Once all instances are completed, some other activity needs to be started. It is important to note that even while some of the instances are being executed or already completed, new ones can be created.
Examples:
- Consider the reviewing process of a paper for a conference when there are are multiple reviews for one paper. The number of reviewers may be fixed initially, say three, but may be increased when there are conflicting reviews or missing reviews. For example, initially three reviewers are appointed to review a paper. However, halfway through the reviewing period, a reviewer indicates that he will not be able to complete the review. As a result, a fourth reviewer (i.e., the fourth instance) is appointed. At the end of the review period, only two reviews are returned. Moreover, the two reviews are conflicting (strong accept versus strong reject). As a result, the PC chair appoints a fifth reviewer (i.e., the fifth instance).
- The requisition of 100 computers involves an unknown number of deliveries. The number of computers per delivery is unknown and, therefore, the total number of deliveries is not known in advance. After each delivery, it can be determined whether a next delivery is to come by comparing the total number of goods delivered so far with the number of the goods requested. After processing all deliveries, the requisition has to be closed.

- For the processing of an insurance claim, zero or more eyewitness reports should be handled. The number of eyewitness reports may vary. Even when processing eyewitness reports for a given insurance claim, new eyewitnesses may surface and the number of instances may change.

Problem: Some workflow engines provide support for generating multiple instances only if the number of instances is known at some stage of the process. This can be compared to a "for" loop in procedural languages. However, these constructs are of no help in processes requiring "while" loop functionality. Note that the comparison with the "while" construct may be misleading since all instances may run in parallel.

Implementation: YAWL directly supports multiple instances. Figure 8.7 shows a process in which there may be multiple witnesses processed in parallel. Composite activity B (*process_witness_statements*) consists of three steps (D, E, F) that are executed for each witness. Figure 8.7 does not show that the number of instances of B may be changed at any point in time, even when the processing of witness statements has already started. This is a setting of B.

FLOWer (see Chapter 15) is one of the few commercial systems directly supporting this pattern. In FLOWer, it is possible to have dynamic subplans. The number of instances of each subplan can be changed at any time (unless specified otherwise).

If the pattern is not supported directly, typical implementation strategies involve a counter indicating the number of active instances. The counter is incremented each time an instance is created and decremented each time an instance is completed. If, after activation, the counter returns to 0, then the construct completes and the flow continues. Consider for example, Figure 8.7. In activity A, the counter is set to some value indicating the initial number of witnesses. While executing instances of composite activity B, new instances may be created. For each created instance, the counter is incremented by 1. Each time F is executed for some instance (i.e., B completes), the counter is decremented. If the value of

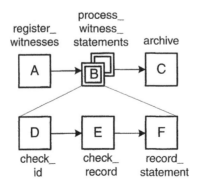

Figure 8.7   Illustration of the multiple instances pattern in YAWL.

the counter equals 0, no more witnesses can be added and C is enabled. Note that the counter only takes care of the synchronization problem. In addition to the counter, the implementation should allow for multiple instances running in parallel.   □

Multiple instances are not only interesting from a control-flow point of view. Note that each instance will have its own data elements but, at the same time, it may be necessary to aggregate data. For example, in Figure 8.7 each witness may have an address; that is, each instance of B has a case attribute indicating the address of the witness. However, in C it may be interesting to determine the number of witnesses living at the same address. This implies that it is possible to query the data of each instance to do some calculations.

8.3.5   State-Based Patterns

In [17], supporting evidence from a number of sources is collected with respect to the time spent waiting as a percentage of the total execution time (i.e., cycle/flow time) averaged over workflow instances in areas dealing with insurance and pension claims, and tax returns. In the five sources mentioned, this average percentage is at least 95% (and in three of these sources at least 99%), which implies that workflow instances are typically in a state awaiting processing rather than being processed. Many computer scientists, however, seem to have a frame of mind, typically derived from programming, in which the notion of state is interpreted in a narrower fashion and is essentially reduced to the concept of data or a position in the queue of some activity. As this section will illustrate, there are real differences between work processes and computing and there are business scenarios in which an explicit notion of state is required.

To illustrate two of the state-based patterns, consider Figure 8.8. This is again a YAWL diagram. However, in contrast to the earlier diagrams, Figure 8.8 explicitly shows the states' in-between activities. Note that YAWL uses a Petri-net-like notation to model states (i.e., places). The initial state of the case (i.e., process instance) is modeled by *start*. The final state is modeled by *end*. If there is a token in *start*, the first activity A can be executed. The last activity I will put a token in *end*. In between A (register) and I (archive) two parallel processes are executed. The upper part models the logistical subprocess, whereas the lower part models the financial subprocess. In the logistical subprocess, there is an exclusive choice (Pattern 4) modeled by B. If the ordered goods are available, B wil put a token in $c5$. If the ordered goods are not available, a replenishment order is planned (token in place $c4$) or the missing goods have already been ordered (token in place $c3$). In the financial subprocess, (i.e., lower part) there is also a choice. After sending the bill (F) there is a choice between the decision to send a reminder (activity H) and the receipt of the payment (activity G). Note that this decision is not made by F; that is, after completing activity F, the choice between H and G is not fixed but depends on external circumstances. For example, H may be triggered by a clock (e.g., after four weeks), whereas G is triggered by the cus-

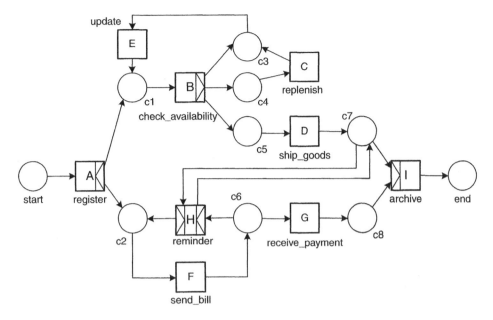

Figure 8.8   An example illustrating state-based patterns.

tomer actually paying for the ordered goods. Note that the choice modeled by c6 is different from the choice modeled by B; in $c6$ there is a "race" between two activities H and G, whereas after executing B, the next activity is fixed by putting a token in $c3$ (E), $c4$ (C), or $c5$ (D). As indicated, the construct involving B corresponds to the traditional *exclusive choice* (Pattern 4) supported by most systems and languages. The construct involving $c6$ corresponds to the *deferred choice* (Pattern 16) described in this section. Figure 8.8 also shows another state-based pattern, the *milestone* (Pattern 18). This is the construct involving H and $c7$. Note that H is an AND-split/AND-join and therefore it can only occur if there is a token in $c7$. This implies that it is only possible to send a reminder if the goods have been shipped. The purpose of the milestone pattern is to be able to test the state in another parallel branch. If H would always occur exactly once, this construct would not be needed, that is, the two arcs representing the milestone could be replaced by a new place connecting D to H. However, for some cases H is not executed at all (that is, the customer is eager to pay), whereas for other cases H is executed multiple times (that is, the customer is reluctant to pay). Therefore, this alternative solution does not work properly. Note that in this case, the deferred choice and milestone are connected. In general this is not the case since both patterns can occur independently. There is a third state-based pattern (Pattern 17, interleaved parallel routing). The pattern is used to enforce mutual exclusion without enforcing a fixed order. A discussion of this pattern is beyond the scope of this chapter. Instead we restrict our attention to the deferred choice.

Pattern 16 (Deferred Choice)

Description: A point in the workflow process where one of several branches is chosen. In contrast to the XOR-split, the choice is not made explicitly (e.g., based on data or a decision) but several alternatives are offered to the environment. However, in contrast to the OR-split, only one of the alternatives is executed. This means that once the environment activates one of the branches, the other alternative branches are withdrawn. It is important to note that the choice is delayed until the processing in one of the alternative branches is actually started, that is, the moment of choice is as late as possible.

Synonyms: External choice, implicit choice, deferred XOR-split.

Examples:

- See the YAWL diagram shown in Figure 8.8. After sending the bill (*F*) there is a choice between the decision to send a reminder (activity *H*) and the receipt of the payment (activity *G*). This decision is not made by *F* but is resolved by the "race" between *H* and *G*, that is, a race between a time trigger (end of a four-week period) and an external trigger (the receipt of the payment).

- At certain points during the processing of insurance claims, quality assurance audits are undertaken at random by a unit external to those processing the claim. The occurrence of an audit depends on the availability of resources to undertake the audit, and not on any knowledge related to the insurance claim. Deferred choices can be used at points at which an audit might be undertaken. The choice is then between the audit and the next activity in the processing chain. The activity capturing the audit triggers the next activity to preserve the processing chain.

Problem: Many workflow management systems support the XOR-split described in Pattern 4 (exclusive choice) but do not support the deferred choice. Since both types of choices are desirable (see examples), the absence of the deferred choice is a real problem. The essence of the problem is the moment of choice, as illustrated by Figure 8.9. In Figure 8.9(a), the choice is as late a possible (i.e., when *B* or *C* occurs), whereas in Figure 8.9(b) the choice is resolved when completing *A*.

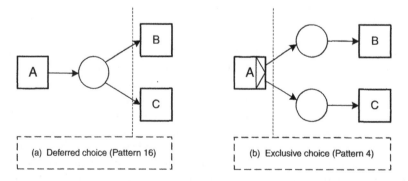

Figure 8.9   The moment of choice in the Patterns 4 (exclusive choice) and 16 (deferred choice).

Implementation:

- COSA is one of the few systems that directly supports the deferred choice. Since COSA is based on Petri nets, it is possible to model implicit choices, as indicated in Figure 8.9(a). YAWL is also based on Petri nets and, therefore, also supports the deferred choice. Some systems offer partial support for this pattern by offering special constructs for a deferred choice between a user action and a time out (e.g., Staffware) or two user actions (e.g., FLOWer).
- Although many workflow management systems have problems dealing with deferred choice, emerging standards in the Web services composition domain have no problems supporting the patterns. For example, BPEL offers a construct called *pick* that directly captures this pattern.
- Assume that the workflow language being used supports cancelation of activities (Pattern 19) through either a special transition (e.g., Staffware) or through an API (most other engines). Cancelation of an activity means that the activity is removed from the designated worklist as long as it has not been started yet. The deferred choice can be realized by enabling all alternatives via an AND-split. Once the processing of one of the alternatives is started, all other alternatives are canceled. Consider the deferred choice between B and C in Figure 8.9(a). This could be implemented using cancelation of activities in the following way. After A, both B and C are enabled. Once B is selected/ executed, activity C is canceled. Once C is selected/executed, activity B is canceled. Note that the solution does not always work because B and C can be selected/executed concurrently.
- Another solution to the problem is to replace the deferred choice by an explicit XOR-split, that is, an additional activity is added. All triggers activating the alternative branches are redirected to the added activity. Assuming that the activity can distinguish between triggers, it can activate the proper branch. Note that the solution moves part of the routing to the application or task level. Moreover, this solution assumes that the choice is made based on the type of trigger.□

8.3.6 Cancelation Patterns

When discussing possible solutions for the deferred choice pattern, we mentioned the pattern *Cancel activity* (Pattern 19). This is one of two cancelation patterns. The other pattern is *Cancel case* (Pattern 20). The cancel activity pattern disables an enabled activity, that is, a thread waiting for the execution of an activity is removed. The cancel case pattern completely removes a case, that is, workflow instance, (even if parts of the process are instantiated multiple times, all descendants are removed). Both constructs are supported by YAWL through a more generic construct that removes all tokens from a given region. A more detailed discussion of the cancelation patterns is outside the scope of this chapter.

8.4 CONCLUSION

This chapter introduced several control-flow patterns that can be used to support modeling efforts, to train workflow designers, and to assist in the selection of workflow management systems. Although inspired by workflow management systems, the patterns are not limited to workflow technology but applicable to process-aware information systems (PAISs) ranging from EAI platforms and Web services composition languages to case-handling systems and groupware. For example, the patterns have not only been used to evaluate several commercial and academic workflow management systems [5, 4] but also several standards including UML [8], BML [24], and BPEL4WS [23]. For more information on these evaluations and interactive animations for each of the patterns we refer the reader to www.workflowpatterns.com.

Throughout this chapter we used YAWL diagrams to illustrate the patterns. YAWL [4] demonstrates that it is possible to support the patterns in a direct and intuitive manner. YAWL is an open-source initiative and supporting tools can be downloaded from www.yawl-system.com. Current research aims at further developing YAWL and developing patterns and pattern languages for other perspectives besides control flow (notably the resource perspective; a collection of data patterns was recently reported in [20]).

Note that the 20 patterns mentioned in this chapter are not complete. Therefore, we invite users, researchers, and practitioners to contribute. Moreover, some systems and languages have limitations not adequately addressed by the patterns. For example, in Staffware it is not permissible to connect an exclusive choice/XOR-split to a wait step (i.e., synchronization/AND-join). Moreover, an exclusive choice in Staffware is always binary, that is, to model a choice involving three alternatives, two exclusive choice elements are needed. For most of these limitations there are simple workarounds. However, this is not always the case as is illustrated by the following example. Consider the Petri net shown in Figure 8.10. This model shows a simple classical Petri net with 8 transitions. First, A is executed, followed by B and E in parallel. B is followed by C. However, F has to wait for the completion of both E and C, and so on. Finally, H is executed and all transitions have been executed exactly once. Although the Petri net is very simple (e.g., it does not model any choices, only parallelism), process algebras like Pi calculus [16] have problems modeling this example. To understand the problem, consider the Petri net shown in Figure 8.10 without the connection between C and F. In that case the sequences $B. C. D$ and $E. F. G$ are executed in parallel in between A and H. In terms of Pi calculus (or any other process algebra), this is denoted as $A. (B. C. D|E. F. G), H$. In this notation, the "." is used to de-

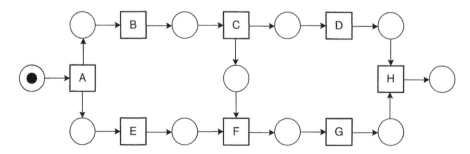

Figure 8.10   How would one model this in terms of Pi calculus?

note sequence and the "|" denotes parallelism. Indeed this notation is elegant and allows for computer manipulation. Unfortunately, such a simple notation is not possible if the connection between C and F is restored. The linear language does not allow for this, whereas for a graph-based language like Petri nets this is not a problem. Note that the claim is *not* that Pi calculus cannot model the process shown in Figure 8.10. However, it illustrates that even powerful languages like Pi calculus have problems supporting certain patterns. This is particularly relevant in the domain of Web services, in which Pi calculus is being put forward as a starting point for developing (future versions of) languages for describing service-based processes.

8.5 EXERCISES

Exercise 1 (identification of patterns in an informal description)
Consider the following informal description of a process for insurance claim handling. When a claim is received, it is first registered. After registration, the claim is classified, leading to two possible outcomes: simple or complex. If the claim is simple, the insurance is checked. For complex claims, both the insurance and the damage are checked independently. After the check(s), an assessment is performed, which may lead to two possible outcomes: positive or negative. If the assessment is positive, the garage is phoned to authorize the repairs and the payment is scheduled (in this order). In any case (whether the outcome is positive or negative), a letter is sent to the customer and the process is considered to be complete. At any moment after the registration and before the end of the process, the customer may call to modify the details of the claim. If a modification occurs before the payment is scheduled, then the claim is classified again, and the process is repeated from that point on. If a modification occurs after the payment is scheduled and before the letter is sent, a "stop payment" task is performed and the process is repeated starting with the classification of the claim.
 Which tasks can be identified in this scenario, and which workflow patterns link these tasks?

Exercise 2 (identification of patterns in an existing model)
Consider the YAWL specification in Figure 8.11. Which patterns occur in this specification and where? For example, the "AND-split" pattern can be found between tasks "register," "send form," and "evaluate."

Exercise 3 (identification of patterns in an existing model)
Consider the UML activity diagram shown in Figure 5.1 (Chapter 5). Which patterns occur in this model and where? Answer the same question for the ARIS function flow in Figure 6.2 (Chapter 6).

Exercise 4 (pattern implementation)
Figures 8.5 and 8.6 feature two YAWL specifications illustrating the multichoice (OR-split) and the synchonizing merge (OR-join) patterns. Translate these YAWL

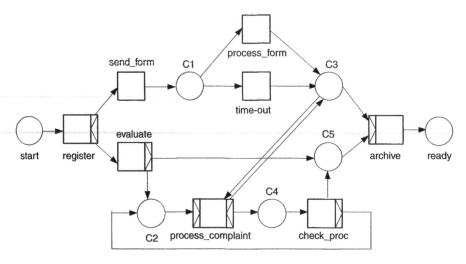

Figure 8.11   YAWL specification for Exercise 2.

specifications into classical Petri nets (see Chapter 7). In other words, expand the YAWL OR-split and OR-join constructs in terms of places and transitions (possibly labeled with empty tasks).

Exercise 5 (pattern implementation)
Figure 8.12 contains a YAWL specification in which the edges are labeled with Boolean expressions $C1$, $C2$, $C3$, and their negations. This specification contains an arbitrary loop in which tasks B and C can be repeated multiple times in alternation until the process completes. A possible execution of this process specification is AB, that is, task A is executed, then predicate $C1$ is true so B is executed, after which predicate $C2$ is false so the process terminates. Other possible executions include AC, ABC, ACB, $ABCB$, $ACBC$, etc.

Some process modeling or process execution languages only provide constructs for structured loops [e.g., constructs of the form `while (boolean expression) {fragment of process to be repeated}`], as in contemporary imperative pro-

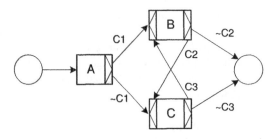

Figure 8.12   YAWL specification for Exercise 5.

gramming languages such as C and Java.[8] How could the specification in Figure 8.12 be expressed in a language that provides "while" loops, conditional statements of the form `if (boolean expression) {fragment of process}`, and simple sequencing between tasks (which can be denoted using a semicolon ";"), but does not support arbitrary loops. Consider each of the following two cases:

1. The language in question supports "break" statements, allowing one to exit a "while" loop in the middle of its body, as in contemporary imperative programming languages such as C and Java.

2. The language in question does not support "break" statements. Hint: You may introduce one or several auxiliary Boolean variable(s). When a predicate is evaluated, it can be assigned to a Boolean variable, and this variable can be used in the Boolean expressions of the "if" and "while" statements.

Exercise 8.6 (evaluation of PAIS development platforms)
Select a tool for PAIS development (see, for example, the tools mentioned in Chapters 1–4). Evaluate the selected tool in terms of the patterns. The evaluation should state, for each of the patterns presented in this chapter, whether the tool provides "direct support" for that pattern or not. If the answer to this question is positive for a given pattern, briefly explain how the pattern is supported. Otherwise, provide (if possible) a workaround solution to capture the pattern in question. Some sample evaluations of tools can be found in [5].

ACKNOWLEDGMENTS

We would like to thank Bartek Kiepuszewski, Alistair Barros, and Petia Wohed for their contribution to the research involving workflow patterns, and Lachlan Aldred for his contribution to the YAWL initiative.

REFERENCES

1. W. M. P. van der Aalst, L. Aldred, M. Dumas, and A. H. M. ter Hofstede. Design and Implementation of the YAWL System. In A. Persson and J. Stirna (Eds.), *Proceedings of the 16th International Conference on Advanced Information Systems Engineering (CAiSE)*, volume 3084 of *Lecture Notes in Computer Science*, pp. 142–159, Riga, Latvia, June 2004. Springer-Verlag.

2. W. M. P. van der Aalst, A. P. Barros, A. H. M. ter Hofstede, and B. Kiepuszewski. Advanced Workflow Patterns. In O. Etzion and P. Scheuermann (Eds.), *Proceedings of the 7th International Conference on Cooperative Information Systems (CoopIS)*, volume 1901 of *Lecture Notes in Computer Science*, pp. 18–29, Eilat, Israel, September 2000. Springer-Verlag.

---

[8]This is the case for example of BPEL, as discussed in Chapter 13.

3. W. M. P. van der Aalst, J. Desel, and E. Kindler. On the Semantics of EPCs: A Vicious Circle. In M. Nüttgens and F. J. Rump (Eds.), *Proceedings of the EPK 2002: Business ProcessManagement using EPCs,* pp. 71–80, Trier, Germany, November 2002. Gesellschaft für Informatik, Bonn, Germany.

4. W. M. P. van der Aalst and A. H. M. ter Hofstede. YAWL: Yet Another Workflow Language. Accepted for publication in *Information Systems,* and also available as QUT Technical report FIT-TR-2003-04, Queensland University of Technology, Brisbane, Australia, 2003.

5. W. M. P. van der Aalst, A. H. M. ter Hofstede, B. Kiepuszewski, and A. P. Barros. Workflow Patterns. *Distributed and Parallel Databases, 14*(1):5–51, 2003.

6. F. Casati, S. Ceri, B. Pernici, and G. Pozzi. Conceptual Modelling of Workflows. In M. P. Papazoglou (Ed.), *Proceedings of the 14th International Object-Oriented and Entity-Relationship Modelling Conference (OOER),* volume 1021 of *Lecture Notes in Computer Science,* pp. 341–354, Gold Coast, Australia, December 1998. Springer-Verlag.

7. J. Dehnert and P. Rittgen. Relaxed Soundness of Business Processes. In K. R. Dittrich, A. Geppert, and M. C. Norrie (Eds.), *Proceedings of the 13th International Conference on Advanced Information Systems Engineering (CAiSE'01),* volume 2068 of *Lecture Notes in Computer Science,* pp. 157–170, Interlaken, Switzerland, June 2001. Springer-Verlag.

8. M. Dumas and A. H. M. ter Hofstede. UML activity diagrams as a workflow specification language. In M. Gogolla and C. Kobryn (Eds.), *Proceedings of the 4th International Conference on the Unified Modeling Language, Modeling Languages, Concepts, and Tools (UML),* volume 2185 of *Lecture Notes in Computer Science,* pp. 76–90, Toronto, Canada, October 2001. Springer-Verlag.

9. L. Fischer (Ed.). *Workflow Handbook 2003,* Workflow Management Coalition, *Future Strategies.* Lighthouse Point, FL, USA, 2003.

10. E. Gamma, R. Helm, R. Johnson, and J. Vlissides. *Design Patterns: Elements of Reusable Object-Oriented Software.* Addison-Wesley, Reading, MA, 1995.

11. S. Jablonski and C. Bussler. *Workflow Management: Modeling Concepts, Architecture, and Implementation.* International Thomson Computer Press, London, 1996.

12. B. Kiepuszewski. *Expressiveness and Suitability of Languages for Control Flow Modelling in Workflows.* PhD thesis, Queensland University of Technology, Brisbane, Australia, 2003. Available via http://www.workflowpatterns.com.

13. B. Kiepuszewski, A. H. M. ter Hofstede, and W. M. P. van der Aalst. Fundamentals of Control Flow in Workflows. *Acta Informatica, 39*(3):143–209, 2003.

14. B. Kiepuszewski, A. H. M. ter Hofstede, and C. Bussler. On structured workflow modelling. In B. Wangler and L. Bergman (Eds.), *Proceedings of the 12th International Conference on Advanced Information Systems Engineering (CAiSE),* volume 1789 of *Lecture Notes in Computer Science,* pp. 431–445, Stockholm, Sweden, June 2000. Springer-Verlag.

15. P. Langner, C. Schneider, and J. Wehler. Petri Net Based Certification of Event driven Process Chains. In J. Desel and M. Silva (Eds.), *Application and Theory of Petri Nets 1998,* volume 1420 of *Lecture Notes in Computer Science,* pp. 286–305, Lisbon, Portugal, June 1998. Springer-Verlag.

16. R. Milner. *Communicating and Mobile Systems: The Pi-Calculus.* Cambridge University Press, Cambridge, UK, 1999.

17. E. A. H. Platier. *A Logistical View on Business Processes: BPR and WFM Concepts (in Dutch)*. PhD thesis, Eindhoven University of Technology, Eindhoven, The Netherlands, 1996.

18. D. Riehle and H. Züllighoven. Understanding and Using Patterns in Software Development. *Theory and Practice of Object Systems, 2*(1):3–13, 1996.

19. P. Rittgen. Modified EPCs and their Formal Semantics. Technical report 99/19, University of Koblenz-Landau, Koblenz, Germany, 1999.

20. N. Russell, A. H. M. ter Hofstede, D. Edmond, and W. M. P. van der Aalst. Workflow Data Patterns. QUT Technical report FIT-TR-2004-01, Queensland University of Technology, Brisbane, Australia, 2004.

21. Eastman Software. *RouteBuilder Tool User's Guide*. Eastman Software, Inc, Billerica, MA, 1998.

22. WfMC. Workflow Management Coalition Terminology and Glossary, Document Number WFMC-TC-1011, Document Status—Issue 3.0. Technical report, Workflow Management Coalition, Brussels, Belgium, February 1999.

23. P. Wohed, W. M. P. van der Aalst, M. Dumas, and A. H. M. ter Hofstede. Analysis of Web Services Composition Languages: The Case of BPEL4WS. In I.Y. Song, S.W. Liddle, T. W. Ling, and P. Scheuermann (Eds.), *Proceedings of the 22nd International Conference on Conceptual Modeling (ER)*, volume 2813 of *Lecture Notes in Computer Science*, pp. 200–215, Chicago, IL, USA, October 2003. Springer-Verlag.

24. P. Wohed, E. Perjons, M. Dumas, and A. ter Hofstede. Pattern-Based Analysis of EAI Languages—The Case of the Business Modeling Language. In O. Camp, J. Filipe, S. Hammoudi, and M. Piatinni (Eds.), *Proceedings of the 5th International Conference on Enterprise Information Systems (ICEIS)*, pp. 174–182, Angers, France, April 2003. Escola Superior de Tecnologia do Instituto Politécnico de Setúbal, Setúbal, Portugal.

PART III

TECHNIQUES

CHAPTER 9

Process Design and Redesign

HAJO A. REIJERS

9.1 INTRODUCTION TO PROCESS (RE)DESIGN

A little over a decade ago, a new way of looking at business emerged in industry. It contrasted sharply with the traditional emphasis on *functional* business areas within companies, such as the procurement, manufacturing, and sales departments. Hammer [5] and Davenport and Short [4] were the first to report on more or less systematic approaches to consider and improve entire *business processes*. The major ingredients in this approach were the application of information technology and a drastic restructuring of the process. The latter part of this approach, *process redesign,* is the subject of this chapter. Clearly, when a process has to be designed from scratch, *process design* would be a more accurate term. However, because this only applies for a minority of occasions—start-ups and entirely new products and services of existing organizations—we will generally speak of process redesign throughout this chapter.

Process redesign has been embraced by companies all over the world, although the success of the approach varies from company to company and from process to process. The drivers behind the popularity of process redesign are manifold. In the first place, companies feel the increasing pressure of a globalizing market. Cost reduction has become necessary to survive and process redesign has turned out be an effective cost cutter. Second, the historically strong position of suppliers in many markets is becoming less dominant compared to that of the customer. To keep customers coming back, companies have to please them by shortening their lead times or by increasing their product quality. These are typical objectives of redesign projects. The last major change driver is the rise of process-aware information systems, which facilitate the execution, analysis, and monitoring of entire business processes. Because a thorough process analysis must precede the implementation of a process-aware information system, it often goes hand in hand with process redesign efforts.

In short, "process-thinking" and process redesign have become mainstream thinking in industry in the 21st century. But surprisingly, it is not very clear how to redesign a process so that it becomes a substantial improvement on the current process layout. At best, most textbooks give directions to manage the organizational risk that is involved with designing or redesigning processes. Even the classic work of Hammer and Champy [6] devotes only 14 out of a total of over 250 pages to this issue, of which 11 pages are used for the description of a case. As Sharp and McDermott [10] commented more recently: "How to get from the as-is to the to-be [in a process redesign] isn't explained, so we conclude that during the break, the famous ATAMO procedure is invoked—And Then, A Miracle Occurs."

This chapter aims to provide more concrete guidance in designing or redesigning business processes. First, an overview will be sketched of existing theory. Then, as a prelude to the discussion of two concrete process redesign methods, the performance indicators that usually drive process redesign projects are described. The process redesign methods are described next; one of them can be characterized as being based on heuristic redesign best practices, whereas the other exploits an information processing perspective on business processes. Finally, the chapter concludes with a short conclusion and a set of exercises.

9.2 METHODOLOGIES, TECHNIQUES, AND TOOLS

Much has been said and written about the subject of process redesign. A plethora of methods, approaches, and guidelines exist, all of which seem to deal with process redesign. The classification of Kettinger et al. [7] is helpful in this respect. There are three levels of abstractions for methods with respect to process redesign: methodologies, techniques, and tools.

A *methodology,* the highest level of abstraction, is defined as a collection of problem-solving methods governed by a set of principles and a common philosophy for solving targeted problems. This is primarily the field of consulting firms who develop proprietary methodologies covering in detail all phases in a redesign project, from the early diagnosis until the implementation and aftercare. Research-oriented methodologies or initiatives for methodologies do exist, but there are relatively few of them. In saying this, we ignore mere lists of activities that should take place within a redesign project without describing in some detail: the activities themselves, the dependencies between these activities, the techniques that should be applied, and the deliverables of the activities. An important aspect to differentiate between various redesign methodologies is whether a *clean sheet* approach is adopted, that is, the process is designed from scratch, or whether an existing process is taken as a starting point, which is gradually refined. There is considerable discussion between practitioners and researchers alike as to which approach is preferable (see O'Neill and Sohal [8]). In general, clean sheet approaches tend to be more risky as they break away from existing, known procedures, but also tend to deliver higher benefits when they succeed, as inefficiencies can be rooted out.

At the next level of abstraction, a *technique* is defined as a set of precisely described procedures for achieving a standard task. Among the often encountered techniques for process diagnosis—one of the phases in a redesign project—are, for example, fishbone diagramming, Pareto diagramming, and cognitive mapping. To support the activity of redesigning, creativity techniques like out-of-box-thinking, affinity diagramming, and the Delphi method (*brainstorming*) are available. For the modeling and/or evaluation of business processes, techniques available use such as flowcharting, IDEF, speech act modeling, data modeling, activity-based costing, time–motion studies, Petri nets, role playing, and simulation.

At the lowest, most concrete level, a *tool* is defined as a computer software package to support one or more techniques. The majority of the business process redesign (BPR) tools focus on the modeling of a business process, be it existing or new. A large number of tools are also available for the evaluation of business process models, in particular through simulation. Fewer tools are available to structurally capture knowledge about the redesign directions or to support existing creativity techniques. Tools are often presented as "intelligent" or "advanced," although hardly any of those actively design business processes.

Obviously, methodologies, techniques, and tools can be linked in different ways. Clearly, a choice of redesign techniques influences the choice for an appropriate redesign methodology and vice versa. In this chapter, the emphasis is on two sets of redesign techniques, either of which may be incorporated in an overall redesign methodology. The actual integration of these techniques in a methodology or the development of tools to support these techniques, however, is not within the scope of this chapter.

Before the two types of techniques are discussed, some background is given on the performance indicators of a process redesign effort.

9.3 BUSINESS PROCESS PERFORMANCE INDICATORS

Before the start of any redesign project, it should be clear which performance aspects of the business process are targeted for improvement. A useful conceptual framework for this purpose is the *devil's quadrangle* [2], which is depicted in Figure 9.1. The framework distinguishes four main performance dimensions along which the effects of redesign can be measured: time, cost, quality, and flexibility. In most circumstances, it would be ideal if a redesign of a business process decreases the time required to handle cases, decreases the required cost of executing the business process, improves the quality of the service delivered, and improves the ability to change the business process to react to variation. The nasty, yet realistic, property of the devil's quadrangle is that, in general, improving upon one dimension may have a weakening effect on another. For example, reconciliation tasks may be added in a business process to improve the quality of the delivered service, but this may produce negative effects on the timeliness of the service delivery. This type of difficult trade-off explains the ominous name of the quadrangle.

For each project and each business process, it will be necessary to formulate exactly how the various performance dimensions of the quadrangle will be measured to establish and determine the goals of the redesign effort. Notwithstanding the popularity of process redesign, different studies have indicated that a large number of

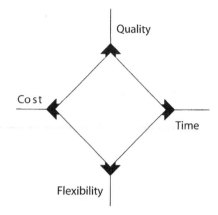

Figure 9.1 The devil's quadrangle.

BPR programs fail. Some failure estimates are up to 70% (e.g., [1, 3]). Although falling short of the intended performance objectives is an obvious mark of failure, it is likely that in many cases *no clear objectives have been formulated at all*. In other words, without setting the performance goals for a redesign effort or without measuring the respective performance indicators both at the *start* and the *end* of a project, it is impossible to determine whether the redesign is a success.

Subsequently, a short overview will be given of some of the most common issues in making the various performance dimensions of the devil's quadrangle operational.

9.3.1 Time

An important performance concept of a business process is its lead time (also known as cycle time, throughput time, etc.). It involves the time that it takes to handle a case from start to end. Although it is usually the aim of a redesign effort to reduce the lead time, there are many different ways of further specifying this aim. For example, one can aim at a reduction of the *average* lead time or the *maximal* lead time over a certain period of time. Both of these entities are absolute measures. It is also possible to focus on the ability to meet lead times that are agreed upon with a client at run time. This is a more relative interpretation of the lead time dimension. Yet another way of looking at the lead time is to try and limit the lead time's variation, so that delivery of service becomes more reliable.

Other aspects of the time dimension come into view when considering the constituents of lead time, which are as follows:

- *Service times*—the time that resources spend on actually handling the case.
- *Queue times*—the time that a case spends waiting in queue because there are no resources available to handle it.

- *Wait times*—all other time a case spends waiting; for example, because synchronization must take place with another process.

In general, there are different ways of measuring each of these constituents. An interesting phenomenon is that the major part of lead time consists of wait and queue time (up to 95%). Decreasing the lead time should, therefore, primarily focus on reducing these components.

9.3.2 Cost

The most common performance targets for redesign projects are of a financial nature. Although the devil's quadrangle mentions "cost," it would also have been possible to put the emphasis on turnover, yield, or revenue. Obviously, an increase of yield may have the same effect on an organization's profit as a decrease of cost. However, redesign is more often associated with reducing cost and not so much with increasing yield.

There are different perspectives on cost. In the first place, it is possible to distinguish between fixed and variable cost. Fixed costs are overhead costs that are (almost) unaffected by the intensity of processing. Typical fixed costs follow from the use of infrastructure and the maintenance of information systems. Variable cost is positively correlated with some variable quantity, such as the level of sales, the number of purchased goods, the number of new hires, etc.

A cost notion that is closely related to productivity is operational cost. Operational costs can be directly related to the outputs of a business process. A substantial part of operational cost (typically 60%) is labor cost, the cost related to human resources in producing a good or delivering a service. Within a redesign effort, it is very common to focus on reducing operation cost, particularly labor cost. The automation of tasks is often seen as an alternative for labor. Obviously, although automation may reduce labor cost, it may cause incidental cost involved with developing the respective application and fixed maintenance cost for the lifetime of the application.

9.3.3 Quality

The quality of a business process can be viewed from at least two different angles: from the client's side and from the worker's side. This is also known as the distinction between external quality and internal quality.

External quality can be measured as the client's satisfaction with either the product or the process. Satisfaction with the product can be expressed as the extent to which a client feels that his specifications or expectations are met by the delivered product. A client's satisfaction with the business process concerns the way it is executed. A typical issue is the amount and quality of the information that a client receives during execution on the progress being made.

The internal quality of a business process involves the condition of working in the business process. Typical issues are: the extent to which a worker feels he or she is in control of the work performed, the level of variation experienced, and whether working in the particular business process is felt to be challenging.

It is interesting to note that there are various direct relations between quality and other dimensions. For example, the external process quality is often measured in terms of time, for example, the lead time.

9.3.4 Flexibility

The criterion most often ignored in measuring the effect of redesign is the involved business process's flexibility. Flexibility can be defined as the ability to react to changes. These changes may concern various parts of the business process as follows:

- The ability of resources to execute different tasks.
- The ability of a business process as a whole to handle various cases and changing workloads.
- The ability of the business's management to change the structure of the process and/or the allocation of resources.
- The organization's ability to change the structure and responsiveness of the business process to meet the wishes of the market and business partners.

Another way of approaching the flexibility issue is to distinguish between run time and build time flexibility. *Run time* flexibility concerns the possibilities of handling changes and variations while executing a specific business process. *Build time* flexibility concerns the possibility of changing the business process structure.

It is important to distinguish the flexibility of a business process from the other dimensions, as will be clear from the discussion of the redesign approach based on best practices in Section 9.4.

9.4 REDESIGNING PROCESSES USING BEST PRACTICES

In this section, a redesign technique is discussed that is based on a large survey of the literature on process redesign (see [9]). From this body of literature, which often expresses the experiences that large companies or consultancy firms have gathered over the years, some 30 *best practices* can be gathered. In the last 20 years, best practices have been collected and applied in various areas, such as business planning, healthcare, manufacturing, and software development. A best practice is often seen as some sort of pattern, expressing the best way to treat a particular problem, which can be replicated in a similar situation or setting (for other types of patterns see Chapter 8). A best practice often needs to be adapted in a skillful way in response to prevailing conditions.

The redesign technique that is based on redesign best practices can now be described as follows. Assuming a number of clearly established performance goals, a team of redesigners uses the list of redesign best practices to evaluate and improve an existing process. Depending on the *fit* of a specific best practice with the performance objectives and the *feasibility* of applying it to the process under consideration, a new, gradually improved version of the process can be derived. This procedure can be continued, repeated, and reiterated until a desired process redesign emerges. Clearly, this approach blends rather well with a redesign methodology in which an existing process is taken as the starting point.

It is noteworthy that the various redesign best practices often lack a quantitative justification in the sources in which they are originally mentioned. They are, nonetheless, "advertised" because they seemed to be working in the past in several situations. In this sense, the presented redesign technique in this section is of a rather heuristic nature.

What follows now is a brief introduction of the various best practices and their (supposed) effectiveness. For each best practice, an acronym will be given (in capitals, between parentheses), followed by its general formulation, desirable effects, and possible drawbacks in terms of the devil's quadrangle. The following classification for discussing the best practices will be used:

- *Task best practices,* which focus on optimizing single tasks within a business process
- *Routing best practices,* which try to improve upon the routing structure of a business process
- *Allocation best practices,* which involve a particular allocation of resources within the business process
- *Resource best practices,* which focus on the types and number of resources
- *Best practices for external parties,* which try to improve upon the collaboration and communication with the client and third parties
- *Integral best practices,* which apply to the business process as a whole

Note that this distinction is not mutually exclusive. In other words, it is to some degree arbitrary to which category a best practice is assigned.

9.4.1   Task Best Practices

Task Elimination (ELIM). The task elimination best practice is to eliminate unnecessary tasks from a business process (see Figure 9.2). A task is said to be unnec-

Figure 9.2   Task elimination.

essary when it adds no value from a client's point of view. Typically, control tasks in a business process do not do this; they are incorporated in the model to fix problems created or not resolved in earlier steps. Control tasks often take the form of iterations or reconciliation tasks. The aims of this best practice are to increase the speed of processing and to reduce the cost of handling a case. An important drawback may be that the quality of the service deteriorates.

Task Composition (COMPOS). The purpose of the task composition heuristic is to combine small tasks into composite tasks and divide large tasks into workable smaller tasks (see Figure 9.3). Combining tasks should result in the reduction of set-up time, that is, the time that is spent by a resource to become familiar with the specifics of a case. By executing a large task that used to consist of several smaller ones, some positive effect may also be expected on the quality of the delivered work. Making tasks too large may result in (a) decreased run-time flexibility and (b) lower quality, as tasks may become unworkable. Both issues are addressed by dividing tasks into smaller ones. Obviously, smaller tasks may result in longer setup times.

Task Automation (AUTO). The task automation best practice is to consider automating tasks (see Figure 9.4). The positive result of automating tasks may be that they can be executed faster, with less cost, and with a better result. An obvious disadvantage is that the development of a system that performs a task may be costly. Generally speaking, a system performing a task is also less flexible in handling variations than a human resource. Instead of fully automating a task, automated support of the resource executing the task may also be considered. This best practice is a specific application of the technology best practice, which will be discussed later on.

9.4.2 Routing Best Practices

Resequencing (RESEQ). The resequencing best practice is to move tasks to more appropriate places (see Figure 9.5). In existing business processes, actual orderings of tasks do not strictly reflect necessary dependencies between those tasks. Therefore, it is sometimes better to postpone a task and carry out another so that, perhaps, its execution may become superfluous. This saves cost. A task may also be moved into the proximity of a similar task, in this way diminishing setup times. Specific applications of the resequencing best practice are the knockout best prac-

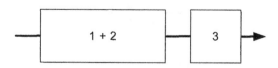

Figure 9.3 Task composition.

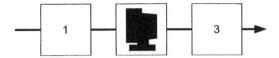

Figure 9.4 Task automation.

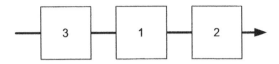

Figure 9.5 Resequencing.

tice, control relocation, and parallelism best practice, which will subsequently be discussed.

Knockout (KO). The knockout best practice is to order knockouts in an increasing order of effort and in a decreasing order of termination probability (see Figure 9.6). A typical part of a business process is checking various conditions that must be satisfied to deliver a positive end result. Any condition that is not met may lead to termination of that part of the business process, the knockout. If there is freedom in choosing the order in which the various conditions are checked, the condition that has the most favorable ratio of expected knockout probability versus the expected effort to check the condition should be pursued, then, the second-best condition is pursued, and so on. This way of ordering checks yields on average of the least costly business process execution. There is no obvious drawback to this best practice, although it may not always be possible to freely order these kinds of checks. Implementing this best practice also may result in all or part of a business process having a longer lead time than one in which all conditions are checked in parallel.

Control Relocation (RELOC). The control relocation best practice means moving controls toward the client (see Figure 9.7). Different checks and reconciliations that are part of a business process may be moved toward the client. By having clients check information themselves with forms or software, the bulk of errors may

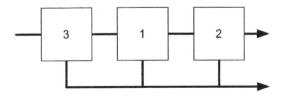

Figure 9.6 Knockout.

Figure 9.7 Control relocation.

be eliminated. This would also improve client satisfaction. A disadvantage of moving a control toward a client is higher probability of fraud, resulting in less yield.

Parallelism (PAR). The parallelism best practice is to consider whether tasks may be executed in parallel (see Figure 9.8). The obvious effect of applying this best practice is that the lead time may be considerably reduced. A drawback of introducing more parallelism in a business process that incorporates possibilities of knockouts is that the cost of business process execution may increase. The management of business processes with concurrent behavior can also become more complex, which may introduce errors (reducing quality) or restrict run time adaptations (reducing flexibility).

Triage (TRI). The main interpretation of the triage best practice is to consider the division of a general task into two or more alternative tasks (see Figure 9.9). Its opposite (and less popular) formulation is to consider the integration of two or more alternative tasks into one general task. When applying the best practice in its main form, it is possible to design tasks that are better aligned with the capabilities of resources and the characteristics of the case. Both of these improve the quality of the business process. Distinguishing alternative tasks also facilitates a better utilization of resources, with obvious cost and time advantages. On the other hand, too much specialization can make processes become less flexible, less efficient, and cause monotonous work with repercussions for quality. This is resolved by using the alternative interpretation of the triage best practice.

A special form of the triage best practice is to divide a task into similar instead of alternative tasks for different subcategories of the case type. For example, a special desk may be set up for clients with an expected low processing time.

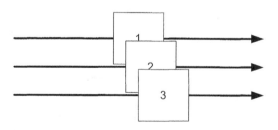

Figure 9.8 Parallelism.

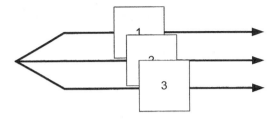

Figure 9.9   Triage.

The triage best practice is related to the task composition best practice in the sense that it is concerned with the division and combination of tasks, but differs from it in the sense that only alternative tasks are considered.

9.4.3   Allocation Best Practices

Case Manager (MAN).   The case manager best practice is to make one person responsible for the handling each case—the case manager (see Figure 9.10). The case manager is responsible for the case, but he or she is not necessarily the only resource that will work on work items for this case. The most important aim of this best practice is to improve upon the external quality of a business process. The business process will become more transparent from the viewpoint of a client if the case manager provides a single point of contact. This positively affects client satisfaction. It may also have a positive effect on the internal quality of the business process, as someone is accountable for correcting mistakes. Obviously, the assignment of a case manager has financial consequences as capacity must be devoted to this job.

Case Assignment (ASSIGN).   The case assignment best practice is to let the same workers perform as many steps as possible in a case (see Figure 9.11). This best practice is different from the case manager best practice. Although a case manager will be responsible for a case, he or she does not have to be involved in executing the business process. By using case assignment in its most extreme form, for each work item a capable person is selected who has worked on the case before. The obvious advantage of this best practice is that this person will already be acquainted with the case and will need less setup time. An additional benefit may be

Figure 9.10   Case manager.

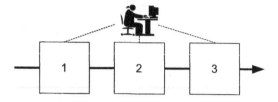

Figure 9.11 Case assignment.

that the quality of service is increased. On the negative side, the flexibility of re-source allocation is seriously reduced. A case may experience substantial queue time when its case manager is not available.

Customer Teams (TEAM). The customer team best practice is to consider com-posing teams of different workers from different departments that will take care of the complete handling of specific sorts of cases (see Figure 9.12). This best practice is a variation of the case assignment best practice. Depending on its exact desired form, the customer team best practice may be implemented by the case assignment best practice. One could also consider a customer team of workers with the same qualifications.

Advantages and disadvantages are similar to those of the case assignment best practices. In addition, teamwork may improve the attractiveness of the work and provide a better understanding of it, which are both quality aspects.

Flexible Assignment (FLEX). The flexible assignment best practice is to assign resources in such a way that maximal flexibility is preserved for the near future (see Figure 9.13). For example, if a work item can be executed by either of two available resources, assign it to the most specialized resource. In this way, the availability of the more general resource to take on the next work item is maximized.

The advantage of this best practice is that the overall queue time is reduced be-cause it is less probable that a case has to await the availability of a specific re-source. Another advantage is that the workers with the highest specialization can be expected to take on most of the work, which may result in a higher quality. The dis-advantages of this best practice can be diverse. For example, work load may be-

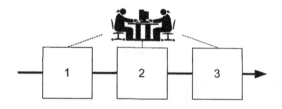

Figure 9.12 Customer teams.

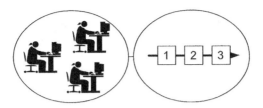

Figure 9.13 Flexible assignment.

come unbalanced, resulting in less job satisfaction. Possibilities for specialists to evolve into generalists are also reduced.

Resource Centralization (CENTR). The resource centralization best practice is to treat geographically dispersed resources as if they are centralized (see Figure 9.14). This best practice is explicitly aimed at exploiting the benefits of process-aware information systems. After all, when a process-aware information system takes care of handing out work items to resources, it has become less relevant where these resources are located geographically. In this sense, this best practice is a special form of the technology best practice. Moreover, it can also be seen as the opposite of the customer teams best practice. The specific advantage of this best practice is that resources can be committed more flexibly, which leads to a better utilization and, possibly, better lead time. The disadvantages are similar to that of the technology best practice.

Split Responsibilities (SPLIT). The split responsibilities best practice is to avoid assignment of task responsibilities to people from different functional units (see Figure 9.15). The idea behind this best practice is that tasks for which different departments share responsibility are more likely to be a source of neglect and conflict. Reducing the overlap in responsibilities should lead to a better quality of task execution. A higher responsiveness to available work items may also be developed, so that clients are served quicker. On the other hand, reducing the effective number of resources that is available for a work item may have a negative effect on its lead time, as more queuing may occur.

Figure 9.14 Resource centralization.

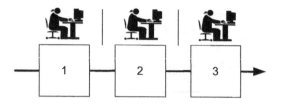

Figure 9.15 Split responsibilities.

9.4.5 Resource Best Practices

Numerical Involvement (NUM). The numerical involvement best practice is to minimize the number of departments, groups, and persons involved in a business process (see Figure 9.16). Applying this best practice should lead to fewer coordination problems. Less time spent on coordination makes more time available for the processing of cases. Reducing the number of departments may lead to less shared responsibilities, with similar pros and cons as the split responsibilities best practice. In addition, smaller numbers of specialized units may prohibit expertise building (a quality issue) and acquiring a routine (a cost issue).

Extra Resources (XRES). The extra resources best practice is to consider increasing the number of resources in a certain resource class if capacity is not sufficient (see Figure 9.17). This straightforward best practice speaks for itself. The obvious effect of extra resources is that there is more capacity for handling cases, reducing queue time. It may also help to implement a more flexible assignment policy. Of course, hiring or buying extra resources has its cost. Note the contrast of this best practice with the numerical involvement best practice. Also note that it deals with the number of people actually involved in a process, not with the priority with which they receive work (as in the flexible assignment best practice).

Specialist–Generalist (SPEC). The specialist–generalist best practice is to consider making resources more specialized or more generalized (see Figure 9.18). Resources may be turned from specialists into generalists or the other way round. A specialist resource can be trained for other qualifications; a generalist may be as-

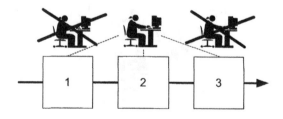

Figure 9.16 Numerical involvement.

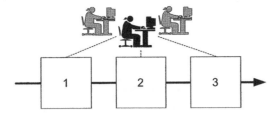

Figure 9.17 Extra resources.

Figure 9.18 Specialist–generalist.

signed to the same type of work for a longer period of time, so that his other qualifications become obsolete. When the redesign of a new business process is considered, application of this best practice comes down to considering the specialist–generalist ratio of new hires.

A specialist builds up a routine more quickly and may have more profound knowledge than a generalist. As a result, he or she works quicker and delivers higher quality. On the other hand, the availability of generalists adds more flexibility to the business process and can lead to a better utilization of resources. Depending on the degree of specialization or generalization, either type of resource may be more costly.

Note that this best practice differs from the triage concept in the sense that the focus is not on the division of tasks.

Empower (EMP). The empower best practice is to give workers most of the decision-making authority and reduce middle management (see Figure 9.19). In traditional business processes, substantial time may be spent on authorizing work that is to be done by others. When workers are empowered to make decisions independently, the result may be smoother operations with lower lead times. The reduction of middle management involvement in the business process also reduces the labor cost spent on the processing of cases. A drawback may be that the quality of the de-

Figure 9.19 Empower.

cisions may be lower and that obvious errors are no longer found. If bad decisions or errors result in rework, the cost of handling a case may actually increase compared to the original situation.

9.4.6 Best Practices for External Parties

Integration (INTG). The integration best practice is to consider the integration with a business process of the client or a supplier (see Figure 9.20). This best practice can be seen as exploiting the supply chain concept known in production. In practice, the application of this best practice may take different forms. For example, when two parties have to agree that the quality of a product they commonly produce is sufficient, it may be more efficient to perform several intermediate reviews than to perform one large review when both parties have completed their part.

In general, integrated business processes should render a more efficient execution, both from a time and cost perspective. The drawback of integration is that dependence grows and, therefore, flexibility may decrease.

Outsourcing (OUT). The outsourcing best practice is to consider outsourcing a business process in whole or in part (see Figure 9.21). Another party may be more efficient in performing the same work, so they might as well perform it. The outsourcing best practice is similar to the business process integration best practice in the sense that it reflects on business processes of other parties.

The obvious aim of outsourcing work is that it will reduce cost. A drawback may be that quality decreases. Outsourcing also requires more coordination efforts and will make the business process more complex.

Interfacing (INTF). The interfacing best practice is to consider a standardized interface with clients and partners (see Figure 9.22). The idea behind this best practice is that a standardized interface will diminish the probability of mistakes, incomplete applications, unintelligible communications, and so on. A standardized interface may result in fewer errors (higher quality), faster processing (reduced time), and less rework (decreased cost). The interfacing best practice can be seen a

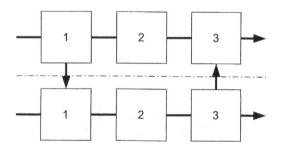

Figure 9.20 Integration.

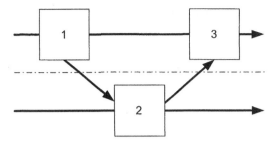

Figure 9.21   Outsourcing.

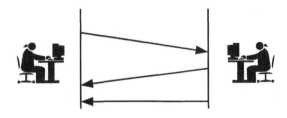

Figure 9.22   Interfacing.

specific interpretation of the integration best practice, with similar advantages and disadvantages.

Contact Reduction (REDUC).   The contact reduction best practice is to reduce the number of contacts with clients and third parties (see Figure 9.23). The exchange of information with a client or third party is always time-consuming. Especially when information exchanges take place by regular mail, substantial wait times may be involved. Each contact also introduces the possibility of introducing

Figure 9.23   Contact reduction.

errors. Reducing the number of contacts may, therefore, decrease lead time and boost quality. Note that it is not always necessary to skip certain information exchanges, but that it is possible to combine them with little extra cost. A disadvantage of a smaller number of contacts might be the loss of essential information, which is a quality issue. Combining contacts may result in the delivery or receipt of too much data, which involves cost.

Note that this best practice is related to the interfacing best practice in the sense that they both try to improve on the collaboration with other parties.

Buffering (BUF). The buffering best practice is to buffer information by subscribing to updates instead of requesting information from an external source (see Figure 9.24). Obtaining information from other parties is a major, time-consuming part of many business processes. By having information directly available when it is required, lead times may be substantially reduced. This best practice can be compared to the caching principle that microprocessors apply. Of course, the subscription fee for information updates may be rather costly. This is especially so when an information source may contain far more information than is ever used. Substantial cost may also be involved with storing all the information.

Note that this best practice is a weak form of the integration best practice. Instead of direct access to the original source of information—the integration alternative—a copy is maintained.

Trusted Party (TRUST). The trusted party best practice is to use results of a trusted party instead of determining information oneself (see Figure 9.25). Some decisions or assessments that are made within business processes are not specific for the business process they are part of. Other parties may have obtained the same information in another context, which, if it were known, could replace the decision or assessment part of the business process. An example is the creditworthiness of a client that bank A wants to establish. If a client can present a recent creditworthiness certificate from bank B, then bank A will accept it. Obviously, the trusted party best practice reduces cost and may even cut back lead time. On the other hand, the quality of the business process becomes dependent upon the quality of some other party's work. Some coordination effort with trusted parties is also likely to be required.

Note that this best practice differs from the outsourcing best practice. When outsourcing, a work item is executed at run time by another party. The trusted party best practice allows for the use of a result from the recent past. It is different from the buffering best practice because the business process owner is not the one obtaining the information.

Figure 9.24 Buffering.

Figure 9.25 Trusted party.

9.4.7 Integral Business Process Best Practices

Case Types (TYP). The case types best practice is to determine whether tasks are related to the same type of case and, if necessary, distinguish new business processes and product types. One should take care of subflows that are not specifically intended to handle the case type of one's umbrella business process (the superflow). Ignoring them may result in less effective management of this subflow and lower efficiency. Applying this best practice may yield faster processing times and lower cost. Distinguishing common subflows of many different flows may also yield efficiency gains. Yet, it may also result in more coordination problems between the business process (lower quality) and fewer possibilities for rearranging the business process as a whole (decreased flexibility).

Note that this best practice is in some sense similar to the triage concept. The main interpretation of the triage concept can be seen as a translation of the case type best practice on a task level.

Technology (TECH). The technology best practice is to try to elevate physical constraints in a business process by applying new technology. In general, new technology can offer all kinds of positive effects. For example, the application of a process-aware information system may result in less time spent on logistical tasks. A document management system will open up the information available on cases to all participants, which may result in a better quality of service. New technology can also change the traditional way of doing business by giving participants completely new possibilities.

Technology-related development, implementation, training, and maintenance efforts are obviously costly. In addition, new technology may instill fear in workers or may result in other adverse effects; this may decrease the quality of the business process.

Exception (EXCEP). The exception best practice is to design business processes for typical cases and isolate exceptional cases from the normal flow. Exceptions may seriously disrupt normal operations. An exception will require workers to get acquainted with a case even though they may not be able to handle it. Setup times are then wasted. Isolating exceptions, for example, by a triage, will make the handling of normal cases more efficient. Isolating exceptions may possibly increase the overall performance as specific worker expertise can be gained by working on the exceptions. By filtering out all exceptions, it may be possible to offer straight-through-processing, that is, completely automated processing. The price paid for isolating exceptions is that the business process will become more complex, possibly decreasing its flexibility. Also, if no special knowledge is developed to handle the exceptions (which is costly), no major improvements are likely to occur.

Case-based Work (CASEB). The case-based work best practice is to consider removing batch processing and periodic activities from a business process. Although business processes in the service industry are essentially case based and make to order, several features may be present in real-life business processes that are in disagreement with these concepts. The most notable examples are (a) the piling up of work items in batches and (b) periodic activities, depending on computer systems that are only available for processing at specific times. Getting rid of these constraints may significantly speed up the handling of cases. On the other hand, efficiencies of scale can be reached by batch processing. The cost of making information systems permanently available may also be costly.

9.5   INFORMATION-BASED BUSINESS PROCESS DESIGN

Instead of focusing on the existing process structure and improving it using a set of best practices, it is also possible to take a step back and consider what the *essence* of a business process is. A new design can then be derived from a clean sheet, taking into account the essential functions of the business process and the performance objectives. An advantage of this approach may be that the new process redesign will not incorporate all the superfluous tasks that have entered the existing process, for example, to respond to historic incidents. Also, the actual ordering of tasks in the current design will not play a role in ordering the tasks in a new design. In practice, business processes are often sequential by nature, which decrease the possibilities for parallelization or resequencing.

In this section, we present a redesign technique that adopts an *information processing perspective* on a business process. It is particularly applicable in the setting of administrative products, such as mortgages, insurance, and permits. Using the desired characteristics of the end product of a business process, an attempt is made to reason backward on what the ideal process should look like to make this end product. The technique that we will discuss fits rather well in a clean-sheet redesign methodology. It resembles the way a so-called bill of material is used to determine assembly lines in manufacturing processes.

To see how an information processing perspective may work for process redesign, consider the processing of insurance claims. The product to be delivered on the basis of an actual claim is basically a decision: either the claim is accepted—followed by a payment—or rejected. All kinds of *information elements* may play a role in making this decision, like the amount of damage, the claim history of the claimant, and the coverage of the insurance. For example, one of the standard conditions of the insurance policy may specify that if (i) the amount of damage is below a certain threshold, (ii) the claimant has not issued a claim for over a year, and (iii) the damage is covered, then the claim is accepted and the damage paid for. This hypothetical condition can be seen as a part of the product specification of the insurance. The information elements can be seen as raw materials or subassemblies for the production of a decision. The business process should "assemble" the decision by distinguishing tasks to retrieve and asses the required information elements, while taking criteria such as average lead time, service level, handling costs, and product quality into account. The latter are typically not characteristics of the product, but performance targets that should be specified at the beginning of the project.

In the Section 5.1 we will discuss rather informally the information-based redesign technique using the hypothetical case of the Air Force Test Agency.

9.5.1 The Air Force Test Agency

Among other duties, the Air Force Test Agency is concerned with testing candidates who have applied for the job of helicopter pilot. The procedure for this test has been conducted in the same form for many years, but Air Force management has recently been confronted with serious budget reductions by the government. Also, there has been an increasing stream of complaints from tested individuals who state that the procedure takes an excessive amount of time considering the mild complexity of the test procedure. This has triggered the initiative for a redesign of the helicopter pilot test procedure. The Air Force has asked a team of graduate students from the nearby university to come up with a redesign of the test procedure.

Instead of trying to improve the existing business process, the redesigners try to understand the process in terms of its information processing. From studying the formal requirements as issued by the Air Force in various regulations, they establish that the suitability of a candidate can be directly determined on the basis of any of the following:

1. By examining the outcomes of a psychological test and a physical test. If both scores are sufficiently high, the candidate can be approved.
2. By considering the outcome of the latest helicopter pilot test, if any, of the candidate: If the test has been executed within the last two years and the candidate was rejected, the candidate must be rejected again.
3. The eyesight quality of the candidate: If this is under a certain threshold, the candidate can be directly rejected.

Just as in the general case, these *business rules* are related. The eyesight test from business rule 3 is but one of the parts of the physical test mentioned in business rule 1. So, if a candidate's eyesight is below the quality threshold, that same candidate can never pass the physical test.

From a further analysis, the redesigners distill some more business rules and conclude that there are six important pieces of information that play a role in the test process:

1. The candidate's suitability to become a helicopter pilot
2. The candidate's psychological fitness
3. The candidate's physical fitness
4. The latest result of the candidate's suitability test in the previous two years
5. The candidate's quality of reflexes
6. The candidate's quality of eyesight

The redesigners specify these information elements as well as their dependencies according to the established business rules in an *information element structure* (see Figure 9.26).

In the information element structure, the individual information elements are depicted as rectangles. The business rules are depicted as directed arrows, combining one or more information elements, of which the values are needed as inputs, to determine the value of another information element—the business rule's output. So, from the three incoming arrows in the information element structure of Figure 9.26, it can be seen that there are three ways to come up with a final decision on a candidate's suitability to become a helicopter pilot (information element *a*). One of these ways involves the combination of the test results on both physical and psychological fitness, that is, information elements *b* and *c*. Four information elements, that is, *b*, *d*, *e*, and *f*, are not output elements of any business rule, which means that they can be established in absence of any further information.

The redesigners proceed to gather additional information on the established business rules. In particular, they want to know what are the constraints, performance characteristics, and execution probabilities of these business rules. They interview the various involved professionals from the test agency and come up with the data shown in Table 9.1.

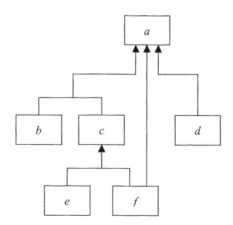

Figure 9.26   Helicopter pilot information element structure.

Table 9.1 Performance parameters

Index	x	$constr(x)$	$cst(x)$	$flow(x)$	$prob(x)$
1	$(a, \{b, c\})$	true	80	1	1, 0
2	$(a, \{d\})$	$*d \in \{$suitable, not suitable$\}$	10	1	0, 1
3	$(a, \{f\})$	$*f < -3, 0$ or $*f > +3, 0$	5	1	0, 4
4	$(b, \varnothing)$	true	150	12	1, 0
5	$(c, \{e, f\})$	true	50	1	1, 0
6	$(d, \varnothing)$	true	10	16	1, 0
7	$(e, \varnothing)$	true	60	4	1, 0
8	$(f, \varnothing)$	true	60	4	1, 0

Each business rule in Table 9.1 is represented as a pair of an information element inputs and a set of information element outputs. So, $(a, \{c, c\})$ refers to the business rule mentioned earlier, in which the outcome of the psychological test (b) can be combined with the outcome of the physical test (c) to determine the candidate's suitability (a). There are no constraints on applying the business rule, aside from the availability of relevant values for b and c. The cost of executing this business rule is roughly 80 Euros, the average lead time is 1 working hour, and it will produce a result with probability 1. The probability, however, of using an earlier test result to come up with a decisive conclusion on a candidate's suitability is only 0.1. For the sake of simplicity, it is assumed that all values are independent of each other.

Based on this analysis, the redesign team comes up with two alternative business process designs. They are represented as workflow (WF) nets in Figure 9.27. To many of the tasks in each of the models, a label is associated that indicates the business rule that is evaluated when executing the specific task. Tasks with no labels are incorporated for routing reasons. See Chapter 7 for a further explanation of the modeling technique.

The redesign team took much care to design both alternatives in such a way that the models addressed the following properties:

- *Soundness* (see Chapter 7),
- *Completeness*—each business rule is incorporated in the model
- *Conformity*—each business rule is associated with a task such that it is guaranteed that all its inputs have been determined in preceding tasks

The process model on the left-hand side of the figure is presented to Air Force management as an alternative with a very favorable expected lead time. After all, much emphasis is put on determining the values of many information elements in parallel, even though some of the information may turn out to be superfluous if the suitability of the candidate can be determined early. The process model on the right-hand side is presented as the alternative with a very favorable expected execution cost. The process is purely sequential and it avoids the computation of information

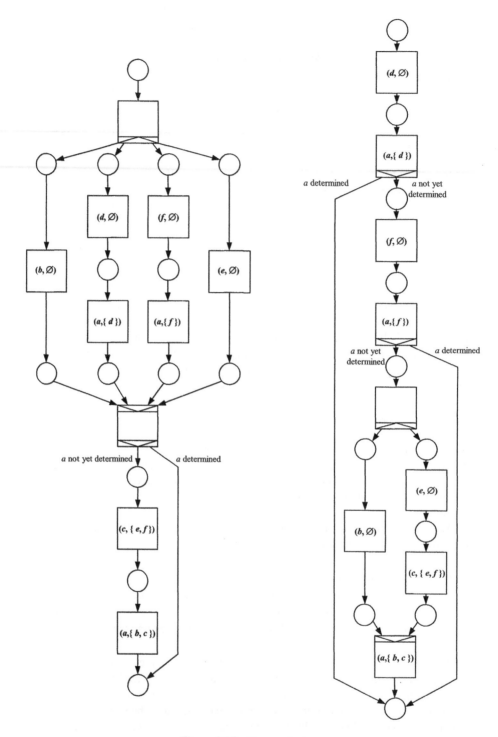

Figure 9.27 Process designs.

that is not strictly necessary. As soon as a decision can be made, the processing stops.

Management is very impressed with results, but dumbfounded at the same time. They have two process designs, but find it difficult to choose between them.

9.5.2 Applicability

Although the Air Force Test Agency case is a simplified one, it contains all essential elements of an information-centered approach to process redesign. The approach has been applied in industry twice, within a financial institution and a social security agency. These typically reflect environments in which the processing of information is the core of the business process. Those who are interested in the descriptions of these cases or a formal description of the approach are referred to [9].

9.6 CONCLUSION

Process redesign is a very popular subject in industry, but concrete redesign techniques are scarce. This chapter gave the essential ideas behind two different redesign approaches. It was shown that process redesign is closely linked to process modeling techniques and simulation.

Regardless of the type of approach that is chosen in a particular situation, it is always wise to explicitly formulate the redesign goals in tangible, measurable terms. Only then it is possible to establish the effectiveness of process redesign after it has been implemented.

Currently, many organizations are enabling their business processes for interaction with their clients, suppliers, and third parties through electronic media such as the World Wide Web, a movement referred to as e-commerce or e-business. Much emphasis is on the construction of flashy Web sites and the creation of contact e-mail addresses. However, if the underlying business processes are to live up to the expectations of doing business in the high-speed age of the electronic highway, this may require considerable redesign of the complex, sequential, paper-based business processes that are essentially behind these Web interfaces. This will ensure a bright future for process redesigners.

9.7 EXERCISES

Exercise 1
Can you explain the three drivers behind the popularity of process redesign?

Exercise 2
Some redesigners favor a so-called "clean sheet" approach to process redesign.
a) What is meant by approach?
b) What are the advantages of this approach?
c) What is the opposite of this approach?

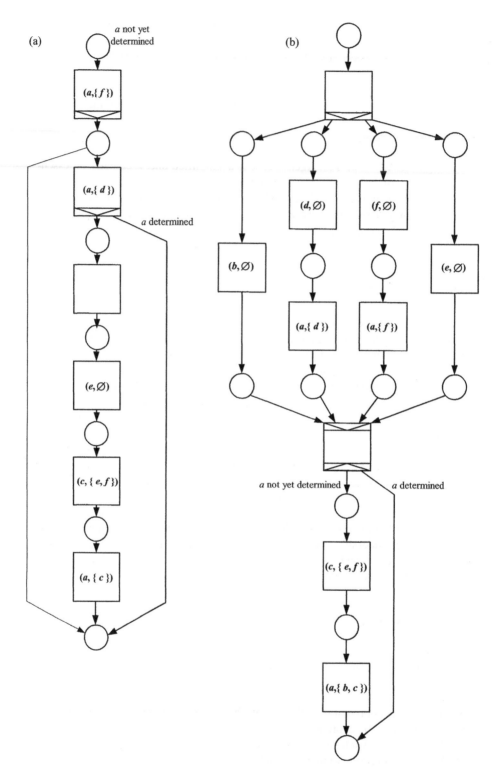

Figure 9.28 Alternative process designs.

Exercise 3
Explain at least one difference between the following redesign best practices in terms of the dimensions of the devil's quadrangle.
a) ASSIGN versus INTF
b) PAR versus KO

Exercise 4
As explained in the Air Force case, the Air Force management finds it difficult to decide between the two alternatives that the redesign team comes up with. Can you offer two possible reasons why this difficulty arises?

Exercise 5
Consider the two design alternatives that the redesign team decides on (see Figure 9.27), with the associated performance parameters (see Table 9.1). You may assume that executing a check will always consume the standard costs, independent of the task's outcome. Moreover, assume an infinity of resources so that no queuing takes place.
a) For both alternatives, determine the expected cost and the expected lead time of an execution of the process.
b) Develop an improved process design for the alternatives in (a) with the lowest expected lead time, such that the expected lead time is even further reduced.
c) Develop an improved process design for the alternatives in (a) with the lowest expected cost, such that the expected cost is even further reduced.

Exercise 6
The Air Force management decides to develop two other process design alternatives themselves, which they consider improvements of the designs presented to them (see Figure 9.28). They are rather happy with themselves; however, can you indicate what is wrong with these designs?

REFERENCES

1. S. Bradley. Creating and Adhering to a BPR Methodology. Gartner Group Report, 1–30, 1994.
2. N. Brand and H. van der Kolk. *Workflow Analysis and Design.* Kluwer Bedrijfswetenschappen, 1995. (In Dutch)
3. J. Champy. *Reengineering Management.* HarperCollins, London, 1995.
4. T. H. Davenport and J. E. Short. The New Industrial Engineering: Information Technology and Business Process Redesign. *Sloan Management Review, 31*(4):11–27, 1990.
5. M. Hammer. Reengineering Work: Don't Automate, Obliterate. *Harvard Business Review,* 70–91, 1990.
6. M. Hammer and J. Champy. *Reengineering the Corporation; A Manifesto for Business Revolution.* Harper Business, New York, 1993.

7. W. J. Kettinger, J. T. C. Teng, and S. Guha. Business Process Change: A Study of Methodologies, Techniques, and Tools. *MIS Quarterly, 21*(1):55–80, 1997.

8. P. O'Neill and A. S. Sohal. Business Process Reengineering: a Review of Recent Literature. *Technovation 19*(9): 571–581, 1999.

9. H. A. Reijers. *Design and Control of Workflow Processes: Business Process Management for the Service Industry.* Springer-Verlag, Berlin, 2003.

10. Sharp and P. McDermott. *Workflow Modeling: Tools for Process Improvement and Application Development.* Artech House Publishers, Boston, 2001.

CHAPTER 10

Process Mining

WIL van der AALST and A. J. M. M. (TON) WEIJTERS

10.1 INTRODUCTION

The basic idea of process mining is to extract knowledge from event logs recorded by an information system. Until recently, the information in these event logs was rarely used to analyze the underlying processes. Process mining aims at improving this by providing techniques and tools for discovering process, control, data, organizational, and social structures from event logs. Fueled by the omnipresence of event logs in transactional information systems (cf. WFM, ERP, CRM, SCM, and B2B systems), process mining has become a vivid research area [4, 5]. In this chapter we provide an overview of process mining techniques and tools and discuss one algorithm (the α algorithm) in detail.

As explained in Chapter 1, many information systems have become process-aware. This awareness can be used in various ways; for example, the process-aware information system may *enforce* a specific way of working but may also just *monitor* the process and suggest alternative ways of working. Workflow management (WFM) systems such as Staffware, IBM MQSeries, COSA, and so on (see Chapter 2) can be used to enforce a specific way of working but may also allow for predefined choices based on human judgment, properties of the case being handled, or a changing context. Case handling (CH) systems such as FLOWer (see Chapter 15) allow for more flexibility by enabling alternative paths that are implicitly defined (e.g. the ability to skip, roll back, or change the order of activities). Both for WFM and CH systems, there is an explicit process model that is *actively* used to support the process. In many other process-aware systems, the process model plays a less explicit role. For example, although ERP (enterprise resource planning) systems such as SAP, PeopleSoft, Baan, and Oracle offer a workflow component, process models are often hard-coded or used in a passive way. SAP supports a wide variety of processes. Parts of these processes are hard-coded in the software, whereas other parts of the process are only described in so-called *reference models*. These reference models describe how people *should* use the system. Although process models in a WFM are used *actively*, these reference models are only used *passively*. Another example is a hospital infor-

Process-Aware Information Systems. Edited by Dumas, van der Aalst, and ter Hofstede
Copyright © 2005 John Wiley & Sons, Inc.

mation system (HIS) supporting clinical guidelines. These guidelines describe the treatment of a patient having a specific health problem and can be used in an active way (e.g., automatically suggest actions to the medical staff) or a passive way (e.g., the medical staff can consult the clinical guideline when needed). Other process-aware information systems such as CRM (customer relationship management) software, SCM (supply chain management) systems, B2B (business to business) applications, and so on may use process models actively or passively and these models may be hard-coded in the software, implicit (as in a CH system), or explicit (as in a WFM system). Despite the different ways in which models are used, most of these systems log events in some way. In this chapter, we do not focus on the design of these models but instead on techniques for *monitoring* enterprise information systems (i.e., WFM, ERP, CRM, and SCM-like systems).

As mentioned, many of today's enterprise information systems store relevant events in some structured form. For example, workflow management systems typically register the start and completion of activities [2]. ERP systems like SAP log all transactions; for example, users filling out forms and changing documents. Business-to-business (B2B) systems log the exchange of messages with other parties. Call center packages and general-purpose CRM systems log interactions with customers. These examples show that many systems have some kind of *event log,* often referred to as "history," "audit trail," "transaction log," and so on [4, 7, 10, 16]. The event log typically contains information about events referring to an *activity* and a *case*. The case (also named process instance) is the "thing" that is being handled, for example, a customer order, a patient in a hospital, a job application, an insurance claim or a building permit. The activity (also named task, operation, action, or work item) is some operation on the case. Typically, events have a *time stamp* indicating the time of occurrence. Moreover, when people are involved, event logs will typically contain information on the person executing or initiating the event—the *originator*. Based on this information, several tools and techniques for process mining have been developed [1, 3, 6, 7, 8, 11, 12, 14, 16, 19].

It is important to note that all enterprise information systems allow for some form of freedom and that the system is not able to control the entire process. Even in a WFM system, there is some degree of freedom; for example, work items are not allocated to a single user but to a group of users, and the routing may be determined by the user or by the arrival of external triggers (e.g., a cancellation by the customer). Note that a WFM cannot completely control its environment; for example, if work is offered to a user, then the user will determine when and how to perform it. Other systems typically offer even more freedom. In many systems, the user can deviate from the predefined process model; for example, in an ERP system the user does not need to follow the reference model completely (it is just a guideline). The fact that all systems allow for some form of freedom makes it interesting to see how people actually work. This motivates the use of process mining techniques as discussed in this chapter.

Process mining is useful for at least two reasons. First of all, it could be used as a tool to find out how people and/or procedures really work, that is, for *process discovery*. Consider, for example, processes supported by an ERP system like SAP (e.g., a procurement process). Such a system logs all transactions but does not (completely) enforce a specific way of working. In such an environment, process mining could be used to gain insight into the actual process. Another example would be the flow of patients in a hospital. Note that in such an environment all activities are logged but information about the underlying process is typically missing. In this context, it is important to stress that management information systems typically provide information about key performance indicators like resource utilization, flow times, and service levels, but *not* about the underlying business processes (e.g., causal relations, ordering of activities). Second, process mining could be used for *delta analysis,* that is, comparing the actual process with some predefined process. Note that in many situations there is a descriptive or prescriptive process model. Such a model specifies how people and organizations are assumed/expected to work. By comparing the descriptive or prescriptive process model with the discovered model, discrepancies between both can be detected and used to improve the process. Consider, for example, the so-called reference models in the context of SAP. These models describe how the system should be used. Using process mining it is possible to verify whether this is the case. In fact, process mining, could also be used to compare different departments/organizations using the same ERP system.

Process mining can be used to monitor coordination in enterprise information systems. Some of the coordination is done by humans, whereas other coordination tasks are done by software. As indicated, similar interaction patterns occur at the level of software components, business processes, and organizations. Therefore, process mining can be done at many levels.

The topic of process mining is related to management trends such as business process reengineering (BPR, see also Chapter 9), business intelligence (BI), business process analysis (BPA), continuous process improvement (CPI), and knowledge management (KM). Process mining can be seen as part of the BI, BPA, and KM trends. Moreover, process mining can be used as input for BPR and CPI activities. Note that process mining is not a tool to (re)design processes. The goal is to understand what is really going on. Despite the fact that process mining is not a tool for designing processes, it is evident that a good understanding of the existing processes is vital for any redesign effort.

The remainder of this chapter is organized as follows. In Section 10.2, we introduce process mining. Using an example, we illustrate the concept of process mining, discuss the information required to do process mining, and show the various perspectives that can be mined (process perspective, organizational perspective, and case perspective). Section 10.3 focuses on the process perspective and provides a concrete algorithm: the α algorithm. In Section 10.4 we discuss some limitations of the α-algorithm and possible solutions. To conclude, we provide exercises in Section 10.6.

10.2   PROCESS MINING: AN OVERVIEW

The goal of process mining is to extract information about processes from transaction logs [4]. We assume that it is possible to record events such that (i) each event refers to an *activity* (i.e., a well-defined step in the process); (ii) each event refers to a *case* (i.e., a process instance); (iii) each event can have a *performer,* also referred to as *originator* (the person executing or initiating the activity); and (iv) events have a *time stamp* and are totally ordered. Table 10.1 shows an example of a log involving 19 events, 5 activities, and 6 originators. In addition to the information shown in this table, some event logs contain more information on the case itself, that is, data elements referring to properties of the case. For example, the case handling system FLOWer logs every modification of some data element.

Event logs such as the one shown in Table 10.1 are used as the starting point for mining. We distinguish three different perspectives: (1) the process perspective, (2) the organizational perspective, and (3) the case perspective. The *process perspective* focuses on the control flow, that is, the ordering of activities. The goal of mining this perspective is to find a good characterization of all possible paths, for example, expressed in terms of a Petri net [15] (see Chapter 7) or event-driven process chain (EPC) [13, 12] (see Chapter 6). The *organizational perspective* focuses on the originator field—which performers are involved and how they are related. The goal is to either structure the organization by classifying people in terms of roles and organizational units or to show relations between individual performers (i.e., build a social network [17]). The *case perspective* focuses on properties of cases. Cases can be characterized by their path in the process or by the originators working on a case. However, cases can also be characterized by the values of the corresponding data elements. For example, if a case represents a re-

Table 10.1   An event log

Case id	Activity id	Originator	Time stamp
case 1	activity A	John	9-3-2004:15.01
case 2	activity A	John	9-3-2004:15.12
case 3	activity A	Sue	9-3-2004:16.03
case 3	activity B	Carol	9-3-2004:16.07
case 1	activity B	Mike	9-3-2004:18.25
case 1	activity C	John	10-3-2004:9.23
case 2	activity C	Mike	10-3-2004:10.34
case 4	activity A	Sue	10-3-2004:10.35
case 2	activity B	John	10-3-2004:12.34
case 2	activity D	Pete	10-3-2004:12.50
case 5	activity A	Sue	10-3-2004:13.05
case 4	activity C	Carol	11-3-2004:10.12
case 1	activity D	Pete	11-3-2004:10.14
case 3	activity C	Sue	11-3-2004:10.44
case 3	activity D	Pete	11-3-2004:11.03
case 4	activity B	Sue	11-3-2004:11.18
case 5	activity E	Clare	11-3-2004:12.22
case 5	activity D	Clare	11-3-2004:14.34
case 4	activity D	Pete	11-3-2004:15.56

plenishment order, it is interesting to know the supplier or the number of products ordered.

The process perspective is concerned with the "How?" question, the organizational perspective is concerned with the "Who?" question, and the case perspective is concerned with the "What?" question. To illustrate the first two, consider Figure 10.1. The log shown in Table 10.1 contains information about five cases (i.e., process instances). The log shows that for four cases (1, 2, 3, and 4) the activities A, B, C, and D have been executed. For the fifth case, only three activities have been executed: activities A, E, and D. Each case starts with the execution of A and ends with the execution of D. If activity B is executed, then activity C is also executed. However, for some cases, activity C is executed before activity B. Based on the information shown in Table 10.1 and by making some assumptions about the completeness of the log (i.e., assuming that the cases are representative and a sufficiently large subset of possible behaviors has been observed), we can deduce the process model shown in Figure 10.1(a). The model is represented in terms of a Petri net. The Petri net starts with activity A and finishes with activity D. These activities are represented by transitions. After executing A, there is a choice between either executing B and C concurrently (i.e., in parallel or in any order) or just executing activity E. To execute B and C in parallel, two nonobservable activities (AND-split and AND-join) have been added. These activities have been added for routing purposes only and are not present in the event log. Note that for this example, we assume that two activities are concurrent if they appear in any order. By distinguishing between start events and completion events for activities, it is possible to explicitly detect parallelism.

Figure 10.1(a) does not provide any information about the organization, that is, it does not use any information concerning the people executing activities. Information

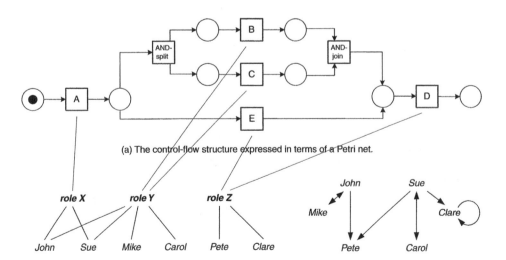

(a) The control-flow structure expressed in terms of a Petri net.

Figure 10.1 Some mining results for the process perspective (a) and organizational perspective (b and c) based on the event log shown in Table 10.1.

about performers of activities, however, is included in Table 10.1. For example, we can deduce that activity A is executed by either John or Sue; activity B is executed by John, Sue, Mike, or Carol; C is executed by John, Sue, Mike, or Carol; D is executed by Pete or Clare; and E is executed by Clare. We could indicate this information in Figure 10.1(a). The information could also be used to "guess" or "discover" organizational structures. For example, a guess could be that there are three roles: X, Y, and Z. For the execution of A, role X is required and John and Sue have this role. For the execution of B and C, role Y is required and John, Sue, Mike, and Carol have this role. For the execution of D and E, role Z is required and Pete and Clare have this role. For five cases, these choices may seem arbitrary, but for larger data sets such inferences capture the dominant roles in an organization. The resulting "activity–role–performer diagram" is shown in Figure 10.1(b). The three "discovered" roles link activities to performers. Figure 10.1(c) shows another view of the organization based on the transfer of work from one individual to another. It does not focus on the relation between the process and individuals but on relations among individuals (or groups of individuals). Consider Table 10.1. Although Carol and Mike can execute the same activities (B and C), Mike is always working with John (cases 1 and 2) and Carol is always working with Sue (cases 3 and 4). Carol and Mike probably have the same role but, based on the small sample shown in Table 10.1, it seems that John is not working with Carol and Sue is not working with Carol.[1]

These examples show that the event log can be used to derive relations between performers of activities, thus resulting in a sociogram. For example, it is possible to generate a sociogram based on the transfers of work from one individual to another, as is shown in Figure 10.1(c). Each node represents one of the six performers, and each arc represents that there has been a transfer of work from one individual to another. The definition of "transfer of work from A to B" is based on whether, in the same case, an activity executed by A is directly followed by an activity executed by B. For example, in both case 1 and 2 there is a transfer from John to Mike. Figure 10.1(c) does not show frequencies. However, for analysis purposes these frequencies can be added. The arc from John to Mike would then have weight 2. Typically, we do not use absolute frequencies but weighted frequencies to get relative values between 0 and 1. Figure 10.1(c) shows that work is transferred to Pete but not vice versa. Mike only interacts with John and Carol only interacts with Sue. Clare is the only person transferring work to herself.

Besides the "How?" and "Who?" questions (i.e., the process and organization perspectives), there is the case perspective that is concerned with the "What?" question. Figure 10.1 does not address this. In fact, focusing on the case perspective is most interesting when data elements are also logged, but these are not listed in Table 10.1. The case perspective looks at the case as a whole and tries to establish relations between the various properties of a case. Note that some of the properties may refer to the activities being executed, the performers working on the case, and the values of various data elements linked to the case. Using clustering algorithms,

---

[1]Clearly the number of events in Table 10.1 is too small to establish these assumptions accurately. However, real event logs will contain thousands or more events.

it would, for example, be possible to show a positive correlation between the size of an order or its handling time and the involvement of specific people.

Orthogonal to the three perspectives (process, organization, and case), the result of a mining effort may refer to *logical* issues and/or *performance* issues. For example, process mining can focus on the logical structure of the process model [e.g., the Petri net shown in Figure 10.1(a)] or on performance issues such as flow time. For mining the organizational perspective, the emphasis can be on the roles or the social network [cf. Figure 10.1(b) and (c)] or on the utilization of performers or execution frequencies. To illustrate the fact that the three perspectives and the type of question (logical or performance oriented) are orthogonal, some examples are given in Table 10.2.

To address the three perspectives and the logical and performance issues, we have developed a set of tools including EMiT [1], Thumb [19], and MinSoN [3]. These tools share a common XML format. Recently, the functionality of these three tools have been merged into the *ProM Framework*. The ProM tool not only supports variants of the α-algorithm; it also supports alternative approaches, for example, approaches based on genetic algorithms. For more details, refer to http://www.processmining.org.

10.3 PROCESS MINING WITH THE α ALGORITHM

In this chapter, we focus on the process perspective. In fact, we consider a specific algorithm: the α algorithm. Before describing the algorithm, we first discuss the input format.

10.3.1 Input

Table 10.1 shows an event log. The basic algorithm only considers the case id and the activity id and not the timestamp and originator of the event. For the α algorithm, the ordering of events within a case is relevant, whereas the ordering of events among

Table 10.2 Some examples of properties that may be investigated using process mining

Perspective	Examples of logical properties	Examples of performance properties
Process perspective	Activity A is always followed by B; activities C and D may be executed in parallel.	The average processing time of activity A is 35 minutes; activity A is executed for 80% of the cases.
Organizational perspective	John and Mary are on the same team; Pete is the manager of department D.	John handles on average 30 cases per day; Mary and Pete work together on 50% of the cases.
Case perspective	Cases of more than 5000 euros are handled by John; activity A is only executed for private customers.	80% of cases of more than 5000 euros are handled within 2 days; the average flow time of cases handled by John and Mary is 2 weeks.

cases is of no importance. In Table 10.1, it is important that for case 1, activity A is followed by B within the context of case 1, and not that activity A of case 1 is followed by activity A of case 2. Therefore, we define an event log as follows. Let T be a set of activities. $\sigma \in T^*$ is an *event trace*—an arbitrary sequence of activity identifiers. $W \subseteq T^*$ is an *event log*—a set of event traces. Note that since W is a set and not a multiset (bag), every event trace can appear only once in a log. In an event log like the one shown in Table 10.1, this is not the case. However, for inferring the structure of a process with the α algorithm, the frequency of an event trace is irrelevant; it does not add information. In more practical mining tools as presented in Section 10.4.2, frequencies become important. If we use this notation to describe the log shown in Table 10.1, we obtain the set $W = \{ABCD, ACBD, AED\}$. Note that cases 1 and 3 have event trace $ABCD$, cases 2 and 4 have trace $ACBD$, and case 5 is the only one having trace AED. Also note that when dealing with noise, frequencies are of the utmost importance (see Section 10.4.2 and [18]). However, for the moment we abstract from noise and simply look at the presence of a trace rather than its frequency.

To find a process model on the basis of an event log, the log should be analyzed for causal dependencies. For example, if an activity is always followed by another activity it is likely that there is a causal relation between both activities. To analyze these relations, we introduce the following notations. Let W be an event log over T, that is, $W \subseteq T^*$. Let $a, b \in T$:

- $a >_W b$ iff there is a trace $\sigma = t_1 t_2 t_3 \dots t_n$ and $i \in \{1, \dots, n-1\}$ such that $\sigma \in W$ and $t_i = a$ and $t_{i+1} = b$
- $a \rightarrow_W b$ iff $a >_W b$ and $b \not>_W a$
- $a \#_W b$ iff $a \not>_W b$ and $b \not>_W a$
- $a \|_W b$ iff $a >_W b$ and $b >_W a$

Consider the event log $W = \{ABCD, ACBD, AED\}$ (i.e., the log shown in Table 10.1). Relation $>_W$ describes which activities appeared in sequence (one directly following the other). Clearly, $A >_W B$, $A >_W C$, $A >_W E$, $B >_W C$, $B >_W D$, $C >_W B$, $C >_W D$, and $E >_W D$. Relation $\rightarrow_W$ can be computed from $>_W$ and is referred to as the *(direct) causal relation* derived from event log W. $A \rightarrow_W B$, $A \rightarrow_W C$, $A \rightarrow_W E$, $B \rightarrow_W D$, $C \rightarrow_W D$, and $E \rightarrow_W D$. Note that $B \not\rightarrow_W C$ because $C >_W B$. Relation $\|_W$ suggests concurrent behavior, that is, potential parallelism. For log W activities, B and C seem to be in parallel, that is, $B \|_W C$ and $C \|_W B$. If two activities can follow each other directly in any order, then all possible interleavings are present and, therefore, they are likely to be in parallel. Relation $\#_W$ gives pairs of transitions that never follow each other directly. This means that there are no direct causal relations and parallelism is unlikely.

10.3.2 The Algorithm

The α algorithm uses notions such as $>_W$, $\rightarrow_W$, $\|_W$, and $\#_W$ to obtain information about the underlying process. The α algorithm represents the discovered process in terms of a Petri net. Let W be an event log over T. $\alpha(W)$ is defined as follows:

1. $T_W = \{t \in T \mid \exists_{\sigma \in W} t \in \sigma\}$ (the set of activities appearing in the log)
2. $T_I = \{t \in T \mid \exists_{\sigma \in W} t = first(\sigma)\}$ (the set of initial activities)
3. $T_O = \{t \in T \mid \exists_{\sigma \in W} t = last(\sigma)\}$ (the set of final activities)
4. $X = \{(A, B) \mid A \subseteq T_W \wedge B \subseteq T_W \wedge \forall_{a \in A} \forall_{b \in B} a \rightarrow_W b \wedge \forall_{a_1,a_2 \in A} a_1 \#_W a_2 \wedge \forall_{b_1,b_2 \in B} b_1 \#_W b_2\}$ (all causality relations)
5. $Y = \{(A, B) \in X \mid \forall_{(A',B') \in X} A \subseteq A' \wedge B \subseteq B' \Longrightarrow (A, B) = (A', B')\}$ (only the minimal causality relations)
6. $P_W = \{p_{(A,B)} \mid (A, B) \in Y\} \cup \{i_W, o_W\}$ (the set of places in the resulting Petri net, $p_{(A,B)}$ is a place connecting transitions in A with transitions in B, i_W is the unique input place denoting the start of the process, and o_W is the unique output place denoting the end of the process)
7. $F_W = \{(a, p_{(A,B)}) \mid (A, B) \in Y \wedge a \in A\} \cup \{(p_{(A,B)}, b) \mid (A, B) \in Y \wedge b \in B\} \cup \{(i_W, t) \mid t \in T_I\} \cup \{(t, o_W) \mid t \in T_O\}$ (the set of connecting arcs in the resulting Petri net)
8. $\alpha(W) = (P_W, T_W, F_W)$ (the resulting Petri net with places P_W, transitions T_W, and arcs F_W).

The α algorithm transforms a log W into a Petri net (P_W, T_W, F_W). The algorithm only uses basic mathematics, the relations $>_W$, $\rightarrow_W$, $\|_W$, and $\#_W$, and the functions *first* and *last* to get the first and last element from a trace.

To illustrate the α algorithm, we show the result of each step using the log $W = \{ABCD, ACBD, AED\}$ (i.e., a log like the one shown in Table 10.1):

1. $T_W = \{A, B, C, D, E\}$
2. $T_I = \{A\}$
3. $T_O = \{D\}$
4. $X = \{(\{A\}, \{B\}), (\{A\}, \{C\}), (\{A\}, \{E\}), (\{B\}, \{D\}), (\{C\}, \{D\}), (\{E\}, \{D\}), (\{A\}, \{B, E\}), (\{A\}, \{C, E\}), (\{B, E\}), (\{D\}), (\{C, E\}), (\{D\})\}$
5. $Y = \{(\{A\}, \{B, E\}), (\{A\}, \{C, E\}), (\{B, E\}, \{D\}), (\{C, E\}, \{D\})\}$
6. $P_W = \{i_W, o_W, p_{(\{A\},\{B,E\})}, p_{(\{A\},\{C,E\})}, p_{(\{B,E\},\{D\})}, p_{(\{C,E\},\{D\})}\}$
7. $F_W = \{(i_W, A), (A, p_{(\{A\},\{B,E\})}), (p_{(\{A\},\{B,E\})}, B) \ldots, (D, o_W)\}$
8. $\alpha(W) = (P_W, T_W, F_W)$ (as shown in Figure 10.2)

It is interesting to note that $\alpha(W)$ shown in Figure 10.2 differs from the Petri net shown in Figure 10.1. This may suggest that the result is not correct. However, from a behavioral point of view, Figure 10.2 and Figure 10.1(a) are equivalent if we abstract from the AND-split and AND-join. Note that every event trace in $W = \{ABCD, ACBD, AED\}$ can be realized by Figure 10.2. Also, every possible firing sequence corresponds to an event trace in W. Therefore, we conclude that although Figure 10.2 and Figure 10.1(a) differ, the α algorithm is able to correctly mine the log shown in Table 10.1.

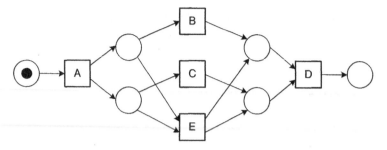

Figure 10.2   Another process model corresponding to the event log shown in Table 10.1. In Section 10.4, we will discuss the problem of "invisible activities"—activities that are not recorded in the log.

10.3.3   How Does it Work?

The fact that the α algorithm is able to "discover" Figure 10.2 based on the event log $W = \{ABCD, ACBD, AED\}$ triggers the question: *How does it work?* To understand the basic idea of the α algorithm, consider Figure 10.3. The algorithm assumes that two activities x and y (i.e., transitions) are connected through some place if and only if $x \rightarrow_W y$ [Figure 10.3(a)]. If activities x and y are concurrent, then they can occur in any order; that is, x may be directly followed by y or vice versa. Therefore, the α algorithm assumes activities x and y are concurrent if and only if $x \parallel_W y$. This is illustrated by Figure 10.3(b). If $x \rightarrow_W y$ and $x \rightarrow_W z$, then there have to be places connecting x and y on the one hand and x and z on the other hand. This can be one place or multiple places. If $y \parallel_W z$, then there should be multiple places to enable concurrency [cf. Figure 10.3(b)]. If $y \#_W z$, then there should be a single place to ensure that only one branch is chosen [cf. Figure 10.3(c)]. Note that in the latter case, y and z never follow one another directly, as expressed by $y \#_W z$ (i.e., $y \nrightarrow_W z$ and $z \nrightarrow_W y$). Figure 10.3(d) shows the AND-join [i.e., counterpart of the AND-split shown in Figure 10.3(b)] and Figure 10.3(e) shows the XOR-join [i.e., counterpart of the XOR-split shown in Figure 10.3(c)].

The basic relations shown in Figure 10.3 are the starting point for the α algorithm. Note that the relations do not always hold, that is, one can think of them as heuristics. For example, it is assumed that the log is *complete* with respect to $>_W$. (Note that $\rightarrow_W$, $\parallel_W$, and $\#_W$ are derived from $>_W$.) This implies that if one activity can be followed by another, this should happen at least once in the log. We will return to the issue of completeness in Section 10.4.2.

10.3.4   Examples

Figure 10.4 is an example of a Petri net in which an AND-split and OR-split are embedded in an AND-split. Given a complete event log, this kind of nesting will not harm the α algorithm in rediscovering the original Petri net.

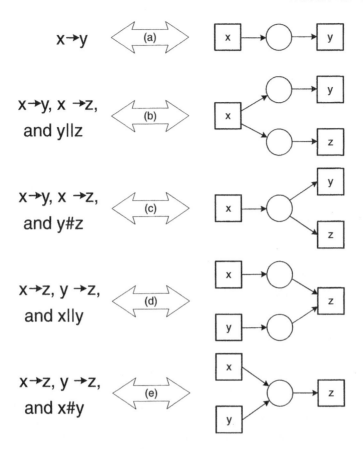

Figure 10.3 Relating the log-based relations $>_W$, $\rightarrow_W$, $\|_W$, and $\#_W$ to basic Petri net constructs.

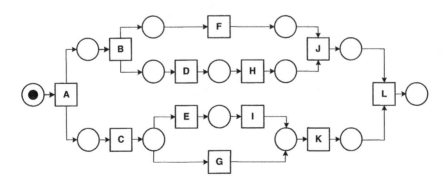

Figure 10.4 An example of a Petri net with nested AND/XOR-splits correctly mined by the α algorithm.

In Figure 10.5, an example of a Petri net with loops is given. Note that an event log with all possible event traces of this Petri net is an *infinite set*. However, a *complete* event log does not need to contain all possible traces. A sufficiently large subset with, on a binary level (i.e. with respect to $>_W$), all possible pairs is sufficient. If we have such a *complete* event log, the α algorithm will, without any problem, correctly rediscover the Petri net of Figure 10.5.

The Petri net of the last example (Figure 10.6) is less abstract and specifies the interactions between a contractor and a subcontractor. First, the contractor sends an order to the subcontractor. Then, the contractor sends a detailed specification to the subcontractor and the subcontractor sends a cost statement to the contractor. Based on the specification, the subcontractor manufactures the desired product and sends it to the contractor. There is no clear owner of the resulting interaction process, but the combined information registered by both parties contains enough information to mine the process (i.e., the process itself and other aspects of it).

10.4 LIMITATIONS OF THE α APPROACH AND POSSIBLE SOLUTIONS

In this chapter, we only consider the basic α algorithm to illustrate the concept of process mining. The α algorithm focuses exclusively on the process perspective (i.e., control flow). As indicated in Section 10.2, process mining can be used to analyze other perspectives (to answer "Who?" and "What?" questions). Despite its focus, the basic α algorithm is still unable to successfully discover some processes. In this section, we identify two classes of problems: logical problems and problems resulting from noise (incorrectly logged data), exceptions (rare events not corresponding to the "normal" behavior), and incompleteness (i.e., too few observations).

10.4.1 Logical Problems

In [6], a formal characterization is given for the class of nets that can be mined correctly. It turns out that assuming a weak notion of completeness (i.e., if one activity

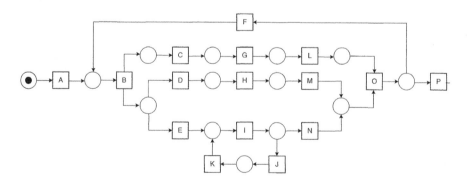

Figure 10.5   Example of a more complex Petri correctly mined by the α algorithm.

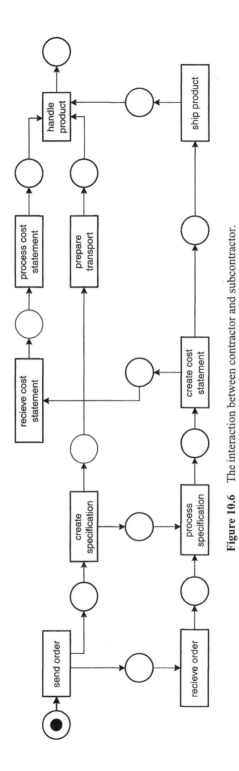

Figure 10.6 The interaction between contractor and subcontractor.

can be followed by another this should happen at least once in the log), any so-called SWF net without short loops and implicit places can be mined correctly. SWF nets are Petri nets with a single source and sink place satisfying some additional syntactical requirements such as the free-choice property. In this chapter, we will not elaborate on formal characterizations of the class of processes that can be successfully mined. Instead, we focus on the practical limitations of the α algorithm, that is, the problems encountered when dealing with invisible activities, duplicate activities, short loops, and so on.

10.4.1.1 *Invisible Activities.*

One of the basic assumptions of process mining is that each event (i.e., the occurrence of an activity for a specific case) is registered in the log. Clearly, it is not possible to find information about activities that are not recorded. However, given a specific language, it is possible to register that there is a so-called "hidden activity." Consider, for example, Table 10.1, where A, B, and C are visible but the AND-split in between A and B and C is not. Clearly, the basic α algorithm is unable to discover activities not appearing in the log. Therefore, the Petri net shown in Figure 10.2 is different from the Petri net shown in Figure 10.1(a). However, both nets are equivalent if we abstract from the AND-split and AND-join. Unfortunately, this is not always the case. Consider, for example, Figure 10.2, where activity E is not visible. The resulting log would be $W = \{ABCD,$ $ACBD, AD\}$ and the α algorithm would be unable to construct the correct model; that is, $\alpha(W) = (\{i_W, o_W, P_{(\{A\},\{B\})}, P_{(\{A\},\{C\})}, P_{(\{B\},\{D\})}, P_{(\{C\},\{D\})}, P_{(\{A\},\{D\})}\}, \{A, B, C,$ $D\}, \{(i_W, A), (A, P_{(\{A\},\{B\})}), \cdots, (A, P_{(\{A\},\{D\})}), (P_{(\{A\},\{D\})}, D), (D, o_W)\})$. The resulting net is shown in Figure 10.7, that is, the original net shown in Figure 10.2 without E but with an additional place connecting A and D. Note that the resulting model does not allow for the event trace AD.

10.4.1.2 *Duplicate Activities.*

The problem of duplicate activities refers to the situation in which one can have a process model (e.g., a Petri net) with two nodes referring to the same activity. Suppose that in Table 10.1 and Figure 10.1, activity E is renamed to B (see Figure 10.8). Clearly, the modified log could be the result of the modified process model. However, it becomes very difficult to automati-

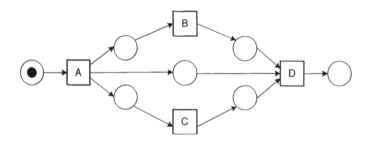

Figure 10.7   If activity E is not visible, the algorithm returns an incorrect model because it does not allow for AD.

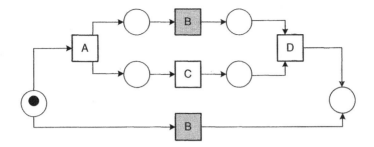

Figure 10.8   A process model with duplicate activities.

cally construct a process model from Table 10.1 with E renamed to B because it is not possible to distinguish the "B" in case 5 from the "Bs" in the other cases. Note that the presence of duplicate activities is related to hidden activities. Many processes with hidden activities but with no duplicate activities can be modified into equivalent processes with duplicate activities but with no hidden activities.

10.4.1.3 *Non-Free-Choice Constructs.* Free-choice Petri nets are Petri nets in which there are no two transitions occupying the same input place but one has an input place that is not an input place of the other [9]. This excludes the possibility of merging choice and synchronization into one construct. Free-choice Petri nets are a well-known and widely used subclass of Petri nets. However, many processes cannot be expressed in terms of a free-choice net. Unfortunately, most of the mining techniques (also those that do not use Petri nets) assume process models corresponding to the class of free-choice Petri nets. Non-free-choice constructs can be used to represent "controlled choices," situations in which the choice between two activities is not determined inside some node in the process model but may depend on choices made in other parts of the process model. Clearly, such nonlocal behavior is difficult to mine with mining approaches primarily based on binary information ($a >_w b$) and may require many observations.

Figure 10.1 is free-choice since synchronization (activity AND-join) is separated from the choice between B and C, and E. Figure 10.9 shows a non-free-choice construct. After executing activity C, there is a choice between activity D and activity E. However, the choice between D and E is "controlled" by the earlier choice between A and B. Note that activities D and E are involved in a choice but also synchronize two flows. Clearly, such constructs are difficult to mine since the choice is nonlocal and the mining algorithm has to "remember" earlier events.

To illustrate that there are also non-free-choice constructs that can be mined correctly using the α algorithm, we consider Figure 10.10. Now, the choice can be detected because of the two new activities, X and Y. Note that X may be directly followed by D but not by E. Hence, the place in between X and D is discovered.

10.4.1.4 *Short Loops.* In a process, it may be possible to execute the same activity multiple times. If this happens, this typically refers to a loop in the corre-

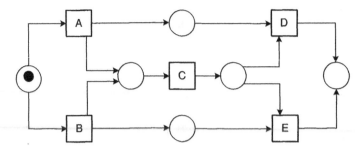

Figure 10.9   A process model with a non-free-choice construct.

sponding model. Figure 10.11 shows an example with a loop. After executing activity B, activity C can be executed arbitrarily many times; possible event sequences are BD, BCD, BCCD, BCCCD, and so on. Loops like the one involving activity C are easy to discover. However, loops can also be used to jump back to any place in the process. For more complex processes, mining loops is far from trivial since there are multiple occurrences of the same activity in a given case. Some techniques number each occurrence; for example, B1 C1 C2 C3 D1 denotes BCCCD. These occurrences are then mapped onto a single activity.

As illustrated by Figure 10.11, there is a relation between loops and duplicate activities. In Figure 10.11, activity A is executed multiple times (i.e., twice) but is not in a loop. Many mining techniques make some assumptions about loops that restrict the class of processes that can be mined correctly.

The logical problems described all apply to the α algorithm. Some of the problems can be resolved quite easily by using a more refined algorithm. Other problems are more fundamental and indicate theoretical limits [6].

10.4.2   Noise, Exceptions, and Incompleteness

The formal approach presented in the preceding section presupposes perfect information: (i) the log must be complete (i.e., if an activity can follow another activity

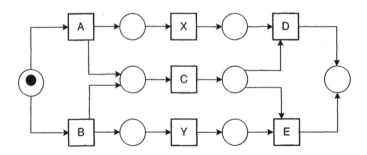

Figure 10.10   A process model with a non-free-choice construct that can be mined correctly.

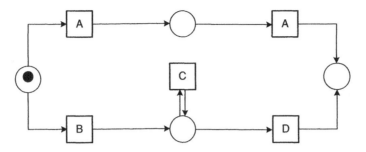

Figure 10.11   A process model with a loop.

directly, the log should contain an example of this behavior) and (ii) we assume that there is no noise in the log (i.e., everything that is registered in the log is correct). However, in practical situations logs are rarely complete and/or noise free. Especially, the differentiation between errors, low-frequency activities, low-frequency activity sequences, and exceptions is problematic. Therefore, in practice, it becomes more difficult to decide if between two activities, say A and B, one of the three basic relations (i.e., $A \rightarrow_W B$, $A \#_W B$, or $A \parallel_W B$) holds. For instance the causality relation as used in the α algorithm ($A \rightarrow_W B$) holds if and only if in the log there is a trace in which A is directly followed by B (i.e., the relation $A >_W B$ holds) and there is no trace in which B is directly followed by A (i.e., not $B >_W A$). However, in a noisy situation one erroneous example can completely mess up the derivation of a correct conclusion. Even if we have thousands of log traces in which A is directly followed by B, then one $B >_W A$ example based on an incorrect registration will prevent a correct conclusion. As noted before, frequency information is not used in the formal approach. For this reason, heuristic mining techniques are developed that are less sensitive to noise and the incompleteness of logs.

As an illustration of a heuristic approach, we briefly discuss the ideas to discover the causality relation as implemented in the heuristic mining tool Little Thumb [18]. In this approach, a frequency-based metric is used to indicate how certain we are that there is truly a causal relation between two events A and B (notation $A \Rightarrow_W B$). The calculated $\Rightarrow_W$ values between the events of an event log are used in a heuristic search for the right relations between events (i.e., $A >_W B$, $A \#_W B$ or $A \parallel_W B$). Below, we first define the $\Rightarrow_W$ metric. After that, we will illustrate how we can use this metric in a simple heuristic in which we search for reliable causal relations (the $A \rightarrow_W B$ relation).

Let W be an event log over T, and $a, b \in T$. Then $|a >_W b|$ is the number of times $a >_W b$ occurs in W, and

$$a \Rightarrow_W b = \left(\frac{|a >_W b| - |b >_W a|}{|a >_W b| + |b >_W a| + 1} \right)$$

First, note that the value of $a \Rightarrow_W b$ is always between -1 and 1. Some simple examples demonstrate the rationale behind this definition. If we use this definition in the situation that, in five traces, activity A is directly followed by activity B but the other way around never occurs, the value of $A \Rightarrow_W B = 5/6 = 0.833$, indicating that we are not completely sure of the causality relation (there are only five observations, possibly caused by noise). However, if there are 50 traces in which A is directly followed by B but the other way around never occurs, the value of $A \Rightarrow_W B = 50/51 = 0.980$ indicates that we are pretty sure of the causality relation. If there are 50 traces in which activity A is directly followed by B and noise caused B to follow A once, the value of $A \Rightarrow_W B$ is still $49/52 = 0.94$, indicating that we are pretty sure of a causal relation.

A high $A \Rightarrow_W B$ value strongly suggests that there is a causal relation between activity A and B. But what is a high value? What is a good threshold to make the decision that B truly depends on A (i.e., $A \rightarrow_W B$ holds)? The threshold appears to be sensitive for the amount of noise, the degree of concurrency in the underlying process, and the frequency of the involved activities.

However, it appears unnecessary to use a threshold value. After all, we know that each noninitial activity must have at least one other activity that is its cause, and each nonfinal activity must have at least one dependent activity. Using this information in a heuristic approach, we can limit the search and take *the best* candidate (with the highest $A \Rightarrow_W B$ score). This simple heuristic helps us enormously in finding reliable causality relations, even if the event log contains noise. As an example, we have applied the heuristic to an event log from the Petri net of Figure 10.1. Thirty event traces are used (nine for each of the three possible traces and three incorrect traces: ABCED, AECBD, AD). We first calculate the $\Rightarrow$-values for all possible activity combinations. The result is displayed in the matrix below.

$\Rightarrow_W$	A	B	C	D	E
A	0.0	0.909	0.900	0.500	0.909
B	0.0	0.0	0.0	0.909	0.0
C	0.0	0.0	0.0	0.900	0.0
D	−0.500	−0.909	−0.909	0.0	−0.909
E	0.0	0.0	0.0	0.909	0.0

As an illustration, we now apply the basic heuristic to this matrix. We can recognize the initial activity A; it is the activity without a positive value in the A column. For the dependent activity of A, we search for the highest value in row A of the matrix. Both B and E are high (0.909). We arbitrarily choose B. If we use the matrix to search for the cause for B (the highest value of the B column), we will again find A as the cause for B. D is the dependent activity of B (D is the highest value of the B row). The result of applying the same procedure on activities B, C, and E is presented in Figure 10.12. Note that only the causal relations are depicted in a so-called dependency graph. The numbers in the activity boxes indicate the frequency of the activity, the numbers on the arcs indicate the reliability of each causal relation, and

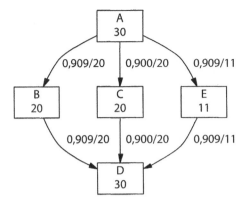

Figure 10.12 A dependency graph resulting from application of the heuristic approach to a noisy log based on the Petri net of Figure 10.1.

the numbers on the nodes indicate the frequencies. In spite of the noise, the causal relations are correctly mined.

The illustrated heuristic procedure is not complete. For example, we need searching procedures for the other basic relations (i.e., $a \#_W b$ and $a \parallel_W b$). Given the correct basic relations, we can use the α algorithm to construct a Petri net. In [18], the experimental results of such an approach to noisy data are presented.

10.5 CONCLUSION

This chapter introduced the topic of process mining by first providing an overview and then zooming in on a specific algorithm for the process perspective (i.e., control flow)—the α algorithm. It is important to realize that this algorithm only tackles one of the cells shown in Table 10.2 (the top-left one). For the other cells, other approaches are needed. However, even within this single cell there are many challenges, as demonstrated in this chapter. The wide applicability of process mining makes is worthwhile to tackle problems such as noise and incompleteness. For more information and to download mining tools, refer to http://www.processmining.org.

ACKNOWLEDGMENTS

The authors would like to thank Boudewijn van Dongen, Ana Karla Alves de Medeiros, Minseok Song, Laura Maruster, Eric Verbeek, Monique Jansen-Vullers, Hajo Reijers, and Peter van den Brand for their ongoing work on process mining techniques and tools at Eindhoven University of Technology. Parts of this chapter have been based in earlier papers written with these researchers.

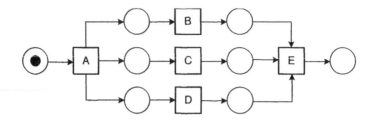

Figure 10.13   A simple parallel Petri-net.

10.6 EXERCISES

Excercise 1
Consider the Petri net shown in Figure 10.13. (1) Determine the event log W with all possible traces. (2) Try to determine a >-complete event log W', with W' a real subset of W ($W' \subset W$).

Excercise 2
Given an event log $W = \{AFBCGD, AFCBGD, AED\}$, use the eight steps of the α algorithm to construct an accompanying Petri net.

Excercise 3
Consider the following log $W = \{ABCDE, ABDCE, ACBDE, ACDBE, ADBCE, ADCBE\}$ originating from the Petri net of Exercise 1. Determinate $A \Rightarrow_W B$ and $B \Rightarrow_W C$.

Excercise 4
Given the following event log $W = [ABCDE, ABDCE, ACBDE, ACDBE, ADBCE, ADCBE, ABCDE, ABDCE\$, ACBDE, ACDBE, ADBCE, ADCBE]$,[2] which originated from the Petri net of Exercise 2. Follow the heuristic of Subsection 10.4.2 to construct a dependency graph as presented in Figure 10.12.

REFERENCES

1. W. M. P. van der Aalst and B. F. van Dongen. Discovering Workflow Performance Models from Timed Logs. In Y. Han, S. Tai, and D. Wikarski (Eds.), *International Conference on Engineering and Deployment of Cooperative Information Systems (EDCIS 2002)*, volume 2480 of *Lecture Notes in Computer Science*, pp. 45–63. Springer-Verlag, Berlin, 2002.

2. W. M. P. van der Aalst and K. M. van Hee. *Workflow Management: Models, Methods, and Systems*. MIT Press, Cambridge, MA, 2002.

3. W. M. P. van der Aalst and M. Song. Mining Social Networks: Uncovering Interaction

---

[2]To express the multiple appearance of traces, we formally have to use the bag or multiset notation instead of the set notation.

Patterns in Business Processes. In J. Desel, B. Pernici, and M. Weske, (Eds.), *International Conference on Business Process Management (BPM 2004),* volume 3080 of *Lecture Notes in Computer Science,* pp. 244–260. Springer-Verlag, Berlin, 2004.

4. W. M. P. van der Aalst, B. F. van Dongen, J. Herbst, L. Maruster, G. Schimm, and A. J. M. M. Weijters. Workflow Mining: A Survey of Issues and Approaches. *Data and Knowledge Engineering, 47*(2):237–267, 2003.

5. W. M. P. van der Aalst and A. J. M. M. Weijters, (Eds.). *Process Mining,* Special Issue of *Computers in Industry, 53,* 3, 2004.

6. W. M. P. van der Aalst, A. J. M. M. Weijters, and L. Maruster. Workflow Mining: Discovering Process Models from Event Logs. *IEEE Transactions on Knowledge and Data Engineering, 16*(9):1128–1142, 2004.

7. R. Agrawal, D. Gunopulos, and F. Leymann. Mining Process Models from Workflow Logs. In *Sixth International Conference on Extending Database Technology,* pp. 469–483, 1998.

8. J. E. Cook and A. L. Wolf. Discovering Models of Software Processes from Event-Based Data. *ACM Transactions on Software Engineering and Methodology, 7*(3):215–249, 1998.

9. J. Desel and J. Esparza. *Free Choice Petri Nets,* volume 40 of *Cambridge Tracts in Theoretical Computer Science.* Cambridge University Press, Cambridge, UK, 1995.

10. D. Grigori, F. Casati, U. Dayal, and M. C. Shan. Improving Business Process Quality through Exception Understanding, Prediction, and Prevention. In P. Apers, P. Atzeni, S. Ceri, S. Paraboschi, K. Ramamohanarao, and R. Snodgrass, (Eds.), *Proceedings of 27th International Conference on Very Large Data Bases (VLDB'01),* pp. 159–168. Morgan Kaufmann, 2001.

11. J. Herbst. A Machine Learning Approach to Workflow Management. In *Proceedings 11th European Conference on Machine Learning,* volume 1810 of *Lecture Notes in Computer Science,* pp. 183–194. Springer-Verlag, Berlin, 2000.

12. IDS Scheer. ARIS Process Performance Manager (ARIS PPM): Measure, Analyze and Optimize Your Business Process Performance (white paper). IDS Scheer, Saarbrücken, Gemany, http://www.ids-scheer.com, 2002.

13. G. Keller and T. Teufel. *SAP R/3 Process Oriented Implementation.* Addison-Wesley, Reading MA, 1998.

14. M. zur Mühlen and M. Rosemann. Workflow-based Process Monitoring and Controlling—Technical and Organizational Issues. In R. Sprague (Ed.), *Proceedings of the 33rd Hawaii International Conference on System Science (HICSS-33),* pp. 1–10. IEEE Computer Society Press, Los Alamitos, CA, 2000.

15. W. Reisig and G. Rozenberg, (Eds.). *Lectures on Petri Nets I: Basic Models,* volume 1491 of *Lecture Notes in Computer Science.* Springer-Verlag, Berlin, 1998.

16. M. Sayal, F. Casati, U. Dayal, and M.C. Shan. Business Process Cockpit. In *Proceedings of 28th International Conference on Very Large Data Bases (VLDB'02),* pp. 880–883. Morgan Kaufmann, 2002.

17. J. Scott. *Social Network Analysis.* Sage, Newbury Park, CA, 1992.

18. A. J. M. M. Weijters and W. M. P. van der Aalst. Workflow Mining: Discovering Workflow Models from Event-Based Data. In C. Dousson, F. Höppner, and R. Quiniou, (Eds.), *Proceedings of the ECAI Workshop on Knowledge Discovery and Spatial Data,* pp. 78–84, 2002.

19. A. J. M. M. Weijters and W. M. P. van der Aalst. Rediscovering Workflow Models from Event-Based Data Using Little Thumb. *Integrated Computer-Aided Engineering, 10*(2):151–162, 2003.

CHAPTER 11

Transactional Business Processes

GUSTAVO ALONSO

> The other terror that scares us from self-trust is our consistency; a reverence for our past act or word, because the eyes of others have no other data for computing our orbit than our past acts, and we are loath to disappoint them.
>
> —Ralph Waldo Emerson

11.1 INTRODUCTION

Since its inception, the transaction concept has played a fundamental role in all forms of information systems. Business processes are no exception and it is not rare to find the term *transactional* attached to that of business processes. Unfortunately, there is quite a lot of confusion about what *transactional* means in this context. This confusion that translates into a rather chaotic set of competing specifications and a wide range of entirely different products that, at least in theory, should be solving the same set of problems. This chapter tries to offer a broader perspective on transactional processes. A perspective that goes beyond existing products and specifications and tackles the essence of the problem and the range of available solutions. To achieve this, we start by clarifying what *transactional* means when applied to business processes and what are the basic techniques needed to support transactional processes. Keeping in mind that the transactional nature of many processes is ad hoc and determined by the programmer, the chapter will also explain how a basic set of design primitives can be used to implement a variety of transactional constructs.

The chapter opens with an overview of the notion of transactional consistency (Section 11.2), starting with the traditional view developed in the realm of databases and showing how this view applies outside the database domain. Next, Section 11.3 explores the concept of atomicity, which is considered to be the transactional property that is most applicable in the area of workflow and business process management. Section 11.4 then describes specific realizations of this concept in the context of system development platforms and standardization initiatives. The chapter closes with an overview of trends (Sections 11.5). Exercises and assignment

Process-Aware Information Systems. Edited by Dumas, van der Aalst, and ter Hofstede
Copyright © 2005 John Wiley & Sons, Inc.

subjects are provided in Section 11.6. Most of these assignments require the reader to refer to external sources and put them in context with respect to the concepts introduced throughout this chapter.

11.2 TRANSACTIONAL CONSISTENCY

The notion of transactional consistency is typically captured by the acronym ACID (atomicity, consistency, isolation, and durability). These four properties cover all possible aspects of transactional consistency. In practice, however, the ACID properties have a meaning only within the context of a transaction model and there are many such models in which some of these properties are ignored. Hence, when talking about transactional consistency, we first need to clarify under what model we are operating and what properties are being enforced. The history of transactional support for different applications is not so much how each property is treated and implemented within each model, but how each transactional model reflects the characteristics of the application.

11.2.1 The ACID Model

The best known and most widely used transaction model is the one used by database management systems to implement concurrency control, recovery, enforcement of constraint, and disaster recovery. This model is discussed in much detail in numerous textbooks [3, 11] so we will concentrate here on a few relevant aspects of this *database transaction model.*

The database transaction model is based on a closed-world assumption. It assumes, in the first place, that transactions can only exchange information by reading or writing to the database. In other words, all communication between transactions happens under the control of the transaction engine. Moreover, transactions are seen as functions that take the database from one state to another. The database state is defined as the values of all the relevant data items and system parameters of the database; this state changes only as a result of the execution of transactions. The correctness criteria and the properties enforced by the database transaction model aim at guaranteeing that the states that are produced by the concurrent execution of transactions are *consistent.* The model also defines what "to be consistent" actually means.

In the database transaction model, the ACID properties can be formulated as follows:

- **Consistency** guarantees that any transaction that (1) starts from a consistent state, (2) executes by itself (no other concurrent transaction in the system), (3) is completely executed (the whole transaction is completed), and (4) executes without failures will produce a consistent state. Consistency in database engines is enforced through *integrity constraints* that check the values written by a transaction and will abort the transaction if any rules are violated.

- **Isolation** guarantees that transactions will behave as if they were alone in the system, even when other transactions are being executed concurrently. Isolation allows running of transactions in parallel while still preserving consistency. In abstract terms, the need for isolation can be explained in terms of consistency (i.e., a transaction must start from a consistent state). If we assume that the database is in a consistent state, executing one transaction at a time guarantees that the database remains consistent. If transactions are executed in parallel, a transaction may actually read an inconsistent state because it has access to the intermediate result of a concurrent transaction. Note that the property of consistency guarantees that a transaction produces consistent results if completely executed. It does not say anything about the state of the database if the transaction is executed only halfway through. If a transaction reads the intermediate state of another transaction rather than a consistent state of the database, the conditions for consistency are not met and that transaction may actually leave the database in an inconsistent state. Isolation is enforced through *concurrency control* mechanisms that enforce a particular ordering in the execution of transactions.

- **Atomicity** is the property that makes sure transactions always leave the database in a consistent state. As the consistency property implies, this can only happen if the transaction is executed in its entirety. Hence, atomicity guarantees that the transaction is either executed to completion (and, hence, leaves the database in a consistent state) or not executed at all (and, hence, the database remains in the initial consistent state). Atomicity is enforced through *recovery* mechanisms that undo transactions that need to be aborted (e.g., because they violate integrity constraints or isolation rules, or because the transaction could not be completed due to a database failure).

- **Durability** is a fuzzier property that, nevertheless, makes a lot of practical sense. It can be interpreted in many different ways but, for most intents and purposes, it can be seen as the guarantee that, if a transaction successfully completes its execution, the changes it has performed will not be lost. Depending on what type of failures are considered, it can be far from trivial to enforce durability, as this might involve using backup systems, replicated databases, and so on.

For our purposes here, the important aspect of these properties is that, as they have been defined, they make sense only within the context of a conventional database engine. Once we are outside databases, many of the ideas built around the ACID principles no longer apply.

11.2.2 Transactions Outside Databases

When we talk about transactional business processes, we are talking about a transactional model that is very different from the one found for database engines. Processes describe operations that are far more complex and varied than the operations considered for the database engine. For instance, an activity within a process may involve an entire transaction or even several transactions against a database. In other cases, the activities may not involve transactions at all, but calls to operations such as sending or receiving messages, printing documents, or getting approval from a manager for a particular business operation. As a result, most of the ACID properties are not considered in models of transactional processes.

In the context of process management, it does not make sense to assume that all communications among processes (or the activities of different processes) are under the control of a single transaction engine or take place only through reading and writing to a common database. Real business processes interact in many different ways, most, if not all, of them outside the control of the system that executed the processes. Hence, in general, isolation is not a good way to make sure things are correct in the context of processes. For one thing, we may not be able to isolate processes from each other. For another, often we do not even want to isolate processes from each other in the way transactions are isolated from each other. Hence, isolation is not a property that is considered in relation with models of transactional processes (although this does not mean that it never makes sense or that it is not possible to so [9]).

For the same reason, consistency is much less clear when processes are involved. Unlike database transactions, processes may terminate in many different ways, all of them correct. For transactions, consistency is enforced by the system through integrity constraints, concurrency control, and recovery mechanisms, all of them identical for all transactions and orthogonal to the transactions themselves. For processes, consistency is typically the responsibility of the process itself. If a process does not reach a given point, then it must clean up whatever needs to be cleaned up to make sure things are consistent (we will see later on in the chapter what this implies). In practice, that means that consistency is not a property enforced by the system or a property of the model, but a characteristic that needs to be programmed by the process designer. Hence, although some of the concepts look very similar, the properties and behavior of the transactional process model and those of the database transaction model are very different. As a result, models for transactional processes do not consider consistency in the same way as database transactions. Consistency typically becomes a programmable feature rather than a system property.

Finally, durability is even fuzzier in transactional processes than it is in database transactions. Typically, durability in the context of processes refers both to the persistence of the process execution (the ability to resume execution from the point where it was interrupted after a failure) and the execution traces that are often required for post-execution analysis and auditing (if necessary, it must be possible to reconstruct the steps taken by the process). Both aspects are very important in practice but differ significantly from the notion of durability considered in databases. As with isolation and consistency, durability is also not part of most models of transactional processes.

In practice, the only property that is really relevant in the context of transactional processes is atomicity, although not for the same reasons as in the database transaction model. Much of the rest of this chapter will be devoted to examining what atomicity means for transactional processes and how it is implemented in practice. But before we get there, it is important to understand how atomicity is implemented in general as, for better or for worse, these same ideas are the basis for implementing atomicity in transactional processes.

11.2.3   Atomicity in Database Transactions

Atomicity in the database transaction model has two sides. One affects transactions executing in a single database and it is enforced using a variety of mechanisms that can *undo* the effects of a transaction at any given point in time. The other affects distributed transactions that involve more than one database. In the latter case, the internal recovery mechanisms of each database are not enough, as the different databases need to be coordinated to make sure that the transaction is executed in its entirety in all databases or not at all in any of the databases. These two sides of atomicity are similar to those used in transactional processes. In the first case (intraprocess atomicity), the concepts are similar, although the mechanisms used to implement them are very different. In the second case (interprocess atomicity), both the concepts and the implementation mechanisms are almost identical.

Atomicity of a Transaction. Transaction recovery is a very complex problem that is still not understood in its entirety. It is far less formalized than concurrency control since it is highly dependent on the underlying implementation of the database. Yet, its basic concepts are relatively easy to understand. To make sure transactions are atomic in their execution, we need to make sure that the transaction can be undone if something goes wrong. Isolation already guarantees that the transaction will see a database that will appear as if no other transaction would be running at the same time. Hence, before the transaction modifies any data item in the database, we make a copy of that item. If the transaction successfully completes, then the new values that it has written are *committed* and we can forget about the old values. If the transaction needs to be canceled, we restore the state of the database before the transaction started making any changes by copying back the values of the data items that were saved before each update took place. Once these values have been copied back, the database appears exactly as it was before the transaction was executed (i.e., the transaction has not been executed at all). The huge advantage of the database model, particularly for relational databases, is that these recovery procedures can be implemented independently of what the transaction does. That is, all transactions follow the same recovery procedure independently of what they modify in the database.

Atomicity of a Distributed Transaction. A distributed transaction is run on several databases. Within each database, recovery takes place as just discussed. However, each database cannot do recovery on its own. All of them need to agree whether to commit or abort the transaction. For this purpose, an *atomic commitment* protocol is needed. The current standard protocol for this purpose is *2 Phase Commit* (2PC). As the name implies, 2PC works in two phases. In the first phase, all *participants* (the databases where the transaction has run) agree on whether the transaction is going to be committed or not. This is commonly done by having a *co-ordinator* requesting and collecting votes. If the coordinator sees that all participants want to commit, then it enters the second phase and sends a message indicating to the participants that they can go ahead and commit the transaction. Simple as it may seem, there are many details and subtleties behind 2PC and much work has been done to understand the problem and come up with optimized solutions [11]. Fortunately, these details are not that important for transactional processes so we will just ignore them in the rest of the chapter and simply use 2PC to imply an atomic commitment protocol like the one just described.

11.3   ATOMICITY

The problem of how to make complex sequences of operations atomic, whether described as a process or not, has been addressed in many different ways. Most of this work was done in the context of *advanced transaction models* [5]. In general, however, the vast majority of these ideas have never been used in real settings. In this section, we cover the most relevant ones and discuss how they have been adapted to modern process modeling languages.

11.3.1   Problem Description

Although the mechanisms used to maintain atomicity within a transactional process are conceptually identical to those used in database transactions, the practical implementation is very different. To start with, it is very rare to be in situations in which the entire process must be atomic. Certainly, we want processes to terminate but we do not necessarily want them to disappear without a trace if they do not succeed.

A simple example can be used to illustrate the differences with database transactions. In a database, the application semantics are ignored. That is, if the transaction encounters any problems, it is simply aborted. Typically, this is done by undoing any changes done by the transaction, a procedure that is done at a very low level (e.g., by replacing images of memory pages) and in a manner that is completely orthogonal to the transaction. At the process level, things are very different. First, the process actually corresponds to the application that submits transactions. It is here where one must react to transactions that abort. It will not do to just kill the process and undo all partial results. Second, even in those cases in which the process must terminate, termination cannot be implemented by undoing the whole process. In the vast majority of cases, the termination can only occur by taking additional steps as part of the process. That is, atomicity is not guaranteed by a low-level mechanism orthogonal to the process but as part of the normal logic of the process. Consider, for instance, a purchase order process. The buyer can cancel the order at any time by sending the appropriate message [see Chapter 9, patterns 19 (cancel activity) and 20 (cancel case)]. If the *cancel* message is received, the process does not just abort and undo all its steps. Canceling the order involves, for example, creating new records that register the cancellation, may involve partial charges if the cancellation arrived late in the process, and requires sending additional cancellation notices to those involved in the purchase (inventory control, production chain, billing, delivery, etc.). In many cases, canceling the order might be as complex as the processing of the purchase order.

The main problem with transactional processes is that the mechanisms necessary to implement atomicity, independently of the type of atomicity chosen, are necessarily programming mechanisms. Thus, the problem of intraprocess atomicity needs to be solved by providing appropriate programming constructs for programmers and process designers to incorporate the necessary transactional guarantees within their processes. In what follows, we outline the most important programming concepts for intraprocess atomicity and how they are used in modern business process modeling languages.

11.3.2 Compensation—Sagas

One of the earliest programming models for dealing with what we now call transactional processes is *sagas* [6]. In their simplest form, sagas assume a process where each step is an independent transaction (in the conventional database sense) that is executed sequentially or in parallel. In the case of a sequential saga, the execution can be represented as the sequence

$$T_1\, T_2 \ldots T_{n-1}\, T_n$$

Atomicity in the context of sagas is achieved through *compensation*. Sagas assume that for each step in the process there is a compensating action. That is, if the effects of the step need to be undone, there is another compensating step that takes care of it. That is, for each step T_i in the normal process, there is a compensating step T_i^{-1}. The compensation can take many forms and sagas do not impose any particular form of compensation. It can be a low-level compensation such as the mechanisms used in databases, or it can be a high-level compensation based on the semantics of the step to be compensated. An example of the latter is a step that places a *purchase order* by sending a particular document to an e-mail address; this step that can be compensated for by sending another e-mail to the same address canceling the purchase order. The sending of the cancellation message is what constitutes the compensating step.

Using sagas, a process designer can ensure the atomicity of a process by indicating a compensating step for every normal step. If at a certain point the process needs to be undone, the compensating steps for all steps that have been completed are executed. The compensating steps are executed in the reverse order of the normal steps to ensure that after the execution of every compensating step, the state of the process is semantically equivalent to having been executed only to that point. That is, to compensate for a saga that has been executed until step k, the execution has the sequence

$$T_1\, T_2 \ldots T_k\, T_k^{-1} \ldots T_2^{-1}\, T_1^{-1}$$

Conceptually, the main contribution of the saga concept to transactional processes are the notion of compensation associated to each step and the undoing of a process by executing the compensating actions in the reverse order. Based on these ideas, many different combinations of this basic model can be derived. For instance, rather than associating with compensating step for every step, a compensating step could be associated to a group of steps. That is, the compensating step undoes not a single step but a group of them. Another example are optimized compensation sequences for Sagas with an arbitrary partial order between the steps. In general, these extension add more flexibility to the use of compensation but do not alter the basic underlying idea behind sagas.

For a programmer, designing a process as a saga facilitates implementing atomicity by simply defining the compensating steps. The underlying execution engine can then keep track of which steps have been executed and, if needed, trigger the execution of the compensating steps in the proper order. This frees the programmer from having to extend the normal control flow with the compensating control flow. The main ideas behind sagas have been incorporated in many programming and specification languages (see Chapter 5 on how UML allows designers to associate compensations with single actions, or Chapter 13 on how compensation is treated in BPEL)

11.3.3 Alternatives—Semiatomicity

Sagas impose a rather rigid structure on the compensation; that is, either the process terminates successfully or it is compensated. This is due to the context in which sagas were proposed—the model of atomicity mirrored that of classic database transactions. In long-lived business processes, it might make sense to be able to undo or roll back part of the process until a given point and then continue from there without compensating the entire process.

This basic idea is taken up by the notion of *semiatomicity* proposed for *flexible transactions* [17]. The idea behind semiatomicity is to extend the basic model of sagas with the possibility of providing alternative execution paths. That is, within a process, there might be different ways to proceed. If one of these options fails, whatever part of this alternative has already been completed is compensated for following an approach identical to sagas. Then, an alternative execution path is selected. From a formal point of view, this model is based on differentiating between three types of steps within a process: *compensatable, pivotal,* and *retriable.* A step is compensatable if it has similar semantics to that in a saga; that is, there is a compensating step for it. A step is retriable if it is guaranteed to succeed if its execution is retried a sufficient number of times. A step is pivotal if it is neither compensatable nor retriable. Based on these three elements, it is possible to construct processes that provide a great deal of flexibility when dealing with failures. The basic rule that needs to be observed is that the resulting combination of steps needs to be such that there is always the possibility of either compensating back to the beginning (as in sagas) or to finish the process by retrying steps. Without getting into the details of the formalism, this implies that once a pivot is executed, there must always be an execution path that ensures the process will terminate (i.e., a path of retriable steps). Similarly, before a pivot is executed, all steps must be compensatable since failure of the pivot implies having to roll back either to the beginning or to a previous alternative path.

These ideas illustrate very well the advantages and disadvantages of advanced transaction models. On the one hand, they apparently provide more flexibility and allow one to automate the navigation over the process by clearly indicating what needs to be done at each point in the execution. For instance, if a pivot fails, then the process needs to be compensated for back to the beginning as in a saga or up to an alternative execution path if any is available. If alternative execution paths are available, it is possible to assign priorities and automatically try each alternative in the proper order until one succeeds. All this can be specified and then navigation proceeds automatically. On the other hand, the programming complexity increases significantly. The designer needs to identify steps as compensatable, retriable and pivotal, combine them accordingly, and establish alternative paths that lead to a well-formed process. Although automatic tools can be used to check the resulting structure, it is the programmer's responsibility to make sure the structure is correct. The problem in doing this is that it might be very difficult to cast the semantics of a process into a correct process structure. In addition, if the process involves steps outside the control of the process designer (e.g., services provided by other companies), it might be impossible to determine the nature of these steps (whether they are compensatable, retriable, or pivotal).

The main contribution of semiatomicity is the idea of using alternative executions paths that can be followed after a given path fails. As with sagas, it is possible to automate the selection of these alternatives, thereby greatly simplifying the design of processes while having more sophisticated recovery and failure handling capabilities. For a programmer, semiatomicity adds to the advantages of sagas the possibility of compensating only parts of a process and of using alternative execution paths to avoid having to discard work already done if something goes wrong at a later point in time.

11.3.4   Mapping to Workflow Processes

Although most advanced transaction models were developed independently of workflow management systems and process modeling languages, the close relation between them was shown long ago [7, 2]. In fact, the three key ideas discussed above—compensation, undoing by reversing the sequence of execution with compensating steps, and alternative paths—can all be easily mapped to almost any form of workflow language. The mapping can be manual or automatic.

In case of manual mapping, it is the process designer who uses constructs in the process representation language to build a process that can recover from failures by implementing either sagas, semiatomicity, or combinations thereof. In the case of manual mapping, the actual model used is not that relevant. The programmer is free to combine many different possibilities. Because of this, the underlying infrastructure and execution engine does not need to be aware of what the programmer intends to do. It just executes what has been specified without being aware of whether a given step is a normal step, a compensating step, an alternative path, and so on.

In the case of automatic mapping, the workflow system can offer one or more transactional models and apply them to the process described by the programmer based on certain rules. In practice, this is feasible only for simple models like sagas, in which, as indicated above, the programmer writes the process and gives compensation steps for each step in the process. Then the system can automatically undo a process by executing compensations in the reverse order of the forward steps. For more complex models, automatic derivation of the compensation procedures is typically not feasible. In addition, any given model restricts the freedom of the programmer in terms of what can be expressed. Thus, automatic mapping is seldom used in practice and only for relatively simple, well-defined cases.

Using a very simple and schematic workflow model, it is possible to show how to implement sagas using standard programming constructs of workflow languages. The idea is to define steps and compensating steps as activities of the process and then include the corresponding control flow based on whether forward activities successfully complete or o not. Figure 11.1 shows how this is done for the simple case of linear sagas. Similar approaches can be taken to implement alternative paths, compensation of subprocesses, and so on.

11.3.5   Atomicity in Process Modeling Languages

Over the years, the concepts discussed above have appeared and disappeared as part of a number of different systems and platforms (e.g., advanced transaction models, workflow systems, composition in object request brokers). As a result, these ideas

LINEAR SAGA

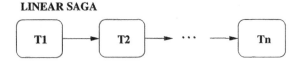

WORKFLOW PROCESS

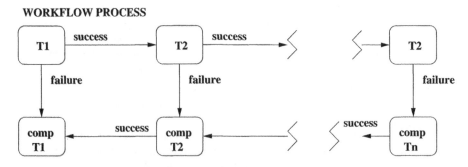

Figure 11.1   A linear saga mapped to a workflow process.

have been assimilated by many developers and programmers and have crystallized into a number of basic primitives that can be found in many process modeling languages. An example is BPEL (also discussed in this book), which contains specific constructs to support these ideas.

In BPEL, the most relevant constructs for transactional purposes are the *fault* and *compensation* handlers, in addition to the standard primitives for implementing different forms of control flow. The fault handlers are a set of steps within the process that will be executed if a failure occurs within a given scope (see Chapter 13 for the definition of scope in BPEL and a detailed description of the types of faults considered in BPEL). Fault handlers are one way to implement alternatives in case there is a problem with a particular service or with a particular execution path within the process. They are, however, not the only way to implement such functionality since similar effects can be achieved by using explicit control flow. Compensation handlers have similar semantics as in sagas and are executed if a particular step or group of steps need to be compensated.

These programming primitives show how the notion of transactional behavior within a process has become a matter of programming the necessary behavior rather than having automatic support. This makes perfect sense since there is no model that fits all possible types of business processes, not even a majority of them. Hence, today support for intraprocess atomicity means supporting the process designer in specifying how the process should behave when different events take place, for example, failures, exceptions, unsuccessful attempts to perform an operation, and cancellations.

11.4   INFRASTRUCTURE FOR IMPLEMENTING ATOMICITY

11.4.1   Atomicity in the Context of Middleware

The notion of atomicity once outside the database revolves around 2PC and how to integrate it with the applications running the transactions. A common simplifying assumption made in textbooks about distributed transactions is that the transaction has a single origin; for example, a program calls first one database, and then another, and so forth. In practice, this is not the always the case. In many architectures where databases are hidden behind higher-level interfaces, transactions are the result of the execution of programs. Distributed transactions arise when a program that itself runs one transaction on one database also calls up a second program that then runs another transaction on another database. In general, these two programs are aware that the other program also calls up a transaction.

This scenario is very common in business processes; in fact, more common than the one in which a single program calls several databases within a single transaction. The practical problem that it creates is, who will be in charge of coordinating the execution of the necessary 2PC protocol. The *Object Transaction Service* (OTS) of CORBA [8] is a good example of how this is done in practice. As shown in Figure 11.2, assume that we have two applications, A and B, each one working on top of a different database. Application A starts executing and since it knows that it will run transactions against its own database, it informs the OTS module that it wants to begin a transaction (Figure 11.2, part 1). The OTS then assigns a transaction identifier and a context to that transaction so that application A can refer to the transaction and the OTS module can keep track of all the operations associated to that transaction. After application A obtains the transactional context from the OTS module, it registers its database with the OTS module (Figure 11.2, part 2). In this way, the OTS module knows where to go when running the 2PC protocol (as it is

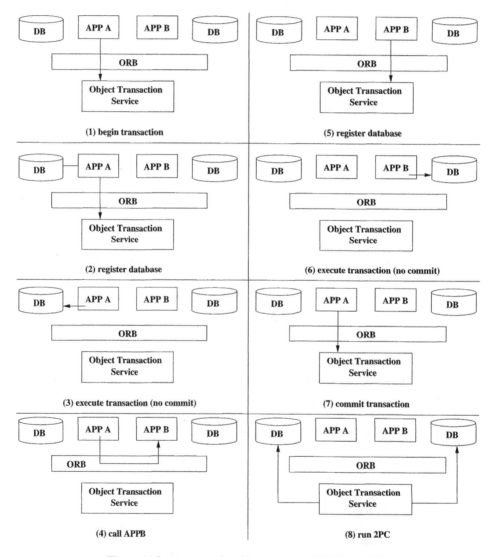

Figure 11.2 A transactional interaction in CORBA's OTS.

the database in charge of the atomicity of the transaction run by application A). After these two steps (obtaining the transactional context and registering the database) are completed, application A can execute its transaction against its database (Figure 11.2, part 3).

Now, assume that application A also wants to call application B and it wants its transaction and whatever B does to be an atomic operation (Figure 11.2, part 4). Let us assume that application A has already executed its transaction on its database and then calls B through whatever interface B makes available. When A calls B, it adds to the call the transactional context it got from the OTS. When B receives the call, it realizes that it is being invoked within the context of a transaction. Since B knows it will run another transaction in its own database, it will repeat the steps of A. First, it will register its database with the OTS module, adding the transactional context it got from A (Figure 11.2, part 5). That way, the OTS module knows that the transaction initiated by A has also ramifications in B and also gets from B the information on how to contact the corresponding database. Once this is done, B can run its transaction against its database and return the results to A (Figure 11.2, part 6). A will then want to finish the transaction and inform the OTS module that the transaction should be committed (Figure 11.2, part 7). The OTS will then commit the transaction by running 2PC among the two databases involved in the exchange (Figure 11.2, part 8). In this case, the OTS module will act as the coordinator and the two databases as the participants. Once the 2PC protocol terminates, the OTS module informs application A of the outcome.

This mechanism is not exclusive to CORBA. With small differences, it is used essentially in all forms of middleware that support transactions and it illustrates very well what is needed for maintaining atomicity of distributed transactions. First, we need a transactional context so that the different participants in the transaction know on behalf of whom they are working. Second, there must be a registration procedure to tell the coordinator who will be in charge of running 2PC once the transaction completes. Third, the participants must support an interface that allows them to run 2PC or a similar protocol under control of the coordinator. Fourth, we need somebody who acts as the coordinator and can both assign transactional contexts and run 2PC. As we will see in what follows, transactional processes use a very similar infrastructure to implement interprocess atomicity.

11.4.2 XA Interface

The *XA interface* specifies the interfaces necessary to implement the type of interaction just described [15]. Although a process designer will typically not interact with an XA interface, it is explained here since it constitutes the conceptual basis (and often the actual physical basis) for transactional specifications targeting business processes.

The model behind the XA interface is shown in Figure 11.3. It contains an *application program,* which is the program that decides to start and end a transaction using the *transaction manager (TM).* The application program communicates with the *resource managers (RM)* (typically databases) to execute operations, but relies on the

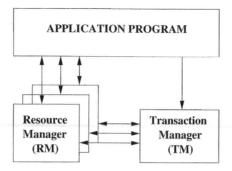

Figure 11.3 The model behind the XA interface.

transaction manager for all aspects of transactional control. The XA interface defines how the transaction manager and the resource managers communicate with each other in order to enforce transactional consistency by running the 2PC protocol.

The most important services of the XA interface are shown in Table 11.1. The interface also contains additional features to deal with threads and other details that are not relevant for the purposes of this chapter. The services shown in the table correspond almost one to one with the ones described for the CORBA OTS: register and unregister a resource manager with the transaction manager, start a transaction, running 2PC, and so on. This is not surprising since the CORBA OTS is designed to work precisely with resource managers that support the XA interface. The important aspect of this interface is that any resource manager supporting it can now participate in the execution of transactions that use the 2PC protocol (the XA interface also defines the different states of the entities involved, the order in which the operations must be invoked, and so on).

11.4.3 WS-Coordination and WS-Transaction

For business processes operating over Web services, there is no direct way to use the XA interface even if the systems involved support such an interface [1]. The

Table 11.1 Services within the XA interface

Name	Description
ax reg	Registers an RM with a TM
ax unreg	Unregisters an RM with a TM
xa open	Initializes an RM for use by an application
xa close	Terminates the use of an RM by an application
xa start	Starts or resumes a transaction associating a transaction ID to it
xa prepare	Tells the RM to prepare to commit a transaction
xa commit	Tells the RM to commit a transaction
xa rollback	Tells the RM to roll back (abort) a transaction

problem is that a Web service front end is needed for the the XA interface to be able to operate using the mechanisms of Web service interactions (SOAP messages). Moreover, since business processes might be distributed across many companies and might be long lived, using 2PC is not always a good idea.

Thus, in the context of Web services, the ideas behind the XA interface have been generalized to operate in that setting without changing the underlying principles. This generalization involves two steps. The first one is related to the fact that distributed business processes like those used in electronic commerce cannot rely on a single transaction manager for executing 2PC or whatever protocol is used. The other step consists of modularizing the transactional protocol itself so that it is not only 2PC that can be run (as it is the case with the XA interface). That way, additional protocols can be defined without having to change the infrastructure to run them.

These two steps have been formalized in two specifications: *WS-Coordination* [12] for the task of the transaction manager, and *WS-Transaction* [13] for the transactional protocols. The WS-Transaction specification has recently been divided into *WS-Atomic Transaction* and *WS-Business Activity*.

WS-Coordination defines a generic infrastructure for the coordination of termination and commitment protocols between Web services (and by extension, between the business processes running atop such Web services). Its main goal is to serve as a generic platform for implementing 2PC and variations of it. The behavior of WS-Coordination is quite similar to that of an XA interface but adapted to Web services. Figure 11.4 shows the structure of the *coordinator* described in the specification. The *activation service* is a Web service that allows an application to create a *context* for a particular interaction. This is similar to obtaining a transaction ID

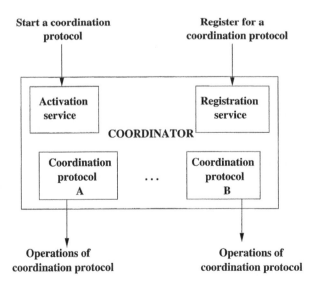

Figure 11.4   Structure of the coordinator in WS-Coordination.

from the TM in the XA interface. In the case of Web services, the context is richer and includes information such as URL of the registration service of that coordinator and the type of protocol being used. Once the application has obtained a context, it includes the context in the header part of SOAP messages to other Web services. A second, remote Web service, upon receiving a request with a context attached to it, will ask for another context from its own coordinator. When it obtains that context, it passes information to its local registration service about the address of the registration service of the invoking Web service. The registration services then talk to each other to synchronize and set up the coordination protocol that has been chosen. This procedure is similar to the registration procedures in the XA interface except that the registration services also need to agree on the protocol they will use and set up the addresses to use for the interactions, which will typically take place as Web service operations. The second application can use the context it has obtained to forward it to other Web services and the procedure will be repeated again. In this way, it is possible to link together several coordination protocols.

WS-Transaction is designed to use WS-Coordination as the underlying infrastructure. WS-Transaction specifies a set of coordination protocols that can be used between Web services. The first of these protocols is now part of the WS-Atomic Transaction specification and it covers interactions similar to 2PC, very much in the same way the XA interface implements 2PC. The other two protocols are part of the WS-Business Activity specification and includes the *business agreement with participant completion protocol* (a simplified form of commitment that supports canceling and compensating), and the *business agreement with coordinator completion protocol* (similar to the previous protocol but with an additional step in the protocol for completing the work that resembles the tentative hold protocol discussed below).

11.4.4   WS-CAF (Web Services Composite Application Framework)

The area of transactional processes in Web services is still a very active arena in which many different groups are competing to dominate the standardization process. WS-Coordination and WS-Transaction have been proposed by a set of vendors, but an alternative, conceptually similar specification has been proposed by another set of vendors. The Web Services Composite Application Framework (WS-CAF) comprises three specifications [14]:

- **Web Service Context (WS-CTX).** WS-CTX is the lowest-level specification and covers basic elements like the generation of context information and the propagation of this information as Web services invoke each other. The idea is that these basic elements and notions will be augmented and tailored by the other two specifications. As such, it supports interfaces that allow an application to create a context, define the demarcation points for an *activity* (e.g., for a transaction), register Web services so that they can participate in an activity with the same context, and propagate the context as needed when invocations are made.

- **Web Service Coordination Framework (WS-CF).** WS-CF specifies a *coordinator,* similar to WS-Coordination, but it is built on top of the WS-CTX specification. The main elements of WS-CF are a coordinator (which extends the notion of activity from WS-CTX to a set of set of tasks with a set of related coordination actions), a *participant* (including the operations involved in the coordination process), and a *coordination service* (defining the behavior of the coordination protocol; instances of the coordination service could be 2PC, sagas, etc.)

- **Web Service Transaction Management (WS-TXM).** WS-TXM is used to define concrete instances of the coordination service of WS-CF (i.e., indicating how the coordinator and the participants must behave) and to augment the distribution context of WS-CTX. Similarly to WS-transactions, WS-TXM provides three different transaction models: *ACID transactions* (following conventional 2PC), *long-running actions* (which include the possibility to compensate), and *business process transactions* (which extend the notion of compensation to interactions based on business processes rather than individual steps). The latter model is included to be able to deal with those processes in which compensation does not necessarily happen in an automatic manner but might involve off-line work to restore a consistent state.

WS-CAF explicitly recognizes the fact that there are two forms of transactional interactions that Web services need to support. One involves conventional middleware platforms and enterprise application integration efforts, in which transactional behavior is covered by 2PC and the mechanisms reflected by the XA interface. The other one involves business processes in which 2PC is not the best way to implement transactional behavior and the coordination procedures are more oriented toward compensation, either following the same model as sagas, semiatomicity, or combinations thereof. WS-Transaction also makes this distinction but in a rather implicit manner. Nevertheless, it is not by accident that both specifications propose three coordination protocols or services: one for 2PC-like interactions (presumably for systems based on the XA interface), one for compensation-based consistency (like sagas), and one for more flexible compensation approaches (like semiatomicity or programmer-defined models).

11.4.5 Tentative Hold Protocol

The *Tentative Hold Protocol* [10] is an interesting case and a contrast to the previous proposals. The protocol addresses more semantic issues than low-level transactional mechanisms and, in fact, heavily depends on the nature of the process and the actual business process being adapted for this type of protocol. The main idea is to allow a nonbinding reservation for a particular item. This reservation is not binding either for the requester or the provider; that is, it can be revoked by sending a message notifying of the fact.

The advantage of the tentative hold protocol lies in allowing one to place a reservation on one or more items while a decision is being made. If the item becomes unavailable, the provider sends a message indicating that this is the case and the requester can then proceed with other options. That way, both sides are informed about the possible future use of that item but without having to block it until the decision is actually made. As an example of where this idea can be applied, consider a company interested in receiving a set of offers for different components that will be used to build a single product. The company may want to obtain offers from different subsets of companies so that there are different combinations of price, quality, delivery dates, and so on, and then eventually decide on a particular combination. The Tentative Hold protocol allows the company in question to place tentative reservations for the different components with different providers without actually committing to purchasing those components. That way, and unlike what happens with conventional transactions, the items are not blocked while a decision is being made. If any of the providers changes the available offer, it notifies the company of the fact, which can then remove that offer from the current options. Once the company decides on a given option, it notifies the providers that the components are no longer on tentative hold.

The Tentative Hold protocol is not so much a protocol for guaranteeing transactional consistency but a practical approach to indicate that, later in time, a transaction may actually arrive that will request a particular item or resource. As such, it can be used in combination with WS-CAF or WS-Coordination/WS-Transactions. First, interest in a given item is expressed using the Tentative Hold protocol and then a transaction, run with any of the approaches described above, will actually request the item in a binding manner. Notice, nevertheless, that the proposals in WS-Transactions (business agreement with completion) and WS-CAF (business process transactions) can be used to implement processes with behavior very similar to that proposed in the Tentative Hold protocol. The difference is that the Tentative Hold protocol, if accepted as a standard, would then provide an accepted framework for tentative reservations. In WS-Transactions and WS-CAF, there is a possibility of implementing a similar behavior but not in a standard manner, as tentative holds would be one of the many possibilities for such business agreements or business process transactions.

11.4.6 Transactional Processes in Electronic Commerce

An often forgotten fact behind the notion of transactional business processes is that the type of transactional characteristics associated with the process depend very much on the programming abstractions used as part of the process. All the ideas and specifications discussed address, almost exclusively, processes in which invocations take place through RPC-like mechanisms (and that includes RMI, J2EE, and most current Web services). Similarly, and as seen in WS-Transactions and WS-CAF, ideas like 2PC are still very much at the top of the list in terms of what it means to have a transactional business process. That has the big advantage that the new specifications fit very well with existing infrastructures (like the XA interface, by now widely used by almost all forms of middleware). Yet, a problem that is becoming increasingly obvious in practice is that business processes that involve remote systems—and not typical middleware or Enterprise Application Integration scenarios—do not use this programming model. Most electronic commerce commerce today is based on the interaction between business processes of different companies through the exchange of documents [4]. In most cases, these exchanges take place through queues and using asynchronous messages. In such systems, the notions of transaction, compensation, and coordination have a very different meaning. In many cases, it is not even clear how to apply transactional notions to pure document exchanges.

Here, it is worth mentioning that some of the ideas developed in the context of message-oriented middleware and persistent queuing [1] can be very useful. A persistent queue is a way of exchanging messages so that instead of going directly from sender to receiver, they are deposited in a queue. That way, the sender and the receiver are completely decoupled from each other as the sender only needs to put messages in the queue and the receiver to read them from that queue. When the interaction is between business processes of different companies, the queuing system typically involves two queues. A sending queue is used to store the messages of the sender until they are successfully transmitted to the receiver. At the receiver side, a queue is used to store all incoming messages until the application retrieves them. As Figure 11.5 shows, these interactions can be made transactional by using simplified versions of 2PC, thereby ensuring that messages are not lost and that operations within each process can actually tie the processing of the messages to actual transactions.

In these systems, the notion of transactional process acquires a different meaning as the transactional parts are the interactions between the participants to ensure the receipt of certain documents. These systems do not use transactional mechanisms within the process itself to enforce consistency. Rather, consistency is part of the process logic itself, as the following example, taking from the xCBL library [16] shows:

The xCBL 4.0 ChangeOrder document is a buyer-initiated document that can be used to change an existing Order already received and responded to by a seller. The document can be used to make changes to header level information, change line items, cancel line items or entire orders, add line items, etc. Note that if an OrderResponse has

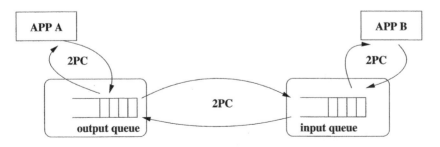

Figure 11.5   Document exchanges using transactional queues.

not been received for a given Order, a ChangeOrder is not necessary (an Order with a purpose of "Replace" should be used). Similarly, if an entire order is to be canceled and an OrderResponse has not been received an Order with a purpose of "Cancellation" can be used. However, if an OrderResponse has been received, a ChangeOrder with a ChangeOrder/ChangeOrderHeader/ChangeType/ChangeTypeCoded = "Cancel" should be used to cancel the entire Order.

The above paragraph describes a standard way of dealing with purchases orders that includes cancellation. It does not say what needs to be done at each end as part of the cancellation. In fact, it is possible that each process, for example, on the seller side, actually reacts to the arrival of such a message by using some sort of compensation mechanism such as those available in BPEL (although it is less than straightforward to express the logic above in BPEL). Yet, what is clear from this description is that there is no coordination protocol between both sides. Cancellation or changes occur as part of the normal processing logic and not as part of a global transaction encompassing the processes of buyer and seller. An interesting problem that will need to be clarified in the short and medium term is how to reconcile specifications like WS-Transactions or WS-CAF (and the models behind them) with processes based on document exchanges like those being developed by xCBL or ebXML [4].

11.5   OUTLOOK

The notion of transactional business processes is anything but new and has been applied in a wide variety of settings: advanced transaction models with more or less conventional transactions, workflow systems, conventional middleware platforms such as transaction processing monitors, modeling languages for processes based on Web services, and so on. In this chapter, we have described the background for most of the ideas used today and discussed the relevance of each one of the existing solutions.

The most important distinction made in the chapter is between mechanisms that are used to guarantee the internal consistency of a given process (typically compensation), and mechanisms used to guarantee the consistency of several business processes that interact with each other (using a variety of models, from 2PCto business transactions). The two aspects are not entirely independent but they should not be confused, as they require and use different mechanisms. Internal consistency has today become a programming problem—how to help a programmer to express complex recovery logic as part of a business process. This question is of vital importance since, as the xCBL example shows (Section 11.4.6), depending on the mechanisms used, it may not be necessary to use distributed coordination protocols. Consistency between processes is today still seen as a problem largely similar to that of transactional consistency using mechanisms like the XA interface. The introduction of other coordination possibilities is interesting and tied to the use of compensation as part of the process logic. Yet, it remains to be seen to what extent these advanced mechanisms will be used in practice and how they can be combined with interactions such as document exchanges.

11.6   EXERCISES AND ASSIGNMENTS

1. Using the process representation format of your choice, describe the behavior of a ChangeOrder document exchange such as the one described in Section 11.4.6.
2. Read reference [2]. Using the process description language of your choice, describe how to implement sagas and other advanced transaction models using that particular format. Is there any construct missing? Are there any programming primitives that would make the whole design easier?
3. Read reference [6]. Extend the ideas in that paper to sagas to allow (a) arbitrary partial orders between the activities and (b) nesting of activities. Can you represent the resulting mechanisms using a process modeling language?
4. Read reference [17]. Using the process description language of your choice, describe how to implement flexible transactions and other advanced transaction models using that particular format. Is there any construct missing? Are there any programming primitives that would make the design easier?
5. Construct (using the process description language of your choice) two processes, one corresponding to the buyer and one corresponding to the seller, that internally use compensation but communicate using the procedure described in the xCBL excerpt of Section 11.4.6.
6. Construct two processes, one for the buyer and one for the seller, that interact through direct invocations (e.g., RPC or Web service operations) and use the Tentative Hold protocol to reserve a particular item before placing the final order.

ACKNOWLEDGMENTS

I would like to thank Claus Hagen, Amaia Lazcano, Win Bausch, Cesare Pautasso, Heiko Schuldt, Daniel Jönnsson and Biörn Biörnstad. All of them Ph.D. students at one time or another at the Department of Computer Science at ETH Zürich and who have worked with me during the last years on topics related to those discussed in this chapter.

REFERENCES

1. G. Alonso, F. Casati, H. Kuno, and V. Machiraju. *Web Services: Concepts, Architectures and Applications.* Springer-Verlag, 2003.
2. G. Alonso, D. Agrawal, A. El Abbadi, M. Kamath, R. Guenthoer, and C. Mohan. Ad-

vanced Transaction Models in Workflow Contexts. In *Proceedings of the 12th International Conference on Data Engineering,* New Orleans, LA, Feb. 26–March 1, 1996.

3. P. A. Bernstein, V. Hadzilacos, and N. Goodman. *Concurrency Control and Recovery in Database Systems.* Addison-Wesley, Reading, MA, 1987.

4. ebXML Technical Architecture Specification v1.04. http://www.ebxml.org/specs/index.htm.

5. A. K. Elmagarmid (Ed.). Transaction Models for Advanced Database Applications. Morgan Kaufmann, 1992.

6. H. Garcia Molina and K. Salem. Sagas. In *Proceedings of ACM SIGMOD,* May 1987, pp. 249–259.

7. F. Leymann. Supporting Business Transactions via Partial Backward Recovery in Workflow Management Systems. In *GI-Fachtagung Datenbanken in Buero Technik und Wissenschaft*—BTW'95. Dresden, Germany. March 1995.

8. Object Management Group. Transaction Service. http://www.omg.org/technology/documents/formal/transaction service.htm.

9. H. Schuldt, G. Alonso, C. Beeri, and H.-J. Schek. Atomicity and Isolation for Transactional Processes. In *ACM Transactions on Database Systems (TODS),* Volume 27, No. 1, March 2002.

10. J. Roberts and K. Srinivasan. Tentative Hold Protocol. W3C Note 28 November 2001. http://www.w3.org/Submission/2001/11/.

11. G. Weikum and G. Vossen. *Transactional Information Systems: Theory, Algorithms, and the Practice of Concurrency Control and Recovery.* Morgan Kaufmann, 2001.

12. F. Cabrera et al. WS-Coordination. http://www-106.ibm.com/developerworks/library/ws-coor/.

13. F. Cabrera et al. WS-Transaction. http://www-106.ibm.com/developerworks/webservices/library/ws-transpec/.

14. Web Services Composite Application Framework (WS-CAF). OASIS WS-CAF Technical Committee. http://www.arjuna.com/standards/ws-caf/.

15. Distributed Transaction Specification: The XA interface. The Open Group. http://www.opengroup.org/onlinepubs/009680699/toc.pdf.

16. XML Common Business Library (xCBL), version 4.0. http://www.xcbl.org/xcbl40/xcbl40.html.

17. A. Zang, M. Nodine, B. Bhargava, and O. Bukhres. Ensuring Relaxed Atomicity for Flexible Transactions in Multidatabase Systems. In *Proceedings of the ACM SIGMOD Conference on Management of Data,* Minneapolis MN, June 1994.

PART IV

STANDARDS AND TOOLS

CHAPTER 12

Standards for Workflow Definition and Execution

JAN MENDLING, MICHAEL zur MUEHLEN, and ADRIAN PRICE

12.1 INTRODUCTION

Process-aware information systems (PAISs) are increasingly used as building blocks of application software architectures as well as embedded components in larger application systems, such as ERP packages. This, ultimately, may lead to the coexistence of several PAISs within an organization. In addition, the use of PAISs and PAIS development tools has been extended to automate business processes beyond organizational boundaries. With an increasing number of trading partners using PAISs, the individual negotiation of system-to-system interoperability aspects becomes cumbersome as the number of interfaces in the network increases geometrically. Accordingly, the use of general standards for process integration purposes becomes economically desirable. Standardization groups in the area of workflow have two purposes. First, the standardization of integration interfaces promises reduced implementation effort and better reuse of workflow components when PAISs have to be integrated with the existing information system infrastructure of an organization. Second, in the field of cross-organizational processes (e.g., in supply chain scenarios), the standardization of invocation interfaces and messaging formats improves the development of plug-and-play solutions and reduces the risk of using proprietary technology for participants in such process chains. To date, several standardization groups dedicated to different aspects of PAISs exist. Other, nonspecialized standardization bodies have published specifications that relate to aspects of PAISs as well. In this chapter, we give an overview of the different standardization bodies relevant to the development of PAISs, and we present two PAIS-related standards in detail: XPDL, which is an XML-based process description language, and Wf-XML, which is an interoperability protocol for process-aware applications.

12.2 STANDARDIZATION BODIES RELEVANT TO PAIS

12.2.1 WfMC

In the early 1990s, the commercial workflow market took a significant upswing. New vendors entered the marketplace, and existing imaging, document management, and messaging software vendors extended their products with workflow capabilities. In mid-1993, a group of large software users organized in the Black Forest Group realized that they would have to make arrangements to deal with the proliferation of multiple workflow management systems in their organizations. They chartered workflow vendors with the task to develop interoperability standards between their systems. Following this challenge, an initial group of vendors and consultants under the leadership of IBM founded the Workflow Management Coalition (WfMC) in August 1993. To date, membership of the WfMC is comprised of more than 250 vendors, users, consultants, and research institutions with an interest in the field of workflow management. WfMC is split into two sections: In Technical Committee meetings, technology representatives, such as the architects of workflow vendors discuss the specification of the different WfMC standards, whereas the External Relations Committee is comprised mostly of marketing and public relations representatives, and discusses the publication and dissemination of standards documents. WfMC standards are treated as recommendations for workflow vendors, but there is no obligation for participating members to actually implement the WfMC specifications. In addition, there is no formal conformance testing in place; vendors' claims of support for WfMC standards rely on self-certification, not on an independent conformance evaluation.

12.2.2 OMG

The Object Management Group (OMG), a standardization body with a focus on software engineering, has published a Workflow Management Facility specification as a component of the Common Object Request Broker Architecture (CORBA). The workflow facility represents a high-level CORBA facility, which allows an object request broker to enable the communication of application components under the control of a process management service. The coordination of business objects using the workflow management facility enables application designers to create workflow-enabled business applications, without the need to implement or integrate a dedicated workflow engine. The creation of the workflow management facility is closely linked to the WfMC.

In 1996, the OMG decided to adopt existing WfMC standards and initiated a fast-track process. However, this move was criticized by some researchers who preferred a less engine-centric and more distributed approach to process automation [19]. As a result, the fast-track process was canceled and a longer request for proposal process was initiated [15]. This request for proposal was answered by two groups: a proposal backed by 19 workflow vendors (all WfMC members, since the WfMC as a body was not eligible to submit a proposal), and a proposal backed by a single vendor and a university (Nortel and the University of Newcastle upon Tyne). After a review and revision period of more than 12 months, the 19 vendor proposal was selected over the alternative submission [17]. The OMG requires vendors who propose a standard to provide an implementation of the standard within a year after the adoption of their proposal [16]. However, the workflow management facility was not implemented by any of the submitters, and the popularity of Web Services Choreography standards (see e.g., [13]) and Wf-XML (see Section 12.5) makes it unlikely that this standard will have any commercial significance in the future.

12.2.3 BPMI

The Business Process Management Initiative (BPMI), founded in 2000, is an industry consortium of approximately 200 companies in the workflow, business process modeling, and systems integration areas. It has the mission of developing an XML-based business process modeling language, a matching notation, and a repository for models specified in this language. The initiative is led by Intalio, which produces a BPMI-compliant business process management system and controls much of the regular BPMI operations. The standards defined by BPMI are the Business Process Modeling Language (BPML) [4] and a matching Business Process Modeling Notation (BPMN) [25]. A query language to control the operations of a PAIS at run time, the Business Process Query Language (BPQL), has been discussed as a BPMI effort but, currently, no specification for such a language exists.

BPML was designed as a modeling language for transactional, discrete business processes. Besides entities such as elementary and complex activities, connectors, and events, the BPML meta-model offers a number of entities for the management of data at run time (e.g., definition of an activity context, which may contain shared data). In addition, elements for exception handling, such as message, time-out, and failure event handlers are provided. Constructs for the modeling of transactional aspects of processes, along the lines of the compensation mechanisms of sagas presented in Chapter 11, are available as well. Based on an early draft of the BPML specification, BPMI members submitted a proposal to the World Wide Web Consortium (W3C), called the Web Services Choreography Interface (WSCI), with the hope of influencing the development of a Web Services Choreography language [5].

Parallel to the development of BPML a working group was established to develop a corresponding visual notation. Two different levels of abstraction of this Business Process Modeling Notation (BPMN) are defined. On the one hand, an execution-level notation will represent the BPML semantics completely. On the other hand, a business-level notation will serve as a "lean" notation for organizational modeling, leaving out details that are relevant for the automation of the modeled processes. Whereas the execution-level notation contains elements such as fault, compensation, transaction, and context, the business-level notation is designed with the intention of providing a comprehensible graphical diagram that allows the grouping of elements through swim lanes or participant lanes, respectively. A first version of the BPMN specification was published in 2004 [25].

12.2.4   OASIS

The Organization for the Advancement of Structured Information Standards (OA-SIS) is an industry group focused on the development of XML-related standards in the area of Web Services and business-to-business processes. Relevant to the development of PAISs, OASIS hosts three working groups that define standards for process interoperability:

- The Asynchronous Services Access Protocol (ASAP) is a protocol for the invocation of long-running services at remote locations, with functionality to query the status of a remote service and manipulate its state [21]. ASAP is a direct descendant of the Simple Workflow Access Protocol, and represents a generic version of the WfMC Wf-XML standard, which is discussed in detail in Section 12.5.

- The Business Process Execution Language for Web Services (BPEL4WS) is an XML-based specification language for business processes that provides powerful constructs for the modeling of a long-running orchestration of Web services [3] (see Chapter 13).

- The ebXML framework consists of standards that describe the structure and access mechanisms of a central registry for Web services, as well as mechanisms to describe these services in a standardized manner. Part of these specifications is a high-level process specification schema standard (BPSS) [23], which describes the interaction between two parties in an ebXML scenario. ebXML is a joint effort of UN/CEFACT and OASIS. UN/CEFACT was the original standardization body for the UN/EDIFACT standard, which found widespread adoption in the commercial world. ebXML is targeted at organizations for which the traditional UN/EDIFACT standard is too expensive to implement.

OASIS is a popular standardization organization for software vendors because its organizational structure allows for the fast setup of working groups and its by-laws provide protection of a vendor's intellectual property (IP) rights, whereas other standards bodies (e.g.,W3C) require submitters of standards proposals to give up the IP rights in their submissions.

12.2.5   W3C

Whereas WfMC, BPMI, and OASIS can be classified as industry consortia, the World Wide Web Consortium (W3C) has the character of a vendor-neutral standards body, similar to the Internet Engineering Task Force (IETF). The W3C is working on the standardization of World-Wide-Web-related technologies and protocols. Most standards that relate to Web services are developed by W3C working groups. The abovementioned WSCI has the status of a W3C note; it is an input document for the W3C working group that develops a Web Services Choreography Definition Language (WS-CDL). Furthermore, W3C has published several standards that are utilized by other standards in the area of PAISs, such as the Simple Object Access Protocol (SOAP),[1] Web Service Description Language (WSDL),[2] and XML.

12.3 WFMC REFERENCE MODEL AND WFMC GLOSSARY

One of the main tasks undertaken by the WfMC is the development of a common understanding of what a workflow management system is and what functionality it should provide. This led to the WfMC Glossary and Reference Model briefly outlined in Chapter 2. In this section, we describe them in more detail.

12.3.1 WfMC Glossary

Due to the increasing number of workflow vendors by the middle of the 1990s, vendor-specific terminology for workflow constructs had led to an inconsistent vocabulary of workflow terms. In order to counter this trend, the first goal of the WfMC was to establish a common terminology for workflow concepts, which led to the publication of the WfMC *Terminology & Glossary* [27]. Today, the WfMC Glossary covers most workflow concepts and gives definitions for terms such as activity, workflow management system, and participant. Although not all workflow vendors use standard terminology, the WfMC vocabulary has found widespread acceptance in practice. It is perceived as a valuable aid for the system selection process, since proprietary terms used by different vendors can be transformed to a common standard, thus enabling a comparison of systems on the basis of a single vocabulary.

Figure 12.1 illustrates the central terms of the WfMC Glossary. The following list gives their definition according to the WfMC Glossary [27].

- A *Business Process* is a "set of one or more linked procedures or activities which collectively realize a business objective or policy goal, normally within the context of an organizational structure defining functional roles and relationships."
- A *Process Definition* is the "representation of a business process in a form which supports automated manipulation, such as modeling, or enactment by a workflow management system. The process definition consists of a network of activities and their relationships, criteria to indicate the start and termination of the process, and information about the individual activities, such as participants, associated IT applications and data, etc."
- A *Workflow* is the "automation of a business process, in whole or part, during which documents, information or tasks are passed from one participant to another for action, according to a set of procedural rules."
- A *Workflow Management System* is a "system that defines, creates and manages the execution of workflows through the use of software, running on one

---

[1]http://www.w3.org/TR/soap/
[2]http://www.w3.org/TR/wsdl/

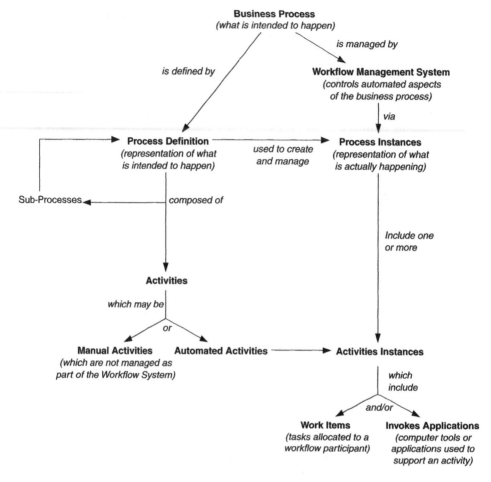

Figure 12.1   Important terms of the WfMC Glossary [27].

or more workflow engines, which is able to interpret the process definition, interact with workflow participants and, where required, invoke the use of IT tools and applications."

- An *Activity* is a "description of a piece of work that forms one logical step within a process. An activity may be a manual activity, which does not support computer automation, or a workflow (automated) activity. A workflow activity requires human and/or machine resources(s) to support process execution; where human resource is required an activity is allocated to a workflow participant."

- An *Instance* (of a process or an activity) is the "representation of a single enactment of a process, or activity within a process, including its associated data. Each instance represents a separate thread of execution of the process or activity, which may be controlled independently and will have its own internal state and externally visible identity, which may be used as a handle, for example, to record or retrieve audit data relating to the individual enactment." The *Process Instance* is the "representation of a single enactment of a process"; and the *Activity Instance* is the "representation of an activity within a (single) enactment of a process, i.e. within a process instance."

- A workflow *Participant* is a "resource which performs the work represented by a workflow activity instance. This work is normally manifested as one or more work items assigned to the workflow participant via the worklist."

- A *Work Item* is the "representation of the work to be processed (by a workflow participant) in the context of an activity within a process instance."

- The *Worklist* is a "list of work items associated with a given workflow participant (or in some cases with a group of workflow participants who may share a common worklist). The worklist forms part of the interface between a workflow engine and the worklist handler."

12.3.2 WfMC Reference Model

The WfMC Reference Model was introduced in 1995 as a means to group typical interfaces of a PAIS according to their purpose and constitutes one of the most influential workflow frameworks to date [8]. A global view of the reference model was given in Chapter 2. Here, we discuss some of the elements of this model in more detail, as they are relevant for the rest of this chapter.

The WfMC reference model is composed of five interfaces, each of which is specified individually through an abstract specification that describes its generic functionality. Depending on the functionality of the interface, one or more interface bindings are provided as illustrations of how an interface can be implemented using particular languages or technologies. All five interfaces are grouped around the execution core of a PAIS, the workflow enactment service, which encapsulates one or more workflow engines. A system may consist of several workflow engines, for example, if it is implemented in a distributed manner. The communication between the workflow enactment service and outside systems is provided through an application programming interface (API), the Workflow API (WAPI). The following sections present the individual interfaces in detail.

12.3.2.1 WfMC Interface 1. The WfMC Interface 1 (Process Interchange) provides a generic process description format, the Workflow Process Definition Language (WPDL). The purpose of WPDL is the exchange of workflow specifications between different workflow systems, or between business process modeling tools and workflow management systems. For this purpose, WPDL provides a common subset of elements found in most workflow management systems. WPDL is specified using an extended Backus–Naur Form (EBNF) and a plain text description, but is independent of a particular system implementation. In 2002, an XML representation called XPDL was published as the successor of WPDL, which has been implemented in a number of open-source workflow projects. It is discussed in detail in Section 12.4.

12.3.2.2 WfMC Interfaces 2 and 3.

The WfMC Interfaces 2 (Workflow Client Applications) and 3 (Invoked Applications) form the core of the WAPI specification. Interface 2 specifies the communication between a workflow engine and client applications that are used by workflow participants to interact with the workflow management system. This concerns mainly the presentation of the work list, the selection of work items, and the notification of workflow participants in the event of overdue activities and/or processes. Interface 3 specifies the API functions for the integration of invoked applications. This relates mainly to the passing of data between the workflow engine and a remote application, and the handling of application return codes. Whereas Interface 2 defines the *invocations of a workflow participant* that demands a work item from a workflow engine for further processing, Interface 3 specifies the *invocations of a workflow engine* that demands another application to execute operations. Although initially separated, the close relationship between the two interfaces ultimately led to a merger of the specifications. This abstract specification defines on one side the operations that external systems can invoke on a workflow engine, and on the other side the operations a workflow management system performs on an outside system. These operations include the instantiation, starting, manipulation, and stopping of workflow instances.

12.3.2.3 WfMC Interface 4.

The WfMC Interface 4 (Interoperability) specifies the communication across different workflow engines in the sense of a process-to-process interaction. The first version of this interface was published in 1996; the current version 2.0 was published in 1999. The specification of Interface 4 consists of an abstract description of interoperability functions (e.g., instantiating a workflow; starting, stopping, and aborting a workflow instance, etc.), as well as bindings to different messaging protocols and transport mechanisms. The first published version allowed for the realization of cross-organizational workflows using e-mail and MIME messages. Operational implementations of this binding were presented in the course of an interoperability challenge by three workflow vendors (DST, FileNet, and Staffware) in March 1999. More recently, the interoperability of PAIS is the focus of several standardization groups in the Web Services Choreography area, and the WfMC has published an interaction protocol that relies on XML message encoding and HTTP as a transport protocol—Wf-XML. This interaction protocol is discussed in more detail in Section 12.5.

12.3.2.4 WfMC Interface 5.

The WfMC Interface 5 (Management and Audit) describes the format of the run time protocol produced by a workflow enactment service, the so-called audit trail [30]. Initially designed for recovery purposes, similar to a database redo/undo-log, the audit trail of workflow systems is becoming increasingly interesting for purposes such as process mining (see Chapter 10) or process monitoring and controlling [31]. The current version of the WfMC Interface 5 (2.0) describes the data format of log entries as well as the state changes responsible for creating these log entries.

12.4   PROCESS DEFINITION IN XPDL

12.4.1   Purpose of XPDL

The XML Process Definition Language (XPDL) has been proposed by WfMC as a standard for interchanging process definitions between process definition tools and workflow management systems (Interface 1). XPDL aims to facilitate the import and export of process definitions by specifying an XML-based interchange format that can be used by process-aware information systems.[3] XPDL has evolved from the Workflow Process Definition Language (WPDL) [26].

When a process definition has to be moved from one workflow management system to another, the process definition needs to be exported to a file conforming to an interchange format that the target system can import. Many vendors offer proprietary interchange formats to support such data transfer between different installations of their software. When process definitions have to be moved between tools of *different vendors,* an interchange format is needed that both tools understand. Furthermore, there may be the desire to represent a process definition using another process modeling technique. *Different techniques* for process modeling (see Chapters 5–8) have individual sets of syntax elements and related semantics. In some cases, it is difficult to define mappings between different techniques; see Dehnert [6] for discussion on mappings between event-driven process chains (EPC) and Petri nets. Moreover, tools serve *different purposes.* There are specialized tools for process modeling via graphical editors, for process simulation using a simulator, for process execution via an execution engine, or for process monitoring and auditing. Each of these specific tools requires different information to be included in the process model.

XPDL aims to serve as an XML-based lingua franca for representation and interchange of process definitions between tools from different vendors, tools using different modeling techniques, or tools that serve different purposes (see Figure 12.2). However, XPDL has been designed around a consensual minimal set of constructs found in the workflow domain. As a consequence, a particular tool will most probably need to store data that is not defined by XPDL. In such a situation, XPDL allows one to define so-called extended attributes. Those can be used to attach arbitrary information to a process or process-related objects. However, these extended attributes will not be understood by other tools. Nevertheless, there is a need for extensibility that is also recognized by academic standardization efforts like Petri Net Markup Language (PNML) [24] or EPC Markup Language [14] to provide for flexible customization and future evolution. Yet, extensibility does not lead to a lingua franca [1]. The minimal-set approach of XPDL contrasts with the workflow patterns and the YAWL initiatives (see [2] and Chapter 8) that aim at identifying a su-

---

[3]See http://www.wfmc.org/standards/XPDL.htm for a list of software supporting XPDL.

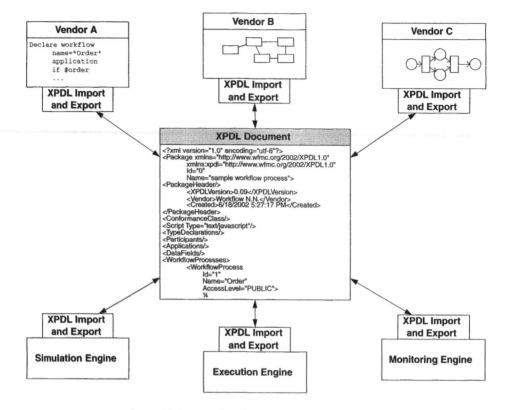

Figure 12.2   Interchanging XPDL process definitions.

perset of control-flow concepts and constructs supported by existing systems. The WfMC is aware of these limitations and encourages vendors to submit proprietary extension attributes to be included in future versions of the XPDL specification.

12.4.2   An Introduction to XPDL

In this section, we assume that the reader has some familiarity with XML and XML Schema. The reader not familiar with these languages is referred to [18]. The syntax of XPDL is defined by an XML Schema[4] document. Its main concepts are the package, applications, participants, data fields and data types, workflow processes, activities, and transitions. Figure 12.3 illustrates the relationships between these entities via a meta-model. In the following, we will first describe elements common to multiple concepts. Afterward, we introduce a use case that will serve as a working example, and then each XPDL concept with its individual elements. We will introduce most of XPDL's elements, but not all. For a complete presentation of XPDL, refer to the specification [29].

[4]http://www.w3.org/TR/xmlschema-1/ and http://www.w3.org/TR/xmlschema-2/.

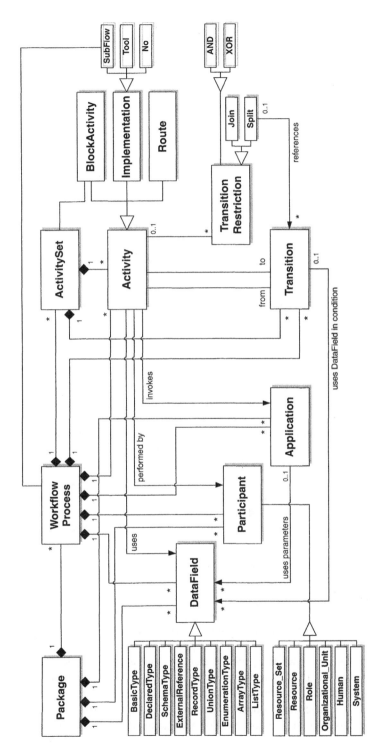

Figure 12.3 XPDL meta-model.

12.4.2.1 Common Elements. There are three general (or common) elements that can be used in conjunction with different XPDL entities: extended attributes, formal parameters, and external references:

- Extended attributes can be used in all XPDL elements. They capture vendor-specific information that cannot be represented by standard elements. Extended attributes are modeled as name–value pairs.
- Formal parameters can be used to represent input and output parameters of workflow processes and applications. A formal parameter has an identifying Id. Its mode attribute defines the semantics of parameter passing.
- External references can be used with data types, participants, and applications. An external reference gives the URI of an external specification in its location attribute. This specification document may be, for example, an XML Schema or a Java class. The xref attribute states the name of the entity in the linked document. In particular, external references can be used to refer to Web services.

12.4.2.2 New Employee Use Case. The XYZ organization requires a workflow process for engaging a new employee. When a new employee is due to join the company, the following actions must be taken by staff in the appropriate departments. The *human resources department* first has to prepare office accommodation and order items from a checklist where necessary. Then, this department has to issue a "preinduction information pack" containing company policy documents. Then, a place on the next scheduled induction course has to be booked. The *IT department* has to procure and configure a computer for the employee. Then, a network user ID and temporary password have to be allocated. Afterward, an e-mail account has to be set up. Then, an extension number has to be assigned and the employee has to be entered into the corporate telephone directory. Finally, access rights have to be granted to the appropriate corporate resources. The finance department has to enter the personnel details of the new employee into the salary system. The process has to be supported by a workflow management system. The workflow management system of the XYZ organization requires the process to be defined as an XPDL document.

12.4.2.3 XPDL Package. The package is the root element of an XPDL document. The package serves as a container to store multiple workflow processes and common information about data fields, application declaration, and participant specification. Accordingly, XPDL can be used to interchange both single process definitions and also a whole repository of process definitions.

Figure 12.4 illustrates the package element with a code snippet from the new employee use case. Every package is identified via a mandatory unique Id attribute and an optional Name attribute. In our example, the name of the process is "New Employee." Packages have a header (element PackageHeader) to capture general information like XPDL version, vendor and name of the tool that generated the

```
<?xml version="1.0" encoding="UTF-8"?>
<Package Id="new-employee"
    Name="New Employee"
    xmlns="http://www.wfmc.org/2002/XPDL1.0"
    xmlns:xpdl="http://www.wfmc.org/2002/XPDL1.0"
    xmlns:xsi="http://www.w3.org/2001/XMLSchema-instance"
    xsi:schemaLocation="http://www.wfmc.org/2002/XPDL1.0
    http://wfmc.org/standards/docs/TC-1025_schema_10_xpdl.xsd">
    <PackageHeader>
        <XPDLVersion>1.0</XPDLVersion>
        <Vendor>Open Business Engine</Vendor>
        <Created>2004-07-25 18:14:19</Created>
    </PackageHeader>
    <RedefinableHeader
        PublicationStatus="UNDER_TEST"/>
    <ConformanceClass
        GraphConformance="NON_BLOCKED"/>
    <Script Type="text/x-xpath"
        Version="1.0"/>
    <WorkflowProcesses>
        <WorkflowProcess AccessLevel="PUBLIC"
            Id="new-employee" Name="New Employee">
        . . .
```

Figure 12.4   Example of an XPDL package.

XPDL file, or creation time. The `RedefinableHeader` element covers general information that may be redefined for each workflow process contained in the package, for example, publication status. The conformance class imposes restrictions on the network structure of the process definition. There are three classes of conformance: NON-BLOCKED (used in the example) indicates that there is no restriction, LOOP-BLOCKED means that the network structure must be acyclic, and FULLBLOCKED means that each split has to be matched with a corresponding join. Split-join blocks may also be nested. The distinction between these conformance classes is important because cycles may result in undefined semantics (see Kindler for a thorough discussion [10]). As a consequence, some workflow vendors do not allow arbitrary cycles in their workflow systems (see, e.g., [11]). The `Script` element identifies a scripting language that is used in expressions of transition conditions. The example process uses XPath[5] (defined in the `Type` attribute). Moreover, a package can contain multiple data fields, participants, applications, and workflow processes (see Figure 12.4). Figure 12.4 does not include data fields, participants, and applications because they are defined local for the workflow process definition of the "new employee use case."

12.4.2.4   *Data Fields and Data Types.*   Data fields represent variables that capture workflow-relevant data. They can be used in the specification of deadline conditions of activities, in transition conditions, or they can be referenced in formal

---

[5]http://www.w3.org/TR/1999/REC-xpath-19991116

parameters that are passed from activities to subflows or applications. Data fields are described within `DataField` elements that can be defined on the package level with global scope or with local scope on the workflow process level.

A data field is identified by a unique `Id` attribute and an `Name` attribute. Furthermore, it has an `IsArray` attribute whose default value is FALSE. Moreover, an `InitialValue` can be defined. A data field has also a `DataType`. XPDL offers nine data types (see Figure 12.5). `ArrayType`, `EnumerationType`, `ListType`, `RecordType`, and `UnionType` are deprecated, that is, they are included only to provide compatibility with WPDL. It is recommended to use `SchemaType` instead. It may contain an XML Schema describing the data. An `ExternalReference` referring to an external document can be used for the same purpose. Furthermore, the `DeclaredType` defines a type by referencing a type declaration within the package element. Finally, an XPDL `BasicType` can be used to specify the data field. There are seven basic types including string, float, integer, reference, datetime, Boolean, and performer. Most of them are self-explanatory but some need further remarks. First, as the reference type is now deprecated, it is recommended to use an external reference instead. Second, the datetime refers to an instance of time, but it does not define a specific date format. Finally, the performer type is a data field that contains a reference to a declared workflow participant.

12.4.2.5 *Participants.*

Participants represent resources that can perform activities of a workflow process. The information relevant to a participant is defined in a Participant element, which is a child element of a package or a workflow process just like the data field element. A participant declared in the package has global scope; others have local scope within the workflow process in which they have been defined. Note that policies for assigning participants to activities are defined in each activity.

Figure 12.6 gives the declaration of the HR department of the new employee use case. A participant is identified by a mandatory unique `Id` attribute ("HR" in the ex-

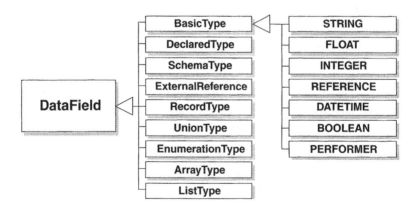

Figure 12.5   Portion of the XPDL meta-model relevant to data fields.

```
<Participant Id="HR" Name="Human Resources Dept.">
    <ParticipantType Type="ORGANIZATIONAL_UNIT" />
</Participant>
```

Figure 12.6   Example of a participant declaration in XPDL.

ample); optionally, a Name attribute can be added. Furthermore, a participant type has to be declared. XPDL offers six predefined types. RESOURCE SET denotes a set of resources. RESOURCE describes a specific resource agent. ROLE defines a functional role that a person can have within an organization. A coordination mechanism may be needed to identify an individual having a role that is assigned to a certain activity. ORGANIZATIONAL UNIT refers to a department or other units within an organization. This type is used in the example for the HR department. HUMAN is an individual user who interacts with a process. Finally, SYSTEM describes an automatic agent.

12.4.2.6 Application. The information needed to invoke operations of an application is included in an Application element, which should appear within the package or a workflow process element. When an application is defined on the package level, it has a global scope and can be referenced from all workflow process definitions. Otherwise, it has local scope within a workflow process. Activities may invoke applications and applications use data.

Figure 12.7 gives an example of an application in the context of the new employee use case. The application is identified by a mandatory, unique Id attribute ("createNetworkUser"). Optionally, a Name attribute can be added. Beyond the optional

```
<Application Id="createNetworkUser"
    Name="Create Network User">
    <Description>Creates a network user. The userID is generated
        from the user name, and the temporary password assigned is
        the userID.</Description>
    <FormalParameters>
        <FormalParameter Id="userName" Mode="IN">
            <DataType>
                <BasicType Type="STRING"/>
            </DataType>
            <Description>The user©s full name</Description>
        </FormalParameter>
        <FormalParameter Id="userID" Mode="OUT">
            <DataType>
                <BasicType Type="STRING"/>
            </DataType>
            <Description>The generated user ID.</Description>
        </FormalParameter>
    </FormalParameters>
</Application>
```

Figure 12.7   Example of a Web service declaration in XPDL.

Description, an application may have FormalParameters to define input and output parameters. The application takes a string "userName" as input, and returns the "userID" of the generated network user, which is also a string. Web services can be defined using the external reference element. For details, see the XPDL specification [29].

12.4.2.7 Workflow Process.

A workflow process is defined by a Workflow-Process element, which is a child element of Package. A workflow process element acts as a container for other elements that are relevant to the description of a workflow process. Similar to the package element, it may contain a ProcessHeader, a RedefinableHeader, DataTypes, and DataFields, as well as Participants and Applications. Furthermore, a workflow process contains multiple ActivitySets, Activities, and Transitions to represent the tasks and control flow of a process definition (see Figure 12.8).

Figure 12.9 gives an example of how a workflow process is defined in XPDL. A workflow process is identified by a mandatory unique Id attribute ("new-employee") and an optional Name attribute ("New Employee"). The AccessLevel attribute set to PUBLIC indicates that the process may be invoked by external applications; if it may only be invoked as a subflow, the access level is set to PRIVATE. When a workflow process serves as an implementation of a subflow activity, it may define multiple FormalParameters.

The control flow of a workflow process definition is defined by activities and control flow transitions between them. In the example, there are five transitions from the activity "Allocate user ID" to other activities. These transitions are referenced in the Split element of the first activity (a5). The type attribute is set to AND; that is, the transitions represent concurrency: each activity a6, a8, a7, a4, and a9 may be executed in parallel. Activity a5 is executed automatically (start and finish mode are set to "automatic") via a procedure of the tool "createNetworkUser." The performer element indicates that the IT department is responsible for this activity.

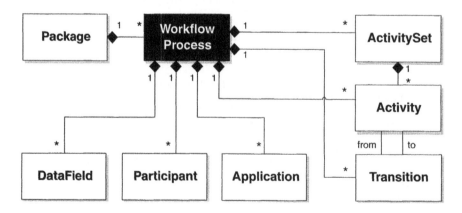

Figure 12.8 Portion of the XPDL meta-model relevant to workflow processes.

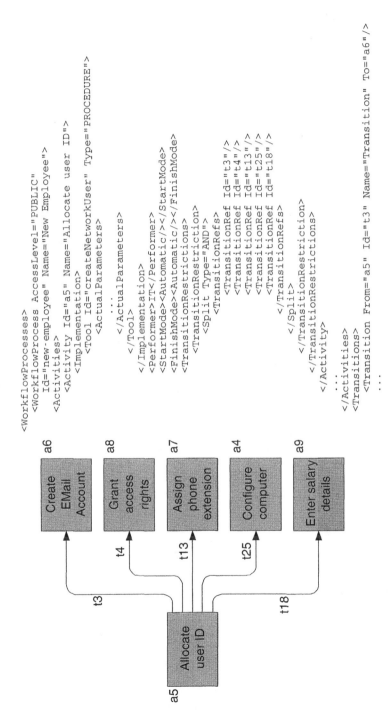

Figure 12.9 Example of a workflow process declaration in XPDL.

297

12.4.2.8 ActivitySet and Activity. The work to be performed in a workflow process is defined by a set of activities (elements `ActivitySet` and `Activity`), which are child elements of the workflow process element). Activity sets may contain multiple activities and transitions that link them. An activity set is identified by an `Id` attribute and can be executed by so-called block activities, which are special kinds of activities explained below. An activity represents a task that is executed within a workflow process. Its execution may involve data fields, participants, and applications. Activities are linked by transitions (see Figure 12.10).

An activity is identified by a mandatory unique `Id` attribute and an optional `Name` attribute (see Figure 12.9). Any activity element may contain multiples of general elements. We give a short description of deadlines, limits, simulation information, start and finish modes, performer information, and transition restrictions. Furthermore, there are elements specific for the different types of activities.

`Deadline` is a general element for the specification of a deadline condition. It includes the name of an exception that is raised on the arrival of the deadline and an `Execution` attribute to describe whether the activity continues after raising the exception (ASYNCHR) or whether the activity is completed abnormally (SYNCHR). The `Limit` element is also related to execution time of an activity. It defines an expected duration that may be used for tool-specific escalation mechanisms. The `SimulationInformation` element is also partly related to time. It can include information about average cost and estimated time subdivided into waiting time, working time, and total duration of an activity. For each activity, a start mode and a finish mode can be defined (elements `StartMode` and `FinishMode`, respectively). Automatic mode denotes that start or end of the activity is controlled by the workflow management system. Manual mode requires user interaction to start or end an activity. Start and finish modes may be defined independently from another. The `Performer` element represents an assignment made by giving a link to a participant who is either declared on the workflow process or on the package level. At run time, this participant assignment may result in an empty set of resources or in mul-

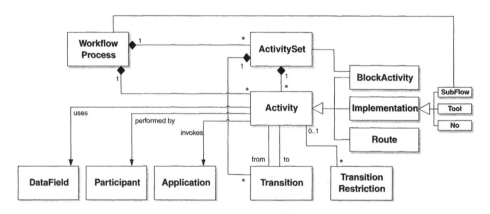

Figure 12.10 Portion of the XPDL meta-model relevant to activities.

tiple resources. XPDL does not make any statement about run time resolution of these cases. Finally, `TransitionRestrictions` is another general element that may be used by all activity types. It defines how multiple transitions are evaluated before starting the activity and after completing the activity (for details, see the Section 12.4.2.9). In the example of Figure 12.9, we have seen a transition restriction of an AND split.

XPDL distinguishes three types of activities: route, block activity, and implementation (see Figure 12.10). An activity element includes one of these three elements. The `Route` element identifies a so-called dummy activity. It has neither a performer nor an application and it does not affect any data field. Its only purpose is to represent complex routing conditions via its transition restrictions. The `Block-Activity` element is related to an activity set. It executes the activities mentioned in the activity set, starting from the first activity. After the exit activity has been reached, the output transitions of the block activity are followed. `Implementation` refers to an activity that is either not automatically executed by the workflow system (denoted by a `No` element), executed by a program or application (`Tool`), or implemented as a workflow subprocess (`SubFlow`). Figure 12.9 contains a tool element used to define an automatic execution.

A `No` implementation represents an activity whose execution is not supported by the workflow management system. This may be a manual activity or a so-called implicit activity. The latter refers to pre- and postprocessing in a workflow process for initializing data or writing data to the archive. Its start and finish modes would be set to automatic. `Tool` describes an activity that is implemented by a tool that may be invoked using the Workflow Client Application API of WfMC's Interface 2. Its `Id` attribute identifies the tool, which may be a `Type` application or procedure. A set of parameters can be defined in the `ActualParameters` element, which is of the type list. Element `SubFlow` corresponds to the third implementation type, covering the case in which the activity is implemented by another workflow process. Its `Id` attribute identifies the invoked workflow process. There is also the option to specify a set of `ActualParameters`. The `Execution` attribute can be set to synchronous (SYNCHR) or asynchronous (ASYNCHR). Synchronous execution refers to a suspension of the activity until the invoked process instance is terminated. In the case of asynchronous execution, the activity is continued after the subflow has been instantiated. If there is a need for a later synchronization with the subflow mechanisms, events have to be used in a vendor-specific way, because they are not defined in XPDL.

12.4.2.9 Transition. Transition is a child element of workflow process and activity set. Transitions define the conditions that enable and disable activities during a workflow execution. They represent control flow via directed edges between activity nodes in a workflow process. XPDL transitions coupled with the split and join elements provide direct support for the first five patterns presented in Chapter 8. A detailed analysis of XPDL in terms of the workflow patterns can be found in [1].

Figure 12.11 gives an example of different transitions. A transition is identified by an `Id` and an optional name attribute. The `From` and the `To` attributes capture the

```
<Transitions>
     <Transition Id="1" From="Act1" To="Act2">
          <Condition Type="CONDITION">salary > 1000</Condition>
     </Transition>
     <Transition Id="2" From="Act1" To="Act3">
          <Condition Type="OTHERWISE"/>
     </Transition>
     <Transition Id="3" From="Act1" To="Act4">
          <Condition Type="EXCEPTION">Expl-Exception</Condition>
     </Transition>
     <Transition Id="4" From="Act1" To="Act5">
          <Condition Type="DEFAULTEXCEPTION"/>
     </Transition>
</Transitions>
```

Figure 12.11   Portion of the XPDL meta-model relevant to transitions.

identifiers of the source activity and the target activity. Each transition contains a
Condition element, which may contain multiple Xpression elements. Such expressions can refer to data fields; they are stated in the scripting language defined
for the package in the script element (i.e., XPath in the new employee use case). A
condition may be of four different types (see Figure 12.11). Type CONDITION indicates that the transition is executed if the condition is true. Type OTHERWISE
defines the transition that is executed when no other conditions are met. Type EXCEPTION defines a transition that is executed in the case when an exception condition is met. Type DEFAULTEXCEPTION specifies the transition that is executed
if an exception occurs but no exception condition is met.

Beyond transition conditions, there can be TransitionRestrictions attached
to activities that influence control flow. The Join element of a transition restriction
defines which multiple incoming transitions are synchronized (AND) and which are
not synchronized (XOR). The Split element specifies which outgoing transitions
are executed. AND denotes that the transitions mentioned in the TransitionRefs
list are executed in parallel, as long as there is no transition condition associated with
the transition that evaluates to false. XOR defines alternative execution paths depending on the conditions of the transitions listed in the TransitionRefs element.

12.4.3   XPDL in Practice

12.4.3.1 Example implementation—The Open Business Engine. The
Open Business Engine is an open source workflow project hosted at SourceForge.[6]
OBE is a flexible, modular, standards-compliant Java workflow engine. It is fully
J2EE compliant and supports several J2EE application servers, operating systems,
and databases. It implements five of the WfMC standards and offers a variety of extensions and enhancements. OBE is equally well suited to embedded or standalone

---

[6]http://sourceforge.net/projects/obe

deployment. OBE is configurable and extensible, and many aspects can be customized. The run-time engine relies upon pluggable services to provide authentication, authorization, persistence, task assignment, inbound event handling, and outbound integration capabilities. OBE provides a comprehensive workflow life-cycle event notification framework to support deep integration with workflow-enabled applications. OBE supports automated, manual, and mixed workflow processes, and has extensible work item allocation and activity completion algorithms. Activities are automated through an extensible system of tool agents, which enable the invocation of external logic defined in Java classes, EJBs, native executables, scripts in arbitrary scripting languages, Web services, and so on. Human interactions are managed through work items, which can be manual or can provide the means to invoke the appropriate software tools. OBE provides a worklist API and worklist clients to manage work items. OBE supports the following WfMC standards:

- Interface 1—XPDL, the XML Process Definition Language
- Interface 2/3—WAPI, the Workflow and Tool Agent APIs
- Interface 4—Wf-XML, Workflow Interoperability[7]
- Interface 5—Audit Audit Data Specification

The OBE project was originated by Anthony Eden. In late 2002, OBE was adopted and, over a period of two years, substantially rewritten by Adrian Price, then of Zaplet, Inc., a Californian software company specializing in collaborative business process management. OBE has been embedded in several commercial applications including Zaplet's own, which is in use in a variety of industries and U.S. government departments, including the CIA and the Terrorist Threat Information Center. The OBE implementation has revealed several weaknesses and ambiguities of XPDL that are described in the following subsection.

12.4.3.2 *XPDL Weaknesses and Ambiguities.* The XPDL 1.0 Final Draft Specification [29] and its associated XML Schema and Sample Workflow Process contain a number of ambiguities and contradictions of which it is important to be aware. These issues exemplify the pitfalls of defining a formal specification without ratifying it through a reference implementation. We discuss them here and, where appropriate, provide guidance for minimizing their impact on the semantic portability of processes expressed in XPDL. The issues fall into four categories: missing defaults, undefined semantics, schema errors and ambiguities, and schema omissions and inconsistencies. In the following, we provide some examples of these issues.[8]

- *Missing Defaults:* These are cases where the XPDL XML schema defines an element or attribute as optional but does not specify the default value. For clarity and portability, it is advisable for XPDL instance documents to specify

---

[7]At the time of writing (September 2004), OBE's Wf-XML support is still under development.
[8]The reader is referred to http://www.openbusinessengine.org/wiki/Wiki.jsp?page=XPDLIssues for a full list.

explicit values for such attributes. An example of this is the `AccessLevel` attribute of the `WorkflowProcess` element. Here, you should specify an explicit value of PUBLIC or PRIVATE. Another example is the `Execution` attribute of the `SubFlow` element where also an explicit value of SYNCHR or ASYNCHR should be specified.

- *Undefined Semantics:* These are cases in which neither the specification nor the schema provide a clear and unambiguous definition of semantics. The `Xpression` element is such an example. The schema permits multiple `Xpression` elements within a `Condition` element of a transition, but the specification does not define the behavior when more than one `Xpression` occurs. For semantic portability, avoid using multiple `Xpression` elements. Another example is the `TransitionRestriction` element. The schema permits multiple occurrences of a `TransitionRestriction` element within a single activity. Each `TransitionRestriction` can contain a `Split`, a Join, or both, but the specification does not define the semantics if more than one split or join is present. To avoid confusion, do not specify more than one split and one join per activity.

- *Schema Errors and Ambiguities:* These are cases in which the schema and the specification reference undefined items, or where the definition is weak or ambiguous. Consider the `Limit` element. The schema incorrectly defines the `Limit` element as type `xsd:string`, whereas, in fact, it is required by the specification to follow the pattern defined for `Duration` (i.e., an integer followed by an optional duration unit Y, M, D, h, m, or s). Furthermore, the schema defines the `Created` element to be `xsd:string`; it is actually a date type but lacking a format definition and, therefore, ambiguous. To ensure portability, do not rely upon the `Created` element. To use it, be aware of your engine's supported date formats.

- *Schema Omissions and Inconsistencies:* These are cases in which the schema uses XML schema ineffectively or inconsistently. The Identifier Length is an example of this case. Unlike WPDL, XPDL does not enforce maximum lengths for any of the identifying strings, but the corresponding identifiers in WfMC Interfaces 2/3 (Workflow/ToolAgent APIs) do. To avoid buffer overrun problems in systems that support WfMC Interfaces 1, 2, and 3, ensure that XPDL identifiers do not exceed 64 characters in length. Moreover, references are used inefficiently. Many elements cross-reference each other, but the schema does not model these relationships using XML schema's keyref feature, nor does it enforce key value uniqueness with the key feature.

12.5 PROCESS INVOCATION USING WF-XML

12.5.1 Purpose of Wf-XML

The initial mandate of the WfMC was to create interoperability between workflow engines of different vendors. Wf-XML is the most recent specification for workflow interoperability, and its functionality extends beyond the coupling of work-flow engines to the remote invocation of a process service by other clients (not necessarily workflow engines). Wf-XML aims to facilitate the remote invocation and manipulation of processes through a lightweight interface that is modeled after the principles of the Hypertext Transfer Protocol. Version 1.1 of Wf-XML relied on a proprietary message format that did not support Web Services [28] but Version 2.0 [20] introduces support for the Simple Object Access Protocol (SOAP) and builds on the Asynchronous Service Access Protocol (ASAP) [21], which is a domain-neutral protocol for the control and monitoring of long-running services. Since ASAP is being standardized within OASIS, but by the same people responsible for Wf-XML, Wf-XML 2.0 can be seen as a specific implementation of ASAP for purposes of workflow management. The following section illustrates Wf-XML Version 2.0.

12.5.2 An Introduction to Wf-XML

The history of Wf-XML goes back to the Simple Workflow Access Protocol (SWAP). This protocol was devised by WfMC members in 1998 as a lightweight alternative to the more complex MIME Interface 4 specification and the OMG Workflow Facility. Although SWAP never became an official standard, its authors pursued the underlying principles in the WfMC Technical Committee, which released the Wf-XML 1.0 specification in 2000 and a revised Version 1.1 in October 2001. In parallel with the development of Wf-XML, SWAP was modified to be applicable to generic long-running services (beyond the workflow domain) and a working group at OASIS was founded to standardize the Asynchronous Service Access Protocol [32]. Wf-XML, like ASAP, describes an interface to an asynchronous process that is based on the architectural principles of Representational State Transfer (REST), a guiding principle for Web applications developed by Roy Fielding [7]. The fundamental concept at the heart of REST is that a client can navigate a service using the standard HTTP commands GET, PUT, POST, and DELETE, without much context information about the service offered. In other words, a client needs the initial address of the service, and can obtain additional interaction possibilities by requesting via GET a copy of the service description. This copy may contain pointers to other services or commands that the client is allowed to use. In essence, a client does not need to know much about the service operations, because these operations are exposed as the interaction between the client and the service progress-es. This loose coupling of client and server forms a contrast to the tight coupling of process partners in approaches such as BPEL4WS, and is one of the distinguishing features of Wf-XML.

ASAP and Wf-XML extend the basic HTTP operations with commands that are more specific to the world of workflows and Web services.[9] In the context of work-flow, the main service is a process factory, which can spawn process instances upon

---

[9]Some REST purists argue that ASAP and Wf-XML violate the purity criteria of the original REST principles, whereas the authors of the specification maintain that this is an acceptable compromise.

request. Note that a process factory is distinct from a workflow management system. A workflow management system can in fact host many process factories, one for each process model that the workflow management system implements. A process factory represents an asynchronous service. An operation of this asynchronous service is started by a request, and the result is later communicated to the client in a separate request. As a service may operate from a few minutes up to a few months, an asynchronous way of interaction is required. ASAP defines operations to monitor, control, and receive notifications about the state of a service. ASAP distinguishes different types of resources. In essence, a *resource* is a service identified by a URI. In ASAP, a Factory resource creates instances of an asynchronous service, and an Instance resource executes the work. Clients may specify an Observer resource, which is able to receive notifications from the instance. Usually, the Observer matches the client that has requested an Instance from the Factory, but a client can also designate other parties as Observers. Wf-XML extends the basic ASAP resource model specifically for the interaction of workflow engines.

In Wf-XML, there are five resource types: The Service Registry, the Factory, the Instance, the Activity, and the Observer. Well-defined service interfaces permit interaction with these resources and their properties. A *property* represents an attribute that captures relevant data of a resource. A *method* is defined by two correlated SOAP messages—a request from the client and a response from the server. The request message includes *context data*. This is XML data that represents the parameters of a method. The response message returns *result data* to the client, that is, XML data created by the successful completion of a method. Beyond these concepts, Wf-XML defines fault messages and a list of error codes. Figure 12.12 gives an overview of Wf-XML services and methods.

12.5.2.1 *Service Registry.* The Service Registry serves as a directory of available processes and allows an administrator to add and remove process definitions to and from the server. It is an extension of Wf-XML and not included in ASAP. Each process definition listed in the registry is represented by a distinct process factory. The Service Registry has the properties *Key,* referring to the URI of the registry, *Name* to carry the name of the registry, a *Description* that is human readable, *Version* of the registry, and *Status*. It offers six access methods. Four of them are dedicated to the search, discovery, and creation of process definitions: *ListDefinition, GetDefinition, SetDefinition,* and *NewDefinition.* Furthermore, the methods *GetProperties* and *SetProperties* allow administrators to query and manipulate the properties of a process definition.

Consider, for example, the *GetDefinition* method illustrated in Figure 12.13. It is initiated by a client who sends a SOAP message to the Service Registry containing a `GetDefinition.Request` element in the message body. In addition, a `ProcessLanguage` can be defined. All Wf-XML-compliant implementations have to support XPDL as a process definition language, and additional process definition languages may be supported in the future. The Service Registry will then return a SOAP message containing a `GetDefinition.Response` element in the message body. This way, the client receives the Wf-XML profile of the requested process

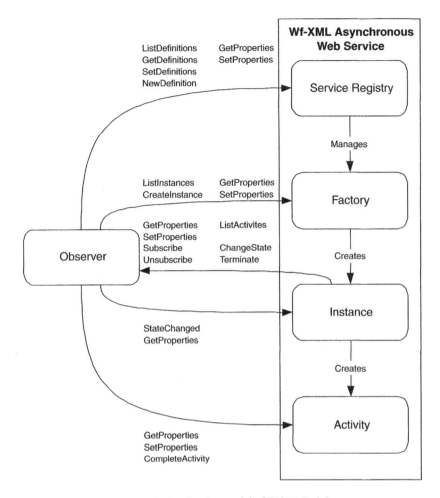

Figure 12.12   Service model of Wf-XML 2.0.

definition (i.e., the address of the process factory and, possibly, context data that is required to create a process instance) and the XML representation of the requested process definition.

12.5.2.2 Factory. The Factory is a dedicated handler for one specific process definition and is an ASAP-compliant service. The Factory provides methods to create new process instances and to search for existing instances. Searching for process instances is an important service because, if the URI of a process instance is lost due to a communication error, the client can query the factory to retrieve the missing address. *Key* identifies the resource via a URI. *Name* provides a human-readable identifier. *Subject* briefly describes the purpose of the process. *Description* offers a long description of the process represented by the Factory. *ContextDataSchema* and

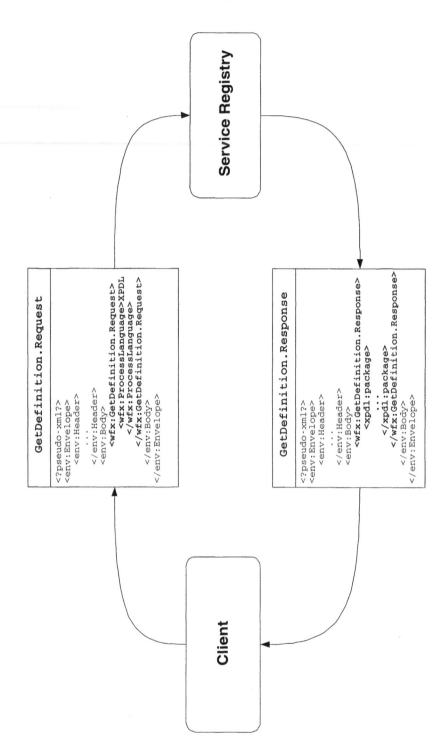

Figure 12.13 GetDefinition Method and related SOAP messages.

ResultDataSchema are XML Schema representations of the data required to instantiate a process and the data that will be returned by the process instance after completion. Finally, *Expiration* defines a minimum period of time the service will be available. There are four methods offered by Factories: *GetProperties* to retrieve process properties, *CreateInstance* to create a process instance using the context data provided in the method request, and *ListInstances* to retrieve all process instances created by this factory. In addition, *SetProperties* can be used to manipulate the attributes of the Factory.

12.5.2.3 Instance. The Instance is a Web service that represents a process instance that performs the requested work. It is defined by ASAP and included in Wf-XML. In the context of Wf-XML, Instances are also called process instances. The methods of this interface permit the monitoring and control of a process instance. Each Instance has the following properties. *Key* identifies the resource via a URI. *State* contains the current state of the process instance (e.g., `open.running` or `closed.completed`). *Name* provides a human readable identifier. *Subject* offers a short description of the process instance, whereas *Description* contains a long description. Furthermore, a list of *ValidStates* is provided, so a client can determine which states a process instance can transition to. For instance, if the list of *ValidStates* contains `closed.cancelled`, a client could stop the instance by sending a state change request. The *FactoryKey* provides a reference to the URI of the instance's factory. *Observer* contains URIs that have been registered as observers. *ContextData* and *ResultData* define data that an instance expects for initialization, which it returns after completion using XML Schemas. Furthermore, a *Priority* can be specified. Modifications and history are stored in *LastModified* and *History,* respectively. Moreover, Instance offers the ASAP methods *GetProperties, SetProperties, Terminate, Subscribe,* and *Unsubscribe,* and a Wf-XML-specific *ListActivities.*

12.5.2.4 Activity. The Activity resource represents a single step in a complex process instance. This resource is a Wf-XML extension of ASAP. It represents a situation in which a process waits for external input. This may be the case when a manual activity is performed by a human user, or if an external service is invoked and needs to complete before the process can continue. An activity has the properties *Key* holding a URI of the activity, *State* to capture its current status, *Name* to store a human-readable identifier, a text *Description, ValidStates* to hold a list of permitted state transitions, *InstanceKey* to refer to the URI of the process instance, *RemoteInstance* to store the URI of the service the activity is waiting for, the *StartedDate* of the activity, a *DueDate* when it is expected to complete, and *LastModified* to represent the date of last modification. Beyond that, it offers the methods *GetProperties* and *SetProperties* for management of properties, and a *CompleteActivity* to be invoked by the performer to signal the completion of the activity.

12.5.2.5 Observer. The Observer is a resource that can receive events about state changes of an activity or a process instance. Wf-XML uses the Observer as defined by ASAP. It has the property *Key* to capture the URI that identifies it. This

URI may be different from the URI of the client that has requested the process instance. The Observer offers the methods *GetProperties*, similar to that of the other services, and *Completed* and *StateChanged* to notify the Observer that the requested process instance has completed or has changed state, respectively.

12.5.3 Wf-XML in Practice

Wf-XML represents a lightweight protocol for the discovery and invocation of processes that are provided by a remote workflow engine. By standardizing the operations that can be invoked on a remote process, it is suitable for the interaction with virtually any processes that an organization might wish to expose, because clients can discover the interaction options available to them as they go along. In contrast, invoking a BPEL4WS process requires the designer of the client application to know exactly the operations that are to be performed on the remote process. This difference is best illustrated by an example. In an insurance claim scenario, a customer might want to create an instance of a damage claim process, schedule a meeting with an appraiser, and be notified about the outcome of the assessment. In a typical Web services implementation, the insurance would expose interfaces that reflect the insurance-specific operations that can be invoked, for example, *createDamageClaim, scheduleAppraiser,* and *queryDamageClaimStatus.* These interfaces have fixed locations for all instances of the damage claim process, and the reference to any particular instance will be given in the body of a SOAP message that is sent to either of these interfaces. In Wf-XML, the damage claim process is represented by a Factory, which in turn would be stored in the Service Directory of the insurer. A client would query the Service Directory for the definition of the *DamageClaim* factory. Using the context data definition received and the URI of the factory, the client then requests a new instance of the *DamageClaim* process, of which the Factory returns the URI to the client. By querying the instance, the client can monitor the status of the damage claim and is notified when it is time to schedule the appraisal (including the URI of the activity that waits for the user input).

Both approaches have advantages and disadvantages. Although Wf-XML requires minimum a priori knowledge about the remote process to be invoked, it relies on more complex document types for data exchange between client and server, and the interaction is difficult to test and debug. Links might get lost at run time and need to be retrieved through the factory. Approaches such as BPEL4WS rely on a tight coupling of client and server, which is easier to test and debug. However, clients need to know exactly which operations a specific process supports, and what the semantics of these operations are. Each approach has its respective place, and it would not be surprising if ASAP/Wf-XML and BPEL4WS coexisted peacefully in the future.

12.6 TRENDS

Technology standardization moves from more elementary technology layers, such as networking protocols and data-encoding formats, to higher-level areas such as messaging patterns and process descriptions. Over the last 10 years, we can observe that standardization has moved up from the lower layers of the technology stack (see Figure 12.14).

In 1994, it was common to have multiple alternatives for the encoding, packaging, transportation, and presentation of data, but today these differences have been widely eliminated through the proliferation of XML, HTTP, and HTML. Current standardization efforts deal with the representation of functional invocation (WSDL), the publication of services (UDDI), and essential service properties such as security and reliability. Although there is consensus in many of these areas, the standardization of process modeling languages, interoperability patterns, and management interfaces is less clear. Half a dozen standardization groups compete within this area for the best solution, and although battles about technical merits and economic benefits are being waged, the dominant winner in the marketplace has yet to be determined.

One of the problematic aspects of PAIS standardization is the lack of agreement on how PAIS standards fit into a larger architectural schema, such as the WfMC reference model. In the course of the Web services movement, individual

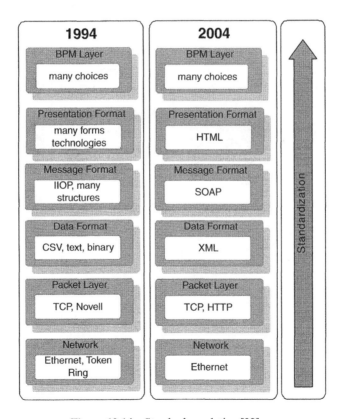

Figure 12.14   Standards evolution [22].

functions that were covered in the WAPI specification have been separated and replaced by more specialized specifications. Figure 12.15 shows a classification of current standardization efforts according to their system coverage (compare [31]).

PAIS-related standards can be grouped into standards that describe the inner workings of a PAIS development tool and standards that describe the externally observable behavior. The former standards describe process modeling capabilities and

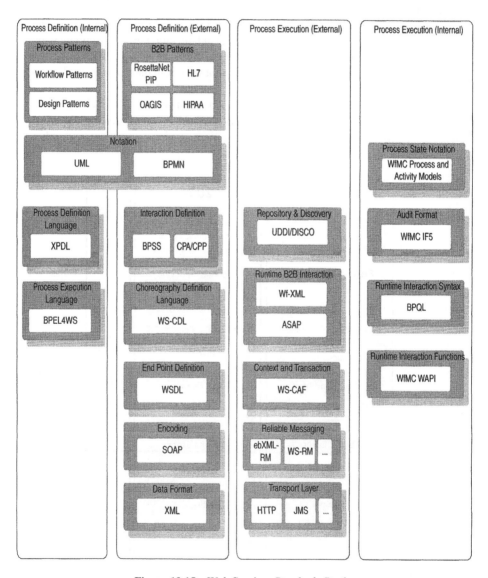

Figure 12.15   Web Services Standards Stack.

patterns; the latter messaging formats and communication protocols. The two columns on the left in Figure 12.15 cover the specification of processes, both the process parts internal to a PAIS, and the publicly visible parts of processes. The two columns on the right contain standards that affect the run-time behavior of PAISs. From the bottom up, the standards start with the most elementary building blocks of modern information system architectures, such as XML for the formatting of data, SOAP for the encoding of messages, and WSDL for the definition of service end-points. In the top-left corner are patterns that can serve as high-level control flow templates for the construction of processes. For the internal specification of processes, the MIT process handbook [12] and the workflow patterns described in Chapter 8 serve as high-level templates. For the specification of interorganizational processes, domain-specific standards such as RosettaNet provide process fragments.

Besides the standards defined by software vendors, a number of standardization initiatives exist that provide process choreography as part of their overall standards framework, but are not necessarily focused on choreography. Examples for these initiatives are ebXML, ACORDE, OAGIS, HIPAA, RosettaNet, SWIFT, and HL7. These groups are typically led by users of the standards, joined in industry coalitions. The integration of the process content of these domain-specific standards with the generic process description standards developed by the likes of OASIS, W3C, and WfMC remains a field for further study.

12.7 EXERCISES

Exercise 1. Workflow Glossary
Look at the WfMC Glossary document [27] that is available on the Web site of the Workflow Management Coalition (http://www.wfmc.org). Give the synonyms mentioned in the glossary for the terms included in Figure 12.1 and give examples of your own to illustrate these terms.

Exercise 2. WfMC Reference Model
Describe the five interfaces of the WfMC Reference Model [8]. Consider which interfaces are relevant for communication *from* outside systems, and which interfaces are relevant for the communication *to* outside systems. For system-to-system communication, when would you use Interface 3, and when Interface 4?

Exercise 3. XPDL—Download Software
For the following exercises you will require an XPDL editor and an XPDL workflow engine. We recommend the following:

- XPDL Editor: JaWE (Java Workflow Editor), download from http://jawe.en-hydra.org/
- XPDL Engine: OBE (Open Business Engine), download from http://source-forge.net/projects/obe

Note that OBE can run stand-alone or under a J2EE application server such as JBoss. The following exercises can be run in stand-alone mode, apart from those that require access to Web services.

Exercise 4. XPDL—New Employee Exercise

Use Case. Reconsider the new employee use case from Section 12.4.2.2. The XYZ organization requires a workflow process for hiring a new employee. The workflow's invocation signature should consist of a single XML document containing the employee details (see below for the schema). When a new employee is due to join the company, the following actions must be taken by staff in the appropriate departments:

- HR Department:

  Prepare office accommodation, ordering items from a checklist where necessary.

  Issue preinduction information pack containing company policy documents.

  Book a place on the next scheduled induction course.

- IT Department:

  Procure and configure a computer for the employee.

  Allocate network user ID and temporary password, `createNetworkUser` `(IN STRING userName, IN STRING department, OUT STRING userID)`

  Set up an email account, `createEMailAccount(IN STRING user-` `Name, IN STRING userID, IN STRING department)`

  Assign an extension number and enter the employee into the corporate telephone directory.

  Grant access rights to the appropriate corporate resources.

- Finance Department:

  Enter personnel details into salary system.

Interfaces

- Input is a string containing an XML document that conforms to the following schema:

```
<?xml version="1.0" encoding="UTF-8"?>
<xsd:schema
    targetNamespace="http://www.xyz.com/2004/Personnel/NewHire"
    xmlns:xsd="http://www.w3.org/2001/XMLSchema"
    xmlns:xyz="http://www.xyz.com/2004/Personnel/NewHire"
    elementFormDefault="qualified"
    attributeFormDefault="unqualified">
    <xsd:element name="new-employee">
        <xsd:complexType>
```

```
            <xsd:attribute name="name" type="xsd:string"
            use="required"/>
            <xsd:attribute name="department" type="xsd:string"
            use="required"/>
                <xsd:simpleType>
                    <xsd:restriction base="xsd:string">
                        <xsd:enumeration value="ENGINEERING"/>
                        <xsd:enumeration value="FINANCE"/>
                        <xsd:enumeration value="HR"/>
                        <xsd:enumeration value="IT"/>
                        <xsd:enumeration value="MARKETING"/>
                        <xsd:enumeration value="SALES"/>
                        <xsd:enumeration value="SUPPORT"/>
                    </xsd:restriction>
                </xsd:simpleType>
            </xsd:attribute>
        </xsd:complexType>
    </xsd:element>
</xsd:schema>
```

- The following procedures will be needed:

  ```
  createNetworkUser(IN STRING userName, IN STRING depart-
  ment, OUT STRING userID)
  createEMailAccount(IN STRING userName, IN STRING userID,
  IN STRING department)
  ```

Hints

- Perform these activities as concurrently as possible, given the logical dependencies between the various tasks (for example, the IT department cannot finish configuring the computer until the network user ID and e-mail addresses
  have been assigned).
- To hold the workflow's XML input document, define a FormalParameter
  containing an ExternalReference with an href of "new-hire.xsd."
- Use Xpath[10] to extract required values from the input document.

Exercise 5. XPDL—Document Approval Chain Exercise
Use Case. The ABC Corporation regularly receives RFQs (requests for quotation)
for custom work from prospective clients. A salesperson is assigned to analyze the
client's requirements and prepare a quotation document. Before the tender document is submitted, a chain of approvals at various levels within the company is required. The following groups or persons are required to approve the document, in
this order:

---

[10]http://www.w3.org/TR/1999/REC-xpath-19991116

- Peer reviewers
- The Vice President of Sales
- The Director of Communications
- The Chief Executive Officer

Each approver can either reject or approve the document, citing a reason. They each have three working days in which to approve the document. If they have not done so by then, the workflow system automatically routes the document to the next approver. If the document is rejected, the workflow system reroutes it to the previous person in the chain.

Whenever such a proposal document is created, the electronic document management system automatically initiates a workflow process, passing on the HTTP URL of the new document and the user ID of the creator. The participants of the document approval process use the corporate standard office software to view and edit the document via WEBDAV.[11] The workflow management system is configured with a tool agent to launch the office software.

Interfaces. The tool agent's invocation interface is: `start(IN STRING docUri)`.

Hints

- Model the peer review stage by assigning the activity to a group of individuals named "Sales." This will create a work item (an assigned activity instance) for each member of the group. Assume that the peer review activity will automatically complete once all the work items are complete (or the activity times out or is manually marked complete).
- The tool agent relies upon the operating system's built-in shell file associations to launch the appropriate application for the document type in question.
- Use a synchronous deadline to implement the three-day time-out periods.
- Check out OBE's pluggable work item assignment strategy, activity completion strategy, and business calendar facilities.

Question: What happens if the time-out period spans a weekend or holiday?

Exercise 6. Wf-XML—New Employee Exercise
Look again at the XML Schema of Exercise 12.7.4. Download an XML editor of your choice and write a SOAP message that is sent to a workflow factory to instantiate a new process. What kind of method has to be used and what kind of message? How is the XML Schema of Exercise 12.7.4 related to the message and to the properties of the workflow factory? Look at the specification of Wf-XML [20] to make sure you use the correct XML name spaces.

---

[11]WebDAV: Web-based Distributed Authoring and Versioning, an HTTP extension protocol supported by word processors.

REFERENCES

1. W. M. P. van der Aalst. *Patterns and XPDL: A Critical Evaluation of the XML Process Definition Language.* QUT Technical report, FIT-TR-2003-06, Queensland University of Technology, Brisbane, 2003.

2. W. M. P. van der Aalst, A. H. M. ter Hofstede, B. Kiepuszewski, and A. P. Barros. Workflow Patterns, *Distributed and Parallel Databases, 14*(1):5–51, 2003.

3. T. Andrews, F. Curbera, H. Dholakia, Y. Goland, J. Klein, F. Leymann, K. Liu, D. Roller, D. Smith, S. Thatte, I. Trickovic, and S. Weerawarana. *Business Process Execution Language for Web Services, Version 1.1.* OASIS, Needham, MA, 2003.

4. A. Arkin. *Business Process Modeling Language (BPML)* Business Process Management Initiative, Alameda, CA, 2002.

5. A. Arkin, S. Askary, S. Fordin, W. Jekeli, K. Kawaguchi, D. Orchard, S. Pogliani, K. Riemer, S. Struble, P. Takacsi-Nagy, I. Trickovic, and S. Zimek. *Web Services Choreography Interface (WSCI) 1.0.* W3C Note 2002-08-08, 2002.

6. J. Dehnert. Making EPCs fit for Workflow Management. In M. Nüttgens and F.J. Rump (Eds.). *Proceedings of the 1st GI Workshop EPK 2002,* pp. 51–69, Gesellschaft für Informatik, Bonn, 2002.

7. R. T. Fielding. *Architectural Styles and the Design of Network-based Software Architectures.* Doctoral Dissertation. Department of Computer Science, University of California, Irvine, CA, 2000.

8. D. Hollingsworth. *The Workflow Reference Model,* Document Number TC00-1003, Document Status—Issue 1.1, 19-Jan-95, Workflow Management Coalition, 1995.

9. D. Hollingsworth. The Workflow Reference Model: 10 Years On. In L. Fischer (Ed.), *Workflow Handbook 2004,* Chapter 20, pp. 295–312, Workflow Management Coalition, 2004.

10. E. Kindler. On the semantics of EPCs: A framework for resolving the vicious circle. In J. Desel, B. Pernici, and M. Weske (Eds.), *Business Process Management: Second International Conference, BPM 2004,* Volume 3080 of *LNCS,* pp. 82–97, Springer-Verlag, Berlin, 2004.

11. F. Leymann and D. Roller. *Production Workflow—Concepts and Techniques.* Prentice-Hall, Upper Saddle River, NJ, 2000.

12. T.W. Malone, W. Crowston, J. Lee, and B. Pentland," Tools for Inventing Organizations: Toward a Handbook for Organizational Processes, *Management Science, 45*(3):425–443, 1999.

13. J. Mendling, G. Neumann, and M.Nüttgens. A Comparison of XML Interchange Formats for Business Process Modelling. In F. Feltz, A. Oberweis, B. Otjacques (Eds.). *Proceedings of the GI Workshop EMISA 2004,* Volume 56 of *LNI,* pp. 129–140, Gesellschaft für Informatik, Bonn, 2004.

14. J. Mendling and M. Nüttgens. Exchanging EPC Business Process Models with EPML. In M. Nüttgens and J. Mendling (Eds.). *Proceedings of the 1st GI Workshop XML4BPM,* pp. 61–79, Gesellschaft für Informatik, Bonn, 2004.

15. Object Management Group. *Workflow Management Facility Request for Proposals.* RFP Document Number cf/97-05-03, OMG, Framingham, MA, May 7th 1997.

16. Object Management Group. *Commercial Considerations in OMG Technology adoption.* Document Number omg/98-03-01, OMG, Framingham, MA, 1998.

17. Object Management Group. *Workflow Management Facility Specification Version 1.2.* Document Number bom/00-05-02, OMG, Framingham, MA, 2000.

18. K. B. Sall. *XML Family of Specifications: A Practical Guide,* Addison-Wesley, Reading, MA, 2002.

19. W. Schulze, M. Böhm, and K. Meyer-Wegener. Services of Workflow Objects and Workflow Meta-Objects in OMG-compliant Environments. In *Proceedings of the 1996 OOPSLA Workshop on Business Object Design and Implementation,* San Jose, CA, 1996. http://jeffsutherland.com/oopsla96/schulze.html.

20. K. D. Swenson, M. D. Gilger, and S. Predhan. *Wf-XML 2.0 XML Based Protocol for Run-Time Integration of Process Engines,* Draft 2003.10.03,Workflow Management Coalition, 2004.

21. K. D. Swenson and M. Silverstein. *Asynchronous Service Access Protocol (ASAP),* Working draft 01, 09 September 2003, OASIS, 2003.

22. K. D. Swenson. Process Management Standards Overview. Presented at the AIIM Conference, New York, April 2004.

23. UN/CEFACT and OASIS. *ebXML Business Process Specification Schema. Version 1.01.* 11 May 2001.

24. M. Weber and E. Kindler. The Petri Net Markup Language. In H. Ehrig, W. Reisig, G. Rozenberg, and H. Weber (Eds.), *Petri Net Technology for Communication-Based Systems—Advances in Petri Nets,* Volume 2472 of *LNCS,* pp. 124–144, Springer-Verlag, Berlin, 2003.

25. S. White. *Business Process Modeling Notation (BPMN) Version 1.0.* May 3, 2004, BPMI.org, San Mateo, CA, 2004.

26. Workflow Management Coalition. *Interface 1: Process Definition Interchange Process Model.* Document Number WfMC TC-1016-P, Document Status—Version 1.1 (Official release). Issued October 29, 1999, Workflow Management Coalition, 1999.

27. Workflow Management Coalition. *Terminology & Glossary,* Document Number WfMC-TC-1011, Document Status—Issue 3.0, Feb 99, Workflow Management Coalition, 1999.

28. Workflow Management Coalition. *Workflow Management Coalition Workflow Standard—Interoperability Wf-XML Binding.* Document Number WfMC-TC-1023, 14-November-2001, Version 1.1,Workflow Management Coalition, 2001.

29. Workflow Management Coalition. *Workflow Process Definition Interface—XML Process Definition Language.* Document NumberWfMC-TC-1025, October 25, 2002, Version 1.0,Workflow Management Coalition, 2002.

30. Workflow Management Coalition. *Audit Data Specification. Version 2.* Document Number WFMC-TC-1015. Workflow Management Coalition, Winchester 1999.

31. M. zur Muehlen. *Workflow-based Process Controlling. Foundation, Design, and Application of Workflow-driven Process Information Systems.* Logos, Berlin 2004.

32. M. zur Muehlen, J. V. Nickerson, K. D. Swenson. Developing Web Services Choreography Standards—The Case of REST vs. SOAP. *Decision Support Systems, 40*(1):9–29, 2005.

CHAPTER 13

The Business Process Execution Language for Web Services

RANIA KHALAF, NIRMAL MUKHI, FRANCISCO CURBERA,
and SANJIVA WEERAWARANA

13.1 INTRODUCTION TO WEB SERVICES

The Web services paradigm [2, 5, 7] has emerged as a response to the shift in the IT landscape away from isolated, tightly coupled, controlled systems to a highly distributed, heterogeneous environment. Web services are based on an extensible, modular set of open XML specifications, enabling different providers to offer software applications as services deployed on the Internet and intranets so that they may be described, accessed, and composed in a loosely coupled manner.

Three Web services specifications initially set the stage: SOAP, WSDL, and UDDI. SOAP provides a standardized XML messaging protocol for interacting with services in a standardized manner, regardless of their internal implementations; WSDL provides a standardized way to describe the functionality provided by a Web service and how and where it may be accessed; and UDDI provides a global registry and defines how to publish, categorize and search for Web services. From there, the Web services stack has expanded to include specifications for defining quality of service policies, addressing schemes, and composition capabilities.

The composition of Web services may be carried out using BPEL4WS, the Business Process Execution Language for Web Services [3], or BPEL for short. In this chapter, we will illustrate the main concepts of the BPEL language and the usage of its constructs to define business processes.

To aid in this task, we will use the example of a business process that models pension disbursement and delivery in a corporation. Each different section will present outtakes from this process to illustrate the particular concept(s) it is addressing. At a high level, this business process is controlled by the human resources department of the company. The process calculates pension payments, manages the funds to be transferred, and arranges for delivery of the correct amount to the eventual recipient. In order to do so, the process also interacts with a bank, an employee database, and a courier service. Our aim here is not to describe a complete working business process, but to use this scenario as a backdrop against which we can develop meaningful examples. The reader is therefore discouraged from assembling the different BPEL code snippets or diagrams into a continuous picture.

13.1.1   Background of WSDL

BPEL's composition model makes extensive use of the Web Services Description Language, WSDL. It is therefore necessary to provide an overview of WSDL before going into the details of BPEL itself. A WSDL description consists of two parts: an abstract part defining the offered functionality, and a concrete part defining how and where this functionality may be accessed. By separating the abstract from the concrete, WSDL enables an abstract component to be implemented by multiple code artifacts and deployed using different communication protocols and programming models.

The abstract part of a WSDL definition consists of one or more interfaces, called *portTypes* in WSDL. PortTypes specify the operations provided by the service, and their input and/or output message structures. Each message consists of a set of parts. The types of these parts are usually defined using XML Schema [8].

The concrete part of a WSDL definition consists of three parts. It binds the portType to available transport protocol and data encoding formats in a set of one or more *bindings*. It provides the location of endpoints that offer the functionality specified in a portType over an available binding in one or more *ports*. Finally, it provides a collections of ports as *services*.

13.2   BPEL4WS

BPEL4WS is a workflow-based composition language geared toward service-oriented computing and layered as part of the Web services technology stack. BPEL composes services by defining control semantics around a set of interactions with the services being composed. The composition is recursive; a BPEL process itself is naturally exposed as a Web service, with incoming messages and their optional replies mapped to calls to WSDL operations offered by the process. Offering processes as services enables interworkflow interaction, higher levels of reuse, and additional scalability.

Processes in BPEL are defined using only the abstract definitions of the composed services, that is, the abstract part (portType/operations/messages) of their WSDL definitions. The binding to actual physical endpoints and the mapping of data to the representation required by these endpoints is intentionally left out of the process definition, allowing the choice to be made at deployment time, design time, or during execution. Added to the use of open XML specifications and standards, this enables two main goals: flexibility of integration and portability of processes.

13.2.1 Abstract and Executable Processes

The language is designed to specify both business protocols and executable process-es. A business protocol, called an "abstract process" in BPEL, specifies the flow of interactions that a service may have with other services. For example, one may accompany a WSDL description with an abstract BPEL process to inform parties using it in what order and in what situations the operations in the WSDL should be called (e.g., a call to a "request for quote" operation must precede a call to a "place order" operation). An "executable process" is similar to an abstract process, except that it has a slightly expanded BPEL vocabulary and includes information that enables the process to be interpreted, such as fully specifying the handling of data values, and including interactions with private services that one does not want to expose in the business protocol. For example, when an order is placed, the executable BPEL process might have to invoke a number of internal applications wrapped as services (e.g., applications related to invoicing, customer relationship management, stock control, and logistics), but these calls should not be visible to the customer and would be omitted from the abstract process the customer sees. In the executable variant, the process can be seen as the implementation of a Web service.

At the time of this writing, most work in BPEL has focused on the executable variant of the language. Additionally, compliance testing between an executable and a corresponding abstract process is still in its early stages.

Referring back to our example, one can see how an abstract process may be used. Assume that the banking industry has an abstract process that standardizes payment interactions with banks as a BPEL abstract process. Then, the process for pension disbursement could use that to model its interactions with the bank according to the banking protocol. Any bank complying with the standardized protocol may then be used with the pension process. On the other hand, the pension process may expose a BPEL abstract process to the courier that consists only of the part of the process flow that the courier is involved with.

13.2.2 The BPEL Process Model

BPEL has its roots in both graph- and calculus-based process models, giving designers the flexibility to use either or both graph primitives (nodes and links) and complex control constructs creating implicit control flow. The two process modeling approaches are integrated through BPEL's exception handling mechanism, detailed in [6] and presented in the subsection on fault handling below.

The composition of services results from the use of predefined interaction activities that can invoke operations on these services and handle invocations to operations exposed by the process itself. The unit of composition in BPEL is the activity. Activities are combined through nesting in complex activities with control semantics, and/or through the use of conditional links. In contrast to traditional workflow systems in which data flow is explicitly defined using data links, BPEL gives activities read/write access to shared, scoped variables. In addition to the main forward flow, BPEL contains fault handling and roll-back capabilities, event handling, and lifecycle management.

Partners and PartnerLinks. A process defines a set of partnerLinks that specify the functionality it offers to and needs from any party it is interacting with. A partnerLink specifies which portType must be supported by each of the parties it connects, and which portType it offers to each of those parties. It is an instance of a typed connector, known as a "partnerLinkType," which specifies a set of roles and the portType (interface) provided by each role. To specify that a set of services are really represented by one party, they may be grouped in a "partner" element. Note that partnerLinkTypes are defined in the WSDL of the process, not in the BPEL itself.

For example, in our pension process, the process interacts with four parties: the Human Resources department, the company's employee database, a bank, and a courier service. Each interaction involves one of the parties (the process or the partner), exposing some function via a portType, and the other party making use of that functionality.

Four partnerLinks are created and named HR, EmployeeDB, Bank, and Courier. The HR department acts as a pure client invoking operations on the process's "pensionPortType." Therefore, the HR partnerLink specifies that the process's role is that of the "pensionDisburser," defined on the partnerLinkType as offering the "pensionPortType." The employee database and the bank are modeled as pure services used by the process. The process invokes their "employeeDataAccessPortType" and "bankPortType." Their corresponding partnerLinks specify only the partner's role on each link. Finally, the courier acts as both a client and a service. It can invoke operations on the pension process to inform it whether the delivery was successful. In this case, we use a partnerLink whose partnerLinkType has two roles, as sender and deliverer. The courier partnerLink then specifies that the process's role (myRole) is that of the "sender" and the partner's role (partnerRole) is that of the "deliverer." This setup is illustrated in Figure 13.1, with only the two-sided courier partnerLink specifically highlighted.

The XML code snippets representing the courier partnerLinkType are shown in Figure 13.2.

Activities. Activities in BPEL have predefined behavior and are split into two main categories: structured and primitive. Structured activities impose behavioral and execution constraints on a set of activities contained within them, such as specifying that they must be executed in sequential or parallel order, or in a loop. At the leaves of all structured activities are primitive activities that perform such actions as sending/receiving messages, invoking other services, waiting, and throwing faults. BPEL's process definition model is not strictly hierarchical since it is possible to define conditional control links between activities. Such links may cross the boundaries of most structured activities. In the following, we first discuss BPEL's supported structured and primitive activities and then the capabilities provided by control links.

Structured Activities. BPEL defines six different structured activities: sequence, while, switch, pick, and flow. The use of these activities and their combinations re-

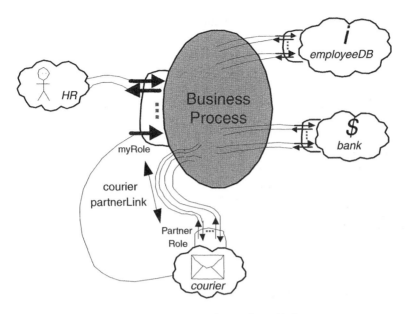

Figure 13.1 A business process interacting with four partners.

sult in enabling BPEL to support most of the workflow patterns described in Chapter 8 and in [14].

The "sequence" activity contains one or more activities that are executed sequentially, and completes when the final activity in the "sequence" has completed. The "flow" specifies that its activities run in parallel, with further control dependen-

```
WSDL snippet:
<plnk:partnerLinkType name="courierPLT">
    <plnk:role name="sender">
        <plnk:portType name="tns:senderPT"/>
    </plnk:role>
    <plnk:role name="deliverer">
        <plnk:portType name="tns:processPT"/>
    </plnk:role>
</plnk:partnerLinkType>

BPEL snippet:
<process ...>
<partnerLinks>
...
<partnerLink name="courier" partnerLinkType="ns1:courierPLT"
            myRole="sender" partnerRole="deliverer"/>
```

Figure 13.2   The courier partnerLink.

cies between them defined using conditional links. It is imperative that the source and target of the link both belong (at arbitrary levels of nesting) inside the flow on which the link is defined. The "while" activity may contain one child activity whose execution is repeated until the specified Boolean condition is no longer true.

The "switch" activity contains an ordered list of one or more conditional branches. Only the activity of the first branch whose condition is true will run. When the selected branch completes, the "switch" activity is complete. The "switch" activity corresponds to the "simple choice" pattern presented in Chapter 8. On the other hand, the "pick" activity corresponds to the "deferred choice" pattern. It is used to express an exclusive choice based on either an incoming message or an alarm. Each branch, therefore, contains either a message handler (onMessage) or an alarm handler (onAlarm) with an activity. As in "switch," only one branch may execute.

The "scope" activity is the unit of recovery and event and exception handling in BPEL, and will be covered in more detail the fault handling subsection.

Primitive Activities. There are three types of interaction activities: invoke, reply, and receive. "OnMessage" is a "receive"-like construct used either in an event handler or the message-waiting part of a "pick" activity, and is treated in much the same manner as a "receive." Interaction activities must specify the partnerLink through which the interaction occurs, the operation involved, the portType in the partnerLink that is being used, and the input and/or output variables. In the case of an invoke activity, the operation and portType that are specified are that of the service being invoked. On the other hand, for "receive" and "reply," the same attributes refer to the operation and portType of the business process that are exposed via that partnerLink.

Specifying which Web service operation to invoke on an "invoke" activity is the BPEL equivalent of assigning an activity implementation to an activity in workflow models such as [11]. An "invoke" blocks while it waits for a response if it is calling a request–response WSDL operation. "Receive" and "reply" are two separate activities that allow message exchanges with external partners. Both specify how they are mapped to the external view of the business process that partners use for interactions with it. The interaction activities also define which variables they will read from or write to. Later, we present BPEL snippets for our scenario that include "invoke" activities for the process to interact with the bank, the employee database, and the courier service. It also includes receive and reply activities to handle requests from the HR department and the courier service.

The relation of the BPEL specification to the WSDL describing the functionality offered by the process comes from the receive and reply activities the process contains. These activities specify which of the process's externally exposed WSDL operations they relate to. A receive/reply pair must map to a request–response operation. In such cases, any control flow between the receive and reply is effectively the implementation of that operation. A receive with no corresponding reply must map to an input-only operation. Multiple receive activities referring to the same operation may be defined at different points in the process. There are three scenarios in which this is desirable: a partner calling the same operation more than once, different partners calling the same operation more than once, and two or more exclusive branches of the process that can handle a call to the operation.

Two questions arise at this point. First, how does one disambiguate which of these receive activities may actually consume the incoming message? Second, if there are multiple reply activities, how does one distinguish which request to reply to? The answers lie in two restrictions. First, BPEL does *not* allow two receive activities to be active (ready to consume messages) at the same time if they have the same partnerLink, portType, operation, and correlation set. Correlation sets are used for routing messages to process instances. If this happens, a built-in fault named "conflictingReceive" must be raised at run time. The second restriction is that BPEL does *not* allow a request to come in to a request–response operation if an active receive is found to consume it but a reply has not yet been sent to a previous request with the same portType, operation, partnerLink, and correlation set. If this occurs, a built-in runtime fault named the "conflictingRequest" fault is thrown. To clarify the difference between the conflictingReceive/Request faults, think of the first as the fault that happens if it is unclear which receive activity to send the input message to. For the second, the process knows which receive to send the message to but knows that it will not be able to match a downstream reply activity to the proper outstanding request.

Aside from the interaction activities, a number of other activities also exist. The "assign" activity is used for the manipulation of process data. It is also used to enable dynamic binding, wherein endpoint information may be copied between variables and the process instance's references to itself and the parties it is interacting with.

The rest of the primitive activities provide additional yet straightforward functionality. These activities enable one to do nothing for a step (<empty/>), wait for a time interval or until a deadline (<wait for/untilExpression?/>), signal a user-defined fault (<throw faultName=? faultVariable?/>), and terminate the process (<terminate/>).

Consider a very basic pension process. HR requests that the pension payment be processed, the employee database is queried for information, the bank is asked to cut a check, and the courier service is asked to deliver it. Once it is delivered, the courier service invokes the process with the delivery confirmation. The BPEL representation of such a process, whose control flow is simply governed by a top-level sequence activity, is represented pictorially in Figure 13.3. Notice the receive/reply mapping to the request–response operation, and that the blocking invokes calling operations on the partners.

The BPEL snippet corresponding to this scenario is given in Figure 13.4, with some sections omitted for clarity and replaced by ellipses. In particular, you may note the reference to "correlations" and "correlation sets." Ignore these for now. Suffice it to say that these are required so the courier's message can be routed back to the correct process instance. The usage of correlation will be discussed in detail further on in this chapter.

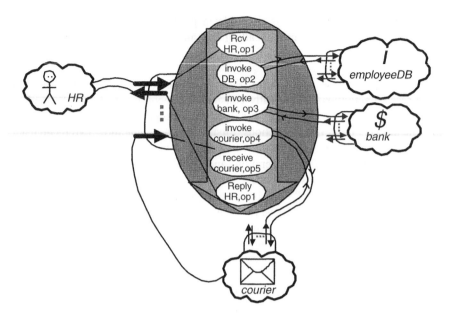

Figure 13.3   A simple pension disbursement and distribution process.

Control Links. Having explained how control is managed in structured activities, we now take a closer look at control links. Control links provide a means to express control flow relations between activities, in addition to the control flow relations captured by structured activities. A control link between two activities denotes a conditional transition.

A BPEL activity requires two things to become active: control from its enclosing structured activity and the satisfactory firing of *all* its incoming links. For example, if the links of the second activity in a sequence have all fired, it cannot run if the first activity in that sequence has not yet completed.

When an activity completes successfully, it activates the control links that originate from it with the value from evaluating each link's *transition condition*. A transition condition is a Boolean expression relating the values of data fields and other states of the process. On the other hand, an activity that is the target of links contains a *join condition,* a Boolean expression in terms of the link values that must be true for the activity to run. The join condition also serves as a synchronization point in the process, since it can only be evaluated once the values of all links coming into the activity are known.

Two link restrictions are important to mention. First, links may only create acyclic graphs. The "while" activity is the only way to create loops. Second, a link may not cross a loop boundary as that would lead to an inconsistent state.

A section of a process more involved than the one in Figure 13.3 is now presented to illustrate the usage of more complex control semantics. Assume that whether the employee's age is past a threshold affects the amount to be paid. Next, assume

```
BPEL snippet:
<process ...>
<partnerLinks> ... </partnerLinks>
<variables>     ... </variables>
<correlationSets> ...  </correlationSets>
<sequence>
   <receive partner="HR" portType="tns:processPT" operation="op1" variable= ... createInstance="yes" />
   <invoke partner="employeeDB" portType= ... operation ="op2" inputVariable=... outputVariable=.../>
   <invoke partner="bank" portType=... operation="op3" inputVariable=... outputVariable=.../>
   <invoke partner="courier" portType=... operation="op4" inputVariable=... outputVariable=...>
      <correlations>...</correlations>
   </invoke>
   <receive partner="courier" portType="tns:processPT" operation="op5" variable=... >
        <correlations>...</correlations>
   </receive>
   <reply partner="HR" portType="tns:processPT" operation="op1" variable =.../>
</sequence>
</process>
```

Figure 13.4   BPEL snippet of simple pension process in Figure 13.3.

that the bank would send a wire transfer if the employee's bank information is available in the database. Only if that is not the case would the process actually ask for a check. In Figure 13.5, this is represented using a "switch" activity.

Again, we include the relevant BPEL snippet in Figure 13.6. Notice how control flow may be defined using either links or structured activities (a "switch"). At the end of a switch or pick, any links leaving unexecuted branches must have their sta-

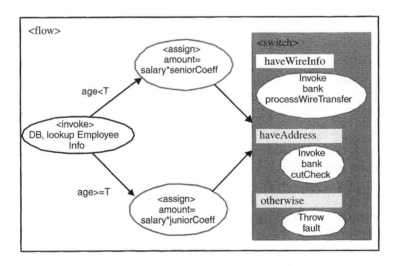

Figure 13.5   Using links and structured activities.

```
BPEL snippet:
<flow suppressJoinFailure = "true">
  <links>
    <link name="senior"/> <link name="junior"/> <link="senior2switch"/> <link="junior2switch"/>
  </links>
  <invoke partner="DB" portType= ... operation ="lookupEmployeeInfo"
                        inputVariable=... outputVariable="employeeInfo">
    <source linkName="senior"
            transitionCondition="bpws:getVariableData(employeeInfo, age)&gt; 65"/>
    <source linkName="junior"
            transitionCondition="bpws:getVariableData(employeeInfo, age)&lt;= 65"/>
  </invoke>
  <assign>
    <copy>
    <from expression="bpws:getVariableData(paymentInfo, amount)*bpws:getVariableData(seniorCoeff)"/
>
    <to variable="paymentInfo" part="amount"/>
    </copy>
    <target linkName="senior"/>
    <source linkName="senior2switch"/>
  </assign>
  <assign>
    <copy>
    <from expression="bpws:getVariableData(paymentInfo, amount)*bpws:getVariableData(juniorCoeff)"/>
    <to variable="paymentInfo" part="amount"/>
    </copy>
    <target linkName="junior"/>
    <source linkName="junior2switch"/>
  </assign>
  <switch>
    <case condition="bpws:getVariableData(employeeInfo, haveWireInfo)">
        <invoke partner="bank" portType= ... operation="processWireTransfer"/>
    </case>
    <case condition="bpws:getVariableData(employeeInfo, haveAddress)">
        <invoke partner="bank" portType= ... operation="cutCheck"/>
    </case>
    <otherwise> <throw ... /> </otherwise>
  </switch>
</flow>
```

Figure 13.6 BPEL snippet of the <flow> in Figure 13.5.

tus set to false so that acitivities downstream that have a control dependency on them do not end up waiting forever. The mechanics of this will become clear later in this chapter, in the section on Dead Path Elimination.

Data—Visibility and Manipulation. Business processes are stateful. The state of the process includes

1. Application data that is consumed by back-end applications driven by the business process.

2. Context data that is of special importance to middleware and may not be visible to applications, such as transaction IDs and session tokens.

A portion of this state is *opaque* to the business process, in the sense that those data values have no relevance for the logic of the business process since they are not examined or manipulated by the process. The bulk of application data falls into this category. Context data that is manipulated by middleware layers separate from the business process (such as security headers) are also opaque. The remainder of the state is used within the process to make data-driven decisions. These data values are visible to and often manipulated by the process logic; they are not relevant to applications. Such data is termed *transparent.* Note that the distinction between opaque and transparent data is a logical one. In practice, a single piece of data, say a Social Security number, may be used by applications and may also be used in the process logic (for correlation for example).

Data Representation. Since BPEL4WS is meant to be used in conjunction with other standards, the concept of a message, prevalent in WSDL, is reused. Thus, a message that flows to or emanates from a BPEL4WS process is in fact defined using WSDL. Interested readers can look at the WSDL specification [4] for a detailed discussion on messages. Briefly, messages consist of a set of named parts, each of which is typed, generally using XML Schema as the type system.

WSDL messages are used in all exchanges between a business process and its constituent back-end services, and also between services provided by business partners and the process itself. Typically, a message is a single representation for both application data and context data. The parts of the message referring to context data are implicit and usually added by middleware logic, in a manner similar to the addition of middleware-specific headers in distributed protocols such as HTTP. These message parts are not meant to be manipulated by applications.

BPEL4WS extends WSDL by allowing the definition of message properties, which are used to represent transparent data.

In Figure 13.7, the HRPensionRqstMsg is shown with three message parts: an employee record, a timestamp, and a delivery acknowledgement. The latter two are context data that are processed by middleware responsible for reliable message delivery. The employee record is application data. This is a complex schema type that consists of a number of elements. Of these, the age and employeeID are both used in our scenario to make data-driven decisions and for correlation, so they are transparent. The other parts of the employee record are opaque.

Message Properties. A message property definition creates a globally unique name and associates it with a schema type. Consider the piece of data representing an employee's age within our scenario. This data might be used within a process specification to make decisions, as we have already seen. An employee's age, in fact, might be carried within many message types. By associating it with a global property name, we give an employee's age, irrespective of what type of message it appears in, the semantic significance necessary for the process logic. All transpar-

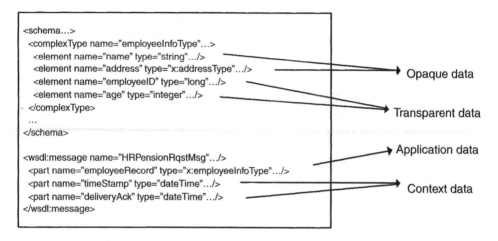

```
<schema...>
 <complexType name="employeeInfoType"...>
  <element name="name" type="string".../>
  <element name="address" type="x:addressType".../>
  <element name="employeeID" type="long".../>
  <element name="age" type="integer".../>
 </complexType>
 ...
</schema>

<wsdl:message name="HRPensionRqstMsg".../>
 <part name="employeeRecord" type="x:employeeInfoType".../>
 <part name="timeStamp" type="dateTime".../>
 <part name="deliveryAck" type="dateTime".../>
</wsdl:message>
```

Opaque data

Transparent data

Application data

Context data

Figure 13.7   Data representation using WSDL messages.

ent data that are manipulated by the process are associated with a property name.
The property name acts as a shorthand for referring to that type of data and, as we
will see, simplifies the syntax for creating expressions for querying and manipulat-
ing the data. Figure 13.8 shows an example declaration of an employeeID property.

Aliasing.   Message properties are merely shorthand for a type. The property values
are instantiated within messages. In order to identify a property value, we need to
map a global property to a portion of a message type. BPEL4WS has defined syntax
for expressing such a mapping, using the notion of aliasing. A message property is
seen as an alias for a message part and location. In Figure 13.8, we have defined a
property alias for the employeeID property. It is mapped to the employeeID element of the employeeRecord part within the HRPensionRqstMsg message. Note
that we can alias the same property more than once, so if the employeeID was carried in another message type, we could use our property as an alias for that data as
well.

Variables.   Variables are the syntactic elements provided by the BPEL4WS language to define containers that can hold data. This data encompasses messages exchanged between the process and its partners as well as intermediate data private to
the process. Variables are typed using WSDL message types, XML schema simple
types, or XML schema elements. Thus the data contained within a variable is a
WSDL message or XML snippet whose type is well defined.

Consider the example below. We define a variable called empData that is typed
using the WSDL message type employeeData, defined above. The *messageType* attribute in the variable definition can be replaced with a *type* attribute or *element* attribute to refer to an XML schema simple type or XML schema element, respectively. Note that if a variable is meant to hold data that is typed using a user-defined

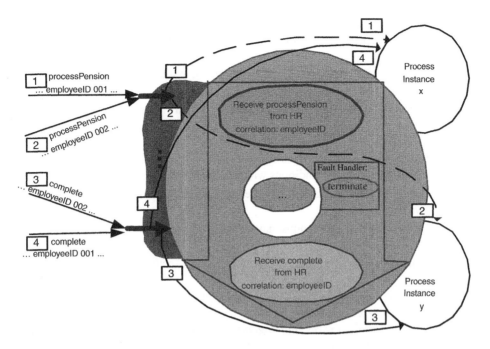

Figure 13.8   Correlating on the employee ID number.

(complex) XML schema type, it is first necessary to define a schema element of that type, and then define a variable in terms of that element.

Variables are defined within a scope, and are said to belong to that scope. Variables belonging to the global process scope are global variables; other variables are local variables. BPEL4WS follows lexical scoping rules; thus, a variable defined in some scope is visible within that scope and all scopes that are nested within it. Additionally, it is possible for a local variable to hide a variable that is defined in an outer scope if the same name is used.

Assignment.   Each "assign" activity consists of a series of separate assignments. Each assignment copies data from one location to another. The source of the copy (specified using the *from* attribute) and its target (specified using the *to* attribute) must be type compatible. The BPEL4WS specification provides a complete set of possible assignment types; here we describe three common uses of assignment:

1. Data construction. Data is assigned from an XPath expression or from a property to part of a variable. Variables and their parts are identified by name. A property value is looked up by examining a variable; it is assumed that the property is aliased to some part of that variable.
2. Copying data. Data is assigned from part of a variable to a part of another variable.

3. Copying references. A partner reference is copied from the reference associated with a particular role of a partner link to another partner link.

Figure 13.6 shows an example of the first type of assignment.

Expressions. BPEL4WS uses expressions of four types to define business processes:

1. Boolean-valued expressions may be used to express transition conditions, join conditions, and conditions within "while" and "switch" activities.
2. Deadline-value expressions may be used within onAlarm event handlers and "wait" activities.
3. Duration-valued expressions may be used within onAlarm event handlers and "wait" activities.
4. General expressions are used within the "assign" activity.

BPEL4WS provides an extensible mechanism for specifying the language that is used to specify expressions. The language used within a particular process definition is declared in the *expressionLanguage* attribute of the *process* element. The language used must have facilities to query data from variables, extract property values from variables, and query the status of links. XPath 1.0 [15] is the default language used. Figure 13.6 shows examples of Boolean-valued expressions (used in a "switch" activity and transition conditions in this case) and general expressions (used in "assign" activities). Figure 13.10 shows a duration-valued expression used within an onAlarm activity.

Process Life cycle. Life-cycle management of a BPEL4WS process consists of two aspects. The first is the creation and destruction of processes, and the second is the routing of messages to the correct running instance of a process.

Recall from Chapter 12, that in Wf-XML, process execution lifecycle management is supported through dedicated operations (e.g., create process instance, change state of running process instance, etc.). In contrast, BPEL handles process life-cycle management and, in particular, process creation, in an explicit manner. A new instance may be created by the activities that receive messages and that are tagged with the ability to create a new instance (creatInstance="yes"). These activities need to come at or near the beginning of the process model such that the process instance can navigate to them once it is created. Upon receiving a message, an existing instance that can consume it is searched for. If none is found and the message can be consumed by such a tagged activity, a new instance of the process is created. Multiple instances of any deployed processes may be running at any time.

Once created, the instance runs based on the prescribed flow of control, stopping at times to wait for incoming messages. Finally, the instance ends when its top-level activity completes successfully. It may also end if the process throws a fault it cannot handle, or if it is explicitly terminated through the "terminate" activity.

The second aspect of of process lifecycle management in BPEL, namely, routing messages to the correct instance, is handled through the built-in "correlation" mechanism, basically using specially marked fields in incoming messages to route messages to existing instances of a business process. These fields are often part of the main application data, such as a person's Social Security number, a purchase order ID, a last name, or a combination of the above. Contrast this to the use of opaque tokens generated and managed by the middleware. "Correlation sets" are defined by naming specific combinations of certain fields in a process's messages. That set may then be referenced from and initialized by receive, reply, invoke, or pick activities.

In order to distinguish between instances to route an incoming message to, the values of its fields denoted by relevant correlation sets are matched against values initialized earlier in each instance's life cycle. The correlation sets are defined using sets of properties and aliased to parts of different messages. If a match is found, the message is routed to that instance. Note that once an instance is created and the creating message reaches its designated receive activity, all other receive activities in that instance that have createInstance set to "yes" lose their creation ability. If an existing instance cannot be found, the process definition is checked for the ability to create one based on the message and activities with createInstance set to "true."

In our example, the courier has to send back a confirmation of having successfully delivered the payment for a particular employee. Assume we run one instance of the process for each employee. In its interactions with the HR department, the

```
WSDL snippet:
 <bpws:property name="employeeID" type="xsd:long"/>
 <bpws:propertyAlias propertyName="tns:emplyeeID" messageType="tns:HRPensionRqstMsg"
                part="employeeRecord" query="/employeeRecord/employeeID"/>


BPEL snippet:
 <process ...>
 <partnerLinks> ... </partnerLinks>
 <variables>    ... </variables>
  <correlationSets>
    <correlationSet name="employeeNum" properties="tns:employeeID"/>
  </correlationSets>
 <sequence>
   <receive partner="HR" portType="tns:processPT" operation="processPension" variable="request"
          createInstance="yes">
     <correlationSets> <correlations set="employeeNum" initiate="yes"/></correlationSets>
   </receive>
    .....
   <receive partner="HR" portType="tns:processPT" operation="complete" variable ="completionInfo">
     <correlationSets> <correlations set="employeeNum" /></correlationSets>
   </receive>
 </sequence>
 </process>
```

Figure 13.9   BPEL and WSDL snippets for Figure 13.8.

```
<scope.../>
 <onMessage partner="HR" portType="tns:processPT"
                operation="cancel" variable
="cancellationInfo">
  <correlationSets>
   <correlations set="employeeNum" />
  </correlationSets>
  <sequence>
   <invoke partner="EmployeeDB"
             portType="employeeDataAccessPortType"
             operation="logCancellation"
             variable="cancellationInfo"/>
    <assign>
     <copy>
      <from expression='true'/>
      <to variable='flagValue' part='state'/>
     </copy>
    </assign>
    <invoke partner="EmployeeDB"
              portType="employeeDataAccessPortType"
              operation="flagRecord"
              variable="flagValue"/>
   </sequence>
  </onMessage>

  <onAlarm for="'P40D'">
   <throw fault="DeliveryConfirmationMissing"/>
  </onAlarm>
  ...
</scope>
```

Figure 13.10   Using event handlers.

process specifies that it will use the employee's ID number as the correlation information. The ID number is part of every message sent by the HR department. HR
sends the first message and an instance is created for Bob's pension. The process
saves his ID number as the value of a correlation set for the instance. This way, all
subsequent messages from HR relating to Bob will be routed to the proper process
instance as shown in Figure 13.8. In this figure, we have abstracted most of the
process model. The instantiating "receive" has a thick border, and one can see messages from HR being routed to the correct instances based on the employee ID
number. Instances of this process will end either once the final activity is reached,
or the terminate activity runs due to a fault in the core of the process instance. The
corresponding BPEL and WSDL snippets are shown in Figure 13.9.

On the other hand, the process specifies that it will use the package tracking
number as the correlation set in its interactions with the courier. This is excluded
from the figure for clarity. The courier returns this as part of its answer to the shipping request. Another correlation set for Bob's process instance gets created. Sub-

sequent calls from the courier with that tracking number will now also be routed to Bob's instance, without the courier knowing about Bob or his ID number.

Event Handlers. Business processes have a need to specify logic to deal with events that take place while the process is running. BPEL4WS provides this function through the use of event handlers. Event handlers are associated with scopes, and specify one or more events followed by an activity. The activity is invoked concurrently when the corresponding event occurs. For that reason, links may not cross the boundaries of event handlers. BPEL4WS allows any type of activity, except compensate activities, to handle events. There are two event types allowed in BPEL4WS:

1. onMessage events are triggered by the arrival of a message. The message that triggers the event is identified by the partner link from which the message arrives, the portType, operation, and optional correlation set. The semantics of the onMessage is similar to the "receive" activity. When the event is triggered, the corresponding activity within the onMessage handler is called up. If this message is part of a synchronous (request–response) operation called by a partner, the event handling logic is expected to have a reply activity in order to fulfill the requirements of the operation. If the incoming message is the input for an asynchronous (one-way) operation, a reply is not needed. onMessage event handlers remain active as long as the scope within which they are defined are active. As long as they are active, they can be triggered multiple times, resulting in the concurrent invocation of the handling code each time. The semantics of simultaneous onMessage events from the same partner, port type, operation, and with the same correlation set are undefined. The reader may recall that receive activities have a similar constraint. Unlike receive activities, however, onMessage events cannot create process instances. An event handler is capable of processing events only once an instance has been created; it is not enabled prior to the creation of an instance.

2. onAlarm events have two forms. The first specifies a duration within the "for" attribute, and allows for signaling of timeouts. In this form, a timer is started when the scope that includes the onAlarm event handler is activated. As soon as the specified duration is reached, the corresponding activity in the handler is executed. In the second form, a time is specified as the value of the "until" attribute. As soon as this time is reached, an event is signaled and the handler is executed. Unlike onMessage events, it must be noted that onAlarm events can be signaled at most one time for an active scope.

In the example in Figure 13.10, we have two event handlers associated with a scope. The first handles the arrival of a cancel message. The intent is to cancel the processing of the identified employee's pension disbursement. The event handler invokes Web services to log the cancellation and add a flag to the employee's record. The second event handler is set to signal an event 40 days after a pension check has been mailed. If the enclosing scope is still active after this time period, the resulting state is an error state, since it indicates that the check has not been cashed. The handler raises a fault to indicate this abnormal state.

Fault Handling and Compensation. Fault handling, especially in long-running business processes, enables a process to recover locally from possible anticipated faults that may arise during its execution. Compensation enables a process to undo already completed actions (see the concept of sagas discussed in Chapter 11). BPEL combines the two in a complementary fashion such that one may undo previously completed actions in a scope before executing a fault handling routine. For example, consider a fault caused by an expired credit card in an ordering process. Compensation could be used to take out a discount that was already given because the initial credit card was enrolled in a special member's program. The fault may then be handled by requesting the information for a new credit card, without having to restart the entire process.

Fault Handling. Fault handling has been common practice in business process modeling and in programming languages for many years. For graph-oriented processes, the Opera [9] system introduced a structured programming approach in which blocks and subprocesses are connected with control links. For strictly structured models like XLANG [13], fault handlers could be attached to a specialized construct that is always strictly nested in the process hierarchy.

BPEL introduces structured fault handling for both structured processes and arbitrary (acyclic) graphs such that the process control structure does not have to follow the strict nesting of the exception handling constructs. BPEL fault handlers may be defined on arbitrarily nested scopes and, most importantly, control links may cross the boundaries of these scopes. However, no links may have their target inside a fault handler if the target lives outside the handler. The converse is not true. A link may leave the fault handler's boundary.

There are three kinds of faults in BPEL4WS. The first two are usually user-defined, whereas the last consists of a set of built-in faults defined in the BPEL specification:

1. Application/service faults, generated by services invoked by the process and defined as part of those services' WSDL definitions

2. Process defined faults, explicitly generated by a process using the "throw" activity

3. System faults, consisting of faults that must be thrown by the process engine when it encounters any of a set of error conditions defined in the specification, such as the conflictingReceive and conflictingRequest faults introduced earlier, datatype mismatches, and asynchronous termination

A fault thrown in a scope is either caught by one of the fault handlers defined on that scope or rethrown to the scope's parent scope. Either way, the scope must stop all activities that have not completed within it, and send out negative status on all unevaluated links starting inside the scope but whose targets lie outside of it. It must also trigger compensation on successfully completed nested scopes in order to undo their actions. If a handler was found to catch the fault, the activity in the exception handler will run. Execution will then continue from the boundary of the scope that handled the fault. Clearly, if no scope in the hierarchy could catch it, then the fault would reach the process root and terminate the entire process.

Note that links leaving from the boundary of a scope that successfully handled an exception are evaluated normally and are not automatically set to negative.

Dead Path Elimination. Incorporating structured process modeling semantics into a graph-based model, as in BPEL, has a number of advantages. Structured control artifacts provide a simple first-class expression of some common control patterns that would otherwise result in a set of complex graph constructs. Structured exception handling provides a clean way to deal with errors, all the while maintaining familiar graph-based approaches.

The operational semantics, however, of the two approaches do not fully align. Specifically, the graph-oriented semantics uses the "dead path elimination" technique (DPE) to automatically skip or disable activities when it is known that they are along a path that will never be reached, for example, if an activity upstream has failed. In such cases, DPE propagates the disabled state transitively across the process graph to all activities waiting on it. This ensures process termination by making sure activities with more than one incoming link do not have to wait forever if it is known that one of the paths being joined will no longer be followed. On the other hand, in purely structured processes, the effects of an evaluation can only propagate vertically through the nesting hierarchy.

To integrate the two process models, BPEL had to solve this mismatch by being able to propagate such information vertically (using structured activity nesting) as well as horizontally (using the links). This was done through the combined usage of fault handling that propagates up the scope hierarchy and the definition of a system fault known as a "joinFailure," to be thrown whenever the joinCondition of an activity is false. A Boolean-valued attribute known as "suppressJoinFailure" was made optional on all activities. From a behavioral perspective, it is equivalent to surrounding the activity with a scope containing an empty fault handler for joinFailures. As a result, DPE semantics could be easily reproduced by specifying that all activities in the process should have this attribute set to "true." This may either be done by specifying it on each activity individually or on the process as a whole. Its default value is "false."

Consider how suppressing the join failure allows dead path elimination to occur. One example has already been presented in Figure 13.6, as you can tell from examining the "flow" activity. If you look closely, you will realize that one of the links leaving the left-most "invoke" activity must fire negatively. The target of that activity will throw a joinFailure during execution. If one had not suppressed the join failure, this would have always caused the entire process to fail.

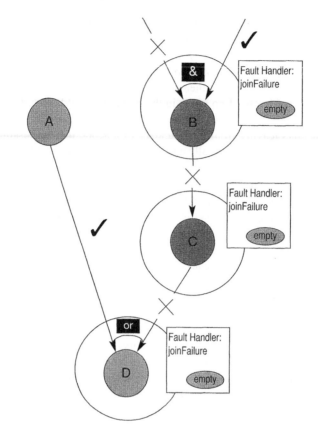

Figure 13.11   DPE semantics shown after expanding the suppressJoinFailure on B, C, and D.

A more interesting example is illustrated in Figure 13.11, in which activity D can continue processing although B has failed. In this figure, activities with suppressJoinFailure set to "true" are illustrated using the "desugared" version, replacing the attribute with a scope with an empty fault handler.

Assume that activity A has completed, sending its link status down to positive/true. Activity B has received the values of its incoming links, so it evaluates the join condition (&) to find that it is false. A "joinFailure" is thrown, disabling B and setting the value of its outgoing link to false. The fault is caught by the handler, activating its "empty" activity, which acts as a no-op. Now, the scope itself completes and the negative link status reaches activity C, which also faults, disables, and sets its link value negative; then the second fault handler runs. Finally, consider activity D. It has to wait until the status of both its links are known before it can react. The link from A is already known to be positive, and after some time the link from C arrives. Its value is negative. Activity D, however, can continue due to its "or" join condition, which is now "true."

Dead-path elimination in BPEL is, therefore, simply a special case of BPEL's fault-handling mechanism. By enabling the use of links in combination with scopes, and allowing these links to cross scope boundaries, the mechanism shown here can be generalized for more advanced behavior that goes beyond "plain vanilla" DPE. For example, putting a link at the boundary of a scope that suppresses a join exception results in anti-DPE behavior; synchronization will occur, but the processing continues normally regardless of the value of the join condition. For a more detailed look into possible uses of the suppressJoinFailure attribute, we refer the reader to [6].

Compensation. Compensation also has a long history in transaction processing and workflow systems (see Chapter 11). The compensation model in BPEL, defined at the scope level, is based on the concept of partial backward recovery introduced by using compensation spheres in [12].

In BPEL, the actions for undoing the work of a completed scope are defined using activities set in a *compensation handler*. For example, the compensating action for a financial debit is a credit. The compensation handler is then attached to the scope. A handler may also be attached to "invoke" activities, resulting in behavior equivalent to wrapping the "invoke" in a scope with that compensation handler attached to it. This shortcut for "invoke" is also true for fault handlers. Once a scope completes, its compensation handler is installed and the data the scope has access to is snapshotted for use when running the handler. The handler remains dormant, activating if compensation is triggered on that scope.

There are two kinds of compensation in BPEL: explicit and default. Explicit compensation is triggered when a "compensate" activity is reached. A "compensate" may only be nested either in fault handlers or other compensation handlers, and it explicitly starts a handler on a named completed scope. On the other hand, default compensation occurs either if a scope with no explicit handler is asked to compensate, or when a fault is thrown. When a fault is thrown to a scope, the scope must (among other things) run the compensation handlers of its direct children, in *reverse order* of completion. If a child has no handler, it performs default compensation itself and the compensation request propagates down the scope hierarchy until it bottoms out at scopes with explicit handlers.

Returning to our example, assume that a fault occurs before the courier has completed shipment, but after the check has been given to the courier to be mailed. So far, the following actions have been completed: employee information look-up from the database, calculation of pension payment, request to bank to cut a check, and sending the check to the courier. Assume the last two are in the same scope. A fault handler on that scope is supposed to try wiring the money instead. Before it can do so, however, it needs to undo the payment and the mailing.

In reverse order, the process first compensates sending the check to the courier by running its handler. The handler asks the courier to return the package. Once the check is returned, handler completes. The next handler to execute will be the compensator of requesting the check from the bank. This handler asks the bank to cancel the check. Finally, the fault handler runs its routine and invokes the bank again, asking for the completion of a wire transfer for the same amount.

13.3 SUMMARY

BPEL's composition model is characterized [15] by its flexible integration model in which services are composed based on their abstract descriptions; recursive composition, in which a process is exposed as (one or more) Web services; the support for both the strictly hierarchical and the conversational patterns of composition; and, finally, its support for life-cycle management using correlation sets. BPEL is also extensible, enabling proprietary or domain-specific extensions to be defined for use in specialized environments.

In this chapter, we have presented the core concepts of the BPEL language after a brief introduction to the Web services paradigm. With roots in prior work on traditional workflow for more tightly coupled systems, BPEL is a composition and workflow language geared specifically to the highly dynamic and heterogeneous networked world of today.

The BPEL4WS language is currently undergoing standardization by the OASIS consortium. The resulting standard will be renamed WS-BPEL.

13.4 EXERCISES

Exercise 1
How are backend applications viewed by BPEL4WS processes? How are BPEL4WS processes viewed by external partners?

Exercise 2
BPEL4WS processes refer to partners only via the abstract function they offer, that is, using portTypes and operations, as opposed to actual deployments of services. How does this impact:
- The software engineering process used to develop an executable business process
- The complexity of developing a software product that supports the BPEL4WS language

Exercise 3
Object-oriented languages such as Java and C++ allow for the specification of nonexecutable program specifications using language constructs such as interfaces, abstract classes, or virtual classes. Compare such object-oriented language features with BPEL4WS's facility to allow the specification of abstract and executable business processes.

Exercise 4
Consider the following fragments of a BPEL4WS process definition:

```
<sequence>
        <receive .../>
```

```
        <invoke .../>
        <reply.../>
</sequence>
```

(a) Rewrite this process definition, replacing the sequence construct with a flow construct containing control links.

(b) Assume the flow construct we defined in (a) has a set of incoming links, and the join condition for the flow activity evaluates to false at runtime. Describe what happens next; specifically, how this affects the activities contained within the flow.

(c) If we had a "while" construct instead of the sequence in the above snippet, could we replace this with a flow as well? If so, demonstrate this; otherwise, explain why not.

Exercise 5

(a) Define the partnerLink between a business process and the external billing service required to capture their interaction. In this relationship, the business process plays the role of the requestor, and the billing service plays the role of the bookkeeper. The requestor role is defined by a single port type, called bs:BillingConfirmationPortType. The bookkeeper is defined by two port types: bs:BillPortType and bs:MonitorBillPortType.

(b) BPEL partnerLinks are limited to describing two-party relationships. Discuss the issues involved in describing n-party relationships.

Exercise 6

Consider the following business process definition:

```
<process...>
        <receive partnerLink="seller" portType="tns:sellerPT"
                            operation="submit"    variable="sellerInfo"    createIn-
stance="yes">
            <correlations>
                    <correlation set="token" initiate="yes"/>
            </correlations>
        </receive>
        <receive partnerLink="buyer" portType="tns:buyerPT"
                            operation="submit"    variable="buyerInfo"    createIn-
stance="yes">
            <correlations>
                    <correlation set="token" initiate="yes"/>
            </correlations>
        </receive>
        ...
</process>
```

Describe what happens when the BPEL4WS runtime engine within which this process is deployed first receives the following messages in this order:

- A message from the buyer, via the buyerPT port type and submit operation. Assume the message contains some data D associated with the token correlation set.

- A message from the seller, via the sellerPT port type and submit operation. Assume the message contains data D, identical to that in the previous message, associated with the token correlation set.

- A message from a buyer, via the buyerPT port type and submit operation. Assume the message contains some data D′ associated with the token correlation set.

- Assume that the buyerInfo and sellerInfo variables each have an integer part called offer and askingPrice respectively. Additionally, assume the existence of a WSDL message type called "result," with a single string part named "outcome." Complete the process definition with logic that compares these two prices. If the offer is higher than or the same as the askingPrice, populate the outcome field of the result message with "success." If the offer is less, populate the outcome message with "failure." Reply to each of the partners with the outcome.

Exercise 7

This exercise involves defining an interaction between two BPEL4WS processes.

(a) Create a business process definition B1 consisting of receive, invoke, and reply activities that are executed in sequence.

(b) Sketch the definition of the portTypes and operations that need to be defined on the external WSDL interface of B1, as viewed by its partners.

(c) Create a business process definition B2, in which B1 is defined as a partner. Define an appropriate partnerLink between B2 and B1. Define an invoke activity in B2's logic, which invokes the request–response operation offered by B1's external WSDL interface.

(d) Comment on whether we can describe B2 as being a subprocess of B1.

Exercise 8

Write a BPEL4WS snippet that does the following:

Declares a variable that is of message type "mycompany:OrderStatus." This message type has two parts. The first part is an integer and is named "orderNumber." The second is a string and is named "status." The process then populates this variable with data as follows [note that to complete parts (a) and (b) you will need to write assign statements involving XPath expressions]:

(a) orderNumber should be set to the "id" property of the variable "OnlineOrder."

(b) Check the value of the part "paymentMade" within the variable "BillingInformation." If the value is true, the status part of the OrderStatus variable should be set to "OK," otherwise it should be set to "Pending."

(c) Assume the assign statement(s) you wrote for parts (a) and (b) above are within the scope of a "while" activity. Describe what happens to the value of the variable at the following points in the life cycle of the "while" activity:

- Before it is enabled
- When it is enabled, but before the assign activities are enabled
- After assignment
- When the "while" activity begin another iteration
- After the "while" activity completes

Exercise 9

Define a business process that is instantiated by the arrival of a message containing a single integer part, representing the number of orders to be processed. The business process includes an event handler that assumes that the employee handling the orders takes 30 minutes to process each order. If the employee has not finished processing orders in the expected time, the handler must send an e-mail message saying "Get moving slowpoke" to the employee (assume there is a partner service that is capable of sending e-mail messages, given the content and destination e-mail address). Use appropriate message definitions and other BPEL4WS constructs in your process definition.

Exercise 10

In the business process we defined in the above exercise, generate a fault called "process:WorkloadExcess" if the number of orders an employee is expected to process exceeds 14. Write a global fault handler to catch this fault and invoke an external Web service that has a port type "AdjustWorkload" with an operation called "reduce." Use appropriate message definitions and other BPEL4WS logic in your process definition.

REFERENCES

1. W. M. P van der Aalst, A. H. M. ter Hofstede, B. Kiepuszewski, and A. P. Barros, "Workflow patterns home page," Available at http://www.tm.tue.nl/it/research/patterns/.

2. G. Alonso, F. Casati, H. Kuno, and V. Machiraju. *Web Services*. Springer-Verlag, 2004.

3. IBM Inc., Microsoft Corp., BEA Inc., SAP AG, and Siebel Systems Inc. *The Business Process Execution Language for Web Services, version 1.1*. Published on the World Wide Web at http://www.ibm.com/developerworks/library/ws-bpel/, 2003.

4. E. Christensen, F. Curbera, G. Meredith, and S. Weerawarana. *Web Services Description Language (WSDL) 1.1*. Published on the World Wide Web by W3C at http://www.w3.org/TR/wsdl, March 2001.

5. F. Curbera, M. Duftler, R. Khalaf, W. Nagy, N. Mukhi, and S. Weerawarana. Unraveling the Web Services Web—An Introduction to SOAP, WSDL, and UDDI. *IEEE Internet Computing* 6(2): 86–93, April 2002.

6. F. Curbera, R. Khalaf, F. Leymann, and S. Weerawarana. Exception Handling in the

BPEL4WS Language. In *Proceedings of the International Conference on Business Process Management,* Eindhoven, The Netherlands, June 2003, pp. 276–290, Springer-Verlag.

7. F. Curbera, R. Khalaf, N. Mukhi, S. Tai, and S. Weerawarana. The Next Step in Web Services. *Communications of the ACM, 46*(10):29–34, 2003.

8. D. C. Fallside. *XML Schema Part 0: Primer.* W3C Recommendation, published on the World Wide Web at http://www.w3.org/TR/xmlschema-0/, 2001.

9. C. Hagen and G. Alonso. Flexible Exception Handling in the OPERA Process Support System. In *Proceedings of the International Conference on Distributed Computing Systems (ICDCS),* pp. 526–533, 1998. Amsterdam, The Netherlands, May 1998. IEEE Computer Society.

10. R. Khalaf, N. Mukhi, and S. Weerawarana. Service-Oriented Composition in BPEL4WS. In *World Wide Web 2003 Conference, Web Services Track,* Budapest, Hungary, May 2003. Amulett Kft Publishers.

11. F. Leymann and D. Roller. Production Workflow: Concepts and Techniques, Prentice-Hall, 2000.

12. F. Leymann. Supporting Business Transactions Via Partial Backward Recovery in Workflow Management Systems. In *Proceedings of Datenbanksysteme in Büro, Technik und Wissenschaft (BTW),* pp. 51–70. Desden, Germany, March 1995. Springer-Verlag.

13. S. Thatte. *XLANG.* Published on the World Wide Web by Microsoft Corp. at http://www.gotdotnet.com/team/xml_wsspecs/xlang-c/default.htm, 2001.

14. P. Wohed, W. M. P. van der Aalst, M. Dumas, and A. H. M. ter Hofstede. Analysis of Web Services Composition Languages: The Case of BPEL4WS. In *Proceedings of the 22nd International Conference on Conceptual Modeling (ER),* pp. 200–215. Chicago IL, October 2003. Springer-Verlag.

15. World Wide Web Consortium. *XML Path Language (XPath), version 1.0.* Published online at http://www.w3.org/TR/xpath , 1999.

CHAPTER 14

Workflow Management in Staffware

CHARLES BROWN

14.1 INTRODUCTION

Carefully designed and robustly implemented process-aware information systems (PAISs) are an essential part of many successful businesses. They help control operational costs and ensure productivity while helping organizations cope with rapid change. They also help in breaking down undesired barriers between departments, applications, and systems by linking all the resources required to perform key business processes and focusing them on end results.

Implementing PAISs, however, is generally a complex and delicate endeavor. The processes supported by a PAIS can be very complex with an incredible number of details that need to be taken into account and should be automated as far as possible. In addition, PAISs usually bring together significant numbers of heterogeneous software applications and organizational resources. Finally, strong requirements are placed on the time frame to complete the implementation and deployment, and it is often expected that deployment occurs with minimal disruptions to day-to-day business operations.

On the other hand, a significant part of the functionality expected from PAISs recurs from one system to another and can, therefore, be packaged into generic tools (or toolsets). As pointed out in Chapter 2, this is the case, in particular, in the area of workflow management. For example, activity dispatching and process monitoring are features required in most (if not all) workflow implementations.

A representative toolset in the area of workflow management is the Staffware[1] suite. This toolset was developed by a company with the same name, until 2004 when it was acquired by TIBCO,[2] a vendor of application integration solutions. It is expected that as a result of this acquisition, some of the functionality of the Staffware suite will be integrated with other TIBCO products. However, the essential parts of Staffware, which are outlined in this chapter, are likely to remain in one

[1]http://www.staffware.com
[2]http://www.tibco.com

Process-Aware Information Systems. Edited by Dumas, van der Aalst, and ter Hofstede **343**
Copyright © 2005 John Wiley & Sons, Inc.

form or another as they correspond to recurrent needs in the context of workflow/business process implementation.

As discussed in Chapter 1, the characteristics of a business process can vary widely, from the one extreme where work is processed without any set structure in a peer-to-peer or ad-hoc manner, such as communication between people via e-mail, to the other extreme being where the processes are highly structured and need to be supported by technology that is robust, scalable and rich in connectivity. The diagram in Figure 14.1 plots a sample list of the PAIS-related processes, including Staffware, with respect to the classification proposed in [3] and briefly presented in Chapter 1.

No one product can be considered appropriate for the whole continuum, so an organization needs to determine the appropriate product for its needs, even if this means selecting more than one. The author has had first-hand experience of unwanted results, on more than one occasion, where a product suited to one end of the range is used for applications with the characteristics of the other end. Because a product's description contains the word "workflow," one should not assume that it will fulfill the needs of an organization across the complete spectrum of its potential requirements.

Figure 14.1 shows that Staffware is aimed at the "production" end of the continuum with the ability to cater to "administrative" processes. It may not be appropriate for the two categories on the left side of the spectrum. It is worth noting, however, that it is more feasible for a production workflow to make available elements of ad-hoc and collaborative functionality than it is for a workflow tool at the left end of the spectrum to emulate the characteristics of a tool designed for production-style processes.

This chapter describes the issues associated with the implementation of workflow/BPM applications in general, with a focus on the use of the Staffware suite of workflow development tools. It looks at the considerations a potential IT decision maker should make in the selection of an appropriate product for a given work-

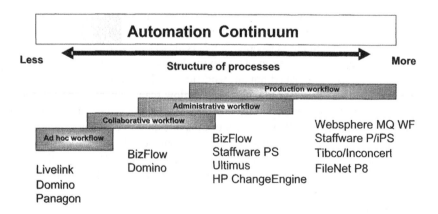

Figure 14.1   The automation continuum.

flow/BPM implementation project. It then explores the attributes and components of such a toolset, using Staffware as the example and the guidelines of the approach or methodology that should be applied. Key in this is a description of the so-called 10 "golden" rules and the presentation of a framework in which to implement solutions to a set of business process requirements. The chapter then finishes with a description of the type and style of resources required for workflow implementation and a concluding summary.

14.2   ARCHITECTURE

14.2.1   The Components of Staffware

As stated above, Staffware is delivered as a toolset, as opposed to a turnkey application. Upon installation, the toolset is used to build an application that can then be used by the business to manage its processes. At the heart of Staffware is a workflow engine (see Chapters 2 and 12) that executes a process application, built using the toolset, that delivers work to participating people and applications via the medium of choice (rich client, Web browser, or mobile). Multiple instances of the engine can be installed to support geographical separation, scalability, and desired technical architecture.

The coarse-grained view of Staffware's architecture is shown in Figure 14.2. It consists of three main components arranged according to a three-tier architectural approach. These components are:

1. Staffware process server(s)
2. Staffware process objects (SPO)
3. Staffware clients (shown with a darker background in Figure 14.2)

The following sections detail each of these three components in turn.

Staffware Process Server(s).   There are, in fact, two process servers that differ in the ability to handle throughput and integration capabilities, and, of course, price. They are called the Process Server (PS) and the iProcess Server (iPS), respectively.

The Process Server, also known as the Process Engine, is designed to cater to the needs of small-to-medium enterprises that are using workflow management to support processes directed mainly at human actors, typically requiring imaging and document management support and relatively modest integration to external applications and other software systems. It consists of a single-threaded "background" process communicating with a number of "foreground" processes managing user log-ins and work queue and work items servers through a FIFO message box (MBOX) architecture. The background process is the "engine" that interprets work arriving from people or other applications, determines the action(s) to be taken and the next participant(s) in the workflow process from the process definition developed as part of the application build and stored in the supporting database, and delivers work item(s) up to the foreground processes.

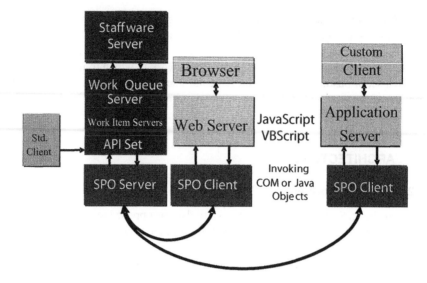

Figure 14.2   Staffware architecture.

The second engine is designed to cater to the needs of large organizations in which a true BPM approach is being taken having the following characteristics:

- High volumes of core business activity
- An end-to-end, straight-through processing model (STP), meaning that several automated steps are chained to achieve a given user goal
- Deep connectivity through a messaging layer

These characteristics do not preclude the iProcess Engine from carrying out those more traditional workflow processes as does the first engine, but such requirements would not be why the iProcess would have been selected. The iProcess Engine adds a multithreaded background set of processes. The number of threads is configurable and can be made to operate across a number of hardware partitions (i.e., on more than one box). This offers virtually linear scalability. Doculabs[3] conducted tests on a 64-CPU Sun Unix server with Staffware configured for eight background processes that accepted log-ons for 30,000 users and processed more than 1.5 million transactions in an hour [2]. Communication with the foreground is carried out by a secure (asynchronous) messaging service such as Oracle Advanced Queueing[4] or Microsoft Message Queuing (MSMQ),[5] depending on the platform. The iProcess Engine provides a step-type process that has been specifically de-

[3]http://www.doculabs.com
[4]http://www.oracle.com
[5]http://www.microsoft.com

signed to support complex integration into middleware, XML, SOAP, and Web services environments; it is not particularly aligned to either .NET and J2EE architectural environments but can operate in both.

Both versions of the Process Engine are supported by a database, which can be either Oracle or SQL Server, depending on platform and preference. The database contains two key elements of data: the current state of work in progress and data associated with each "case" travelling through a process and its audit information. The database also tracks triggers for events such as the expiry of a deadline, initiating escalations as required. Figure 14.3 gives a view of the components of the iProcess Engine.

Staffware Process Objects. The Staffware Process Objects (SPO) layer sits between the Server and any application serving up the user interface or providing the link to an integrated system. It consists of two components: the SPO Server, which sits alongside the Staffware server and thus has to be on the same platform as Staffware; and the so-called SPO Client that sits alongside the client application, such as a Web server or other application server. This can be on a completely separate platform utilizing a different operating system, if required. The SPO Client can also sit on the desktop, if directly supporting an application running on the desktop.

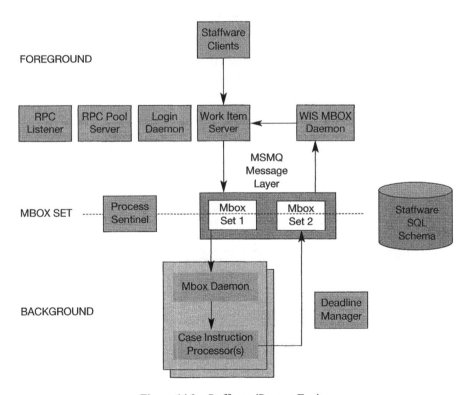

Figure 14.3 Staffware iProcess Engine.

The Process Objects layer provides access to the functions of the Process engines through an object model utilizing any programming language that can invoke COM and COM+ objects or Java classes. The use of the model provides a means of rapidly developing code to allow the user to log on, view work queues, open work items in another user interface such as a Web browser, or other custom application.

The potential functionality provided through the object model outstrips that offered by the standard client; however, both access the server through the Staffware Application Programming Interface (API) layer, so, to the server, both the standard client and the SPO layer look the same, and both means can be used simultaneously by the one user, which has its advantages.

The root of the Process Objects model is the concept of Enterprise User (SWEntUser), which can log on to one or many Staffware Nodes (SWNode) and retrieve information about work queues (SWWorkQ) and the work items (SWWorkItem) that they contain. The user can also retrieve the state of a case (SWCase) associated with a workflow procedure (SWProc). The outline of the Process Objects model is depicted in Figure 14.4. In addition to the objects provided in this figure, there are many other objects that provide access to audit information, form definition, user and group attributes, among others.

Staffware Clients. Staffware provides two "out-of-the-box" clients: the traditional client-server or "rich" client and a "web" client that operates utilizing the SPO architecture described above in both COM and Java forms.

"Rich" Client. This client application provides users with a richly functional view of their workflow delivered tasks, in a manner that is configurable to a degree, but

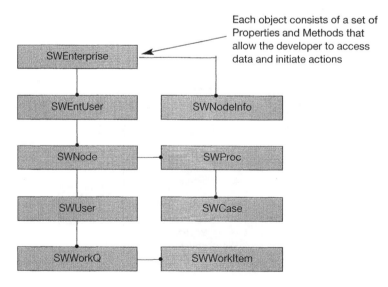

Each object consists of a set of Properties and Methods that allow the developer to access data and initiate actions

Figure 14.4   High-level Process Object view.

nonetheless is a "one size fits all" approach. However, it provides for the vast majority of functionality the normal user requires to participate in a workflow process, provides support for silent roll-out, and has been tested many times for compatibility within an SOE. It remains the quickest and simplest way to get users up and running using applications developed in Staffware, despite the preference of most architects for a browser-based solution. It also provides, as an install option, a set of administrative functions for managing Staffware instances (see Figure 14.5).

"Thin" Client. Staffware's "out-of-the-box" thin client offers less functionality that the client/server offering but, nevertheless, still provides for the basic needs of the workflow participant. However, as most organizations have fairly set standards as to the look and feel of applications delivered over their intranet/extranet, it is likely that, at the very least, the style sheet would need to be modified and logos changed on some of the key pages.

The thin client generates the forms from the standard Staffware rich client forms, as these definitions are accessible through SPO. Although out-of-the-box, it does not support some of the more complex features of the standard Staffware form such as complex conditional statements, form commands, and table look-ups (see Figure 14.6).

The section on methodology below will describe the practical impacts of the above and will identify how most organizations deploying the technology are catering to the needs of a rich user interface at lowest cost of development and ownership.

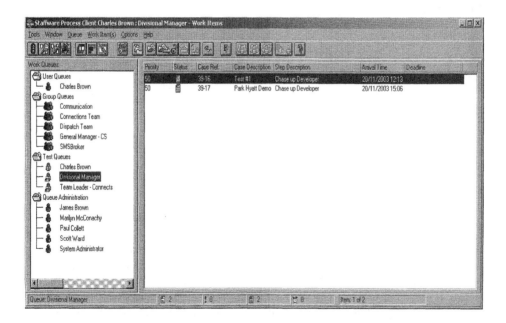

Figure 14.5 Staffware "rich" client.

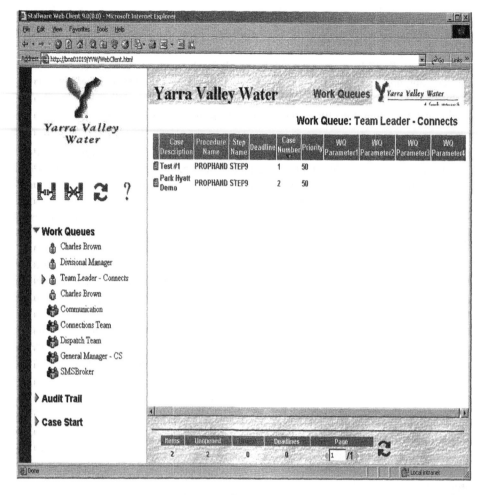

Figure 14.6 Modified Staffware thin client.

14.3 DEVELOPMENT TOOLSET

14.3.1 Process Definer

The Staffware Process Definer (SPD) has one primary objective—to make the analysis and documentation of business processes as simple and intuitive as possible. It is aimed at a nontechnical audience that understands their business. This primary orientation is crucial in eliciting the business community's involvement procedure definition. Failure to do this could seriously jeopardize the success of the final project. The SPD builds on the familiar flowchart metaphor to show, in an unambiguous manner, the flow of work for a particular business process.

The SPD stores the rules defined graphically, in effect creating a "program" that can then be deployed across a wide-ranging hardware architecture.

A major advantage of the SPD is its two-way link to the Staffware Server. This means that the Graphical Toolset may be used to create and maintain live workflows. This is in stark contrast to some other flowchart tools that can only be used to define a procedure in the first instance with no ability to reload an operational workflow procedure. This assists in the cycle of:

- Analyzing the business
- Defining the business process
- Automating the business process
- Collecting metrics on the operational business process
- Feeding the data from the live environment back for further analysis

A business process is made up of one or more Staffware procedures. Procedures are defined, using the SPD, by creating a flowchart diagram, consisting of steps linked together with lines (denoting transitions) and interspersed with diamonds (conditional splits as defined in Chapter 8) and other control-flow constructs. Once the diagram is complete, it is "saved" (like a Word document) and may be run in either the Test or Released environments.

There are a number of different kinds of steps.

- Normal Steps, for normal user interaction and for Staffware Brokers
- Automatic Steps, which run programs on the server
- Event Steps which are used to coordinate with external events
- Open Client Steps, used mainly to deliver work via third-party systems such as Microsoft Exchange
- EAI Steps for enterprise application integration (EAI)
- Subprocedure Steps called both statically and dynamically

These are connected in relation to:

- Lines
- Routers
- Complex routers
- Condition boxes
- Rendezvous points/waits
- Stop signs

The SPD is also used to create Fields, Staffware Scripts and EIS Reports.

There is no technical limit to the size of a Staffware procedure, although we suggest that Processes of more than about 50 steps should be broken down into several subprocedures for ease of use.

Point-and-click subprocedure functionality allows SPD users to drill down from one procedure to the next (subprocedure), selecting the data to be passed to and from, and assigning a deadline to an entire subprocedure. This brings subprocedure definition and use entirely into the realm of the business analyst. It also facilitates the creation of corporate process libraries, through which common processes are made reusable across the business.

When a "normal step" is defined, the SPD provides links to the Staffware User (Group, Role) database, the Staffware Graphical Forms Definer (to create Staffware or VBA Forms), and Deadline and Priority functions.

Readability of the SPD is enhanced by annotations (notes that may be placed anywhere on the chart), user-definable icons for each step and a setup option covering line colours, size, and so on.

14.3 Integration Tools

Integration from Staffware to External Applications. A well-rounded process management product must integrate fully with the other systems that are used in the business in order to optimize productivity, data accuracy, and consistency. Integration must be possible via automated server processes as well as via interactive applications running on the desktop. Staffware has a comprehensive set of tools to facilitate this integration.

Server-based integration—or, more accurately, the automation of a step in a process, without any user intervention—is most typically achieved by components in the Staffware Process Integrator module of the Staffware Process Suite, namely Staffware Broker applications and EAI Steps.

Staffware Broker applications are half "out of the box," because Staffware provides the Staffware Integration Broker as framework in which these applications can run. It is then necessary to develop broker tasks (implemented as Dynamic Link Libraries or DLLs) that can run in this framework. Many template broker applications are provided in the Staffware Process Definers Kit, but these will not include any code that implements third-party API's because of copyright issues.

Figure 14.7 illustrates how a broker interacts with a work item placed in a queue to which a broker has access to carry out a call to a middleware layer.

EAI Steps provide a server-based mechanism to invoke third-party applications via an adapter. Staffware currently provide Oracle, SQL Server, Java, Tuxedo, TCL, XML, and COM+ adapters. This range will be extended according to customer demand.

Figure 14.8 illustrates how a number of EAI steps within a process orchestrate the sequence of integration points between a number of systems required to support the workflow application. The figure shows a process (or procedure) model at the top that contains "EAI steps" that link the process with external systems (namely MapInfo,[6] Trim,[7] and SAP[8]) through "adapters." The key point here is that

[6]http://www.mapinfo.com
[7]http://www.towersoft.com/ap/
[8]http://www.sap.com

Work Queue Manager

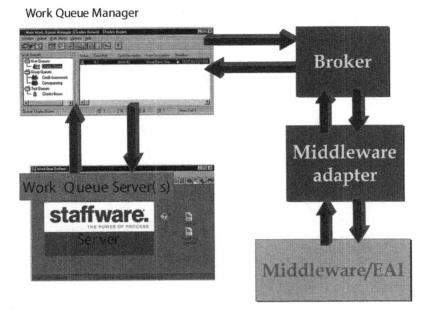

Figure 14.7   Staffware broker.

Staffware manages this system interaction with a very visible tool that a business analyst can use, external to both the applications themselves and the middleware layer itself.

The products identified above are for illustrative purposes only, and do not constitute a limitation or any prescribed architecture.

Integration from External Applications to Staffware. It is a typical requirement to be able to integrate with Staffware from external applications, either to start a new case of a process or to resume an existing case that is suspended and awaiting further information. The information that starts or resumes a case may come from many sources; for example, when a report is generated by a mainframe, when a message is written to an MQ Series message queue, when a Web page is completed, when an e-mail is delivered to a specific mailbox, or when a piece of correspondence is scanned and indexed.

Staffware can interface to all these sources of information via custom integration components, referred to as Workflow Triggers. These components typically implement Staffware Process Objects in order to communicate with Staffware. Cases of a process can also be started or events in cases can be triggered using a command line interface to a server utility program and a DDE interface to a client utility program.

In addition to triggering Staffware in this way, it is also sometimes required to embed Staffware functionality (such as access to Work Items in Work Queues, access to audit trails, or user administration) into a custom GUI application. This inte-

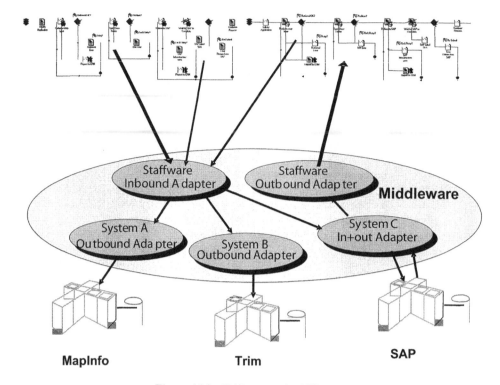

Figure 14.8   EAI steps and middleware.

gration is made possible through Staffware Process Objects, which give access to all Staffware client functionality.

14.4   METHODOLOGY

This section does not constitute a specific methodology to be followed to the letter but, rather, a set of guidelines to help the implementer of workflow/BPM systems avoid some of the pitfalls and be aware of issues that are likely to arise before, during, and after development. In respect of this section, the assumption is made that the traditional business analysis tasks of documenting the requirements, determining the as-is status and its move to the to-be status, and a definition of the scope have been done using whatever tools the organization is familiar with.

14.4.1   The Ten "Golden" Rules

Over many years of developing and implementing workflow/BPM solutions, a certain pattern for maximizing the possibility of a successful result has emerged. This pattern has been codified into a set of "golden" rules. These rules form a useful set of guidelines. Note that they are a set and have a collective strength, so taking any one rule on its own may not be particularly valuable.

Rule 1—Keep it Simple. Experience shows that the success of an IT project is inversely related to the complexity of the project. The greater the number of components in a project, the greater will be the amount of effort needed to complete the project and the greater will be the risk that the project will (a) go over budget and (b) fail to meet requirements. This is the golden rule. It is particularly valid for workflow projects, because in most cases the organization deploying a workflow system has no prior experience of workflow technology (which directly contributes to risk).

Rule 2—Deliver Quick Wins. This rule is another way of expressing the golden rule. A Staffware workflow system is designed primarily to automate processes, not to manage documents or data. Unlike documents or data, an organization's processes will change on a regular basis, according to the pace of change of the market in which the organization operates. Staffware makes the definition and change of business processes a relatively easy task. It is fruitless to spend many months analyzing, defining, and implementing the perfect process, because a dynamic organisation may have changed its processes (as a result of the introduction of new products or services) before its original (and now incorrect) processes have been automated. It is far better to define a narrow project scope and focus on key business problems in order to deliver short-term benefits.

Rule 3—Prepare for Change. Workflow is a unique technology in the way that it forces people to change the way they work. This change particularly affects those people who are responsible for managing work (identifying the types of work to be done, allocating the work to an appropriate work group, making sure the work gets done, and monitoring productivity), because many of the management tasks will be performed automatically by the workflow system. To try to define a perfect automated workflow procedure based on the requirements of the people who operate the current manual process is like trying to draw a picture with a blindfold on. Only by implementing and using a workflow system will a business really understand how it wants to use a workflow system.

Rule 4—Define the Scope. The previous rules may seem to encourage a relaxed approach to scoping a workflow project, but a well-defined scope is actually as important for a workflow project as it is for any other IT project, and the previous rules are intended to assist in making the scope realistic and effective.

Rule 5—Define the Goal. There is a tendency when implementing a workflow system to make the automation of an existing process a goal in its own right. This may be acceptable when the primary goal is to improve the efficiency of an existing process and efficiency gains are expected by virtue of replacing manual tasks (such as the allocation of work) with automatic processes. A business may, however, have many other reasons for implementing a workflow system; for example, to make processes easier to change, to deskill processes so that resource costs are reduced, and to reduce elapsed times from start to end of a process in order to improve customer service. If the goals are not defined, they are unlikely to be achieved.

Rule 6—Define a Plan. Once you know the scope of a workflow project (rule 4) and you have defined the goal of the system to be implemented (rule 5), the task of defining a project plan is greatly simplified, but the importance of defining a project plan is by no means reduced. Because Staffware encourages an iterative, prototyping approach to development of workflow procedures, and because many workflow projects are set up as pilots, before attempting "the real thing," there is danger that project planning will be considered unnecessary. This is, of course, not the case. The well-known axiom applies to all types of workflow projects as much as it does to any other IT project: "If you fail to plan, you plan to fail."

Rule 7—Involve the Business, Appropriately. This is another rule that is common to all IT projects but which has particular relevance to workflow projects because of the profound impact that workflow will have on the way people work. If the people involved in current processes do not participate in the design of workflow procedures, those procedures are unlikely to support the day-to-day processing requirements, and when the workflow procedures are implemented, existing users may actively seek ways to make them fail. However, if the goal of a workflow system is to radically change existing ways of working, which may also imply the need for fewer people with less skill, overreliance on assistance from current users may be considered to be not only inappropriate but also unethical.

Rule 8—Measure the Results. Even when rule 5 is remembered, and the goal of a workflow system is the preeminent factor in the design of the system, rule 8 is often forgotten. If the goal is to increase productivity by 50%, but the productivity before and after implementation of the system is not measured, the success of the system cannot be quantified. Consequently, it is difficult to learn lessons that provide feedback and influence the implementation of the next automated business process.

Rule 9—Cultivate a Business Sponsor. A business sponsor—perhaps a better name is a "business champion"—is essential to see through the implementation of a workflow system. During the course of this implementation, job roles will be challenged, power bases will be threatened, and substantial change will be introduced; in short, many potentially terminal challenges will arise. Without an influential, committed business champion who keeps focused on the goals of the workflow system, the chance of failure is high.

Rule 10—Use Known and Proven Components. It is wise to avoid building any reliance into a workflow system on a component that has never been used before, or never been used in the same way before. This applies as much to the

Staffware toolkit itself, which offers increasing functionality over time, as it does to the middleware components with which Staffware will be integrated. In order to minimize the risk of unproven components, they should, if possible, be evaluated in a technical test project before they become a critical component of a production system.

14.4.2 Rapid Application Development

The concept of rapid application development (RAD) in the Staffware environment consists of a development approach that involves the swift development and evolution of a process prototype using the Staffware Process Definer. This phase of the workflow application development life cycle is relatively short, and in an environment in which the business analysts/workflow developers have some experience, this should take no more than a month. Experience has shown that it generally takes three evolutions of the prototype to get it to the stage where the business process owners can state that they are happy that the scope, the participants and their structure, the process flow, and the functionality offered (but not yet built) are more or less correct. This is sufficient to commence the next phase—applying the prototype to the process framework.

The development of the prototype should go hand in hand with the development of a functional specification that lays out, in business terms, a written definition of each step in the process and the actions arising, the participants reporting, and other administrative requirements and how those functions are met.

The RAD phase must incorporate the following characteristics:

High business involvement. The prototypes should be reviewed through a series of workshops at which the active attendance of the key stakeholders in the business process is essential. They must feel they have ownership of what is being developed.

Three D's: Design, Develop, and Discuss. This phase consists of close interaction between the development team and the business analysts/developers on an iterative basis.

Get it wrong quickly. As per rule 3 above.

Integration is not advised (but can be mapped). The prototype will not contain any real integration but will be simulated, or at least flagged, in order that the workshop attendees can see where in the process such integration takes place.

Three stages of development. These are

Stage 1 (the how). Map the business process flow. How does a piece of work travel through the process? What are all the steps required to take a case form beginning to end?

Stage 2 (the what and who). What data is required at each step of the process? What people are needed to process each step?

Stage 3 (the when). Escalation and deadline settings. Where are the integration points?

14.4.3 Reference Process Framework

The Risks of Large-scale Workflow Implementations. For some significant users of the Staffware product suite, there are multiple, concurrent workflow application developments across many parts of the organization in progress and planned at any one time. If a framework approach is not used, the following risks are taken:

- Each team evolves its own way of development, reinventing the wheel
- Integration methodology is uncontrolled
- There is severely diminished scope for reuse
- Business cases are harder to justify and costs are higher
- Realization of fewer benefits for the business
- A potential management nightmare!

The Staffware Universal Process Framework (UPF). The concept of the UPF is a multitiered design to fit a wide range of typical workflow applications. Its purpose is to construct a means of providing a structure within which (most) administrative and production workflow processes can be developed to reduce the effort required during the build phase and cater to the need to standardize the design approach.

The implementation of a process within the UPF is carried out after the development of the final prototype, developed during the rapid application development phase described above. The Universal Process Framework (see Figure 14.9):

- Is supplied as a generic framework that is tailored to the needs of the organization.
- Is based on and can utilize the complete functionality of the Staffware toolset.
- Contains process maps, common brokers/utilities, and data definitions that provide a horizontal framework applicable to the majority of Workflow implementations.
- Provides highly functional process infrastructure for accelerated implementation of Workflow projects.
- Is a robust, production-ready environment with built-in exception and error handling routines.

A description of each level in the UPF structure follows.

Coordinator Process. The single top-level process that coordinates all incoming events, starts new cases, and triggers to existing cases. This process includes elements that vary by channel; for example, an incoming scanned document may be handled slightly differently from an incoming e-mail.

Line of Business Process. This is a customer-specific process intended to be transparent to a subject matter expert and able to be specified and created by a business analyst who has had Staffware training. This process can be a user interactive,

Customer
Interaction

Coordinator
Process

Line of
Business
Processes

Case Worker
Processes

System Interface
sup-process

Back End
Systems

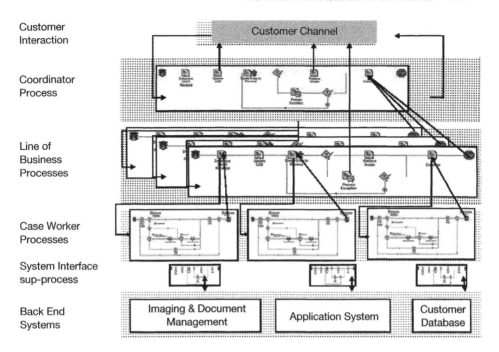

Figure 14.9   The structure of the Universal Process Framework.

a straight-through process, or a mixture of both. Essentially, the line of the business process reflects the business view of the overall process that is being automated, and would be based on the final prototype developed during the rapid application development phase described above. It consists of a series of calls to the Case Worker processes (see below), with business logic determining routing and actions following the return from each Case Worker.

Case Worker Process.   This is a process structure designed to support user and integration activities. It is a subprocess with a series of user interaction options that provides the Line of Business Process a measure of ad-hoc processing. Where users are involved, a work item is delivered to the appropriate recipient, identifying what it is and the function to be carried out. Under rules defined at build time, the user can choose to undertake a number of functions:

- Process (release)
- Defer (keep)
- Diarize for a period of time
- Initiate document creation and transmission
- Escalate to a superior
- Route to a colleague
- Close

The above are provided with the template; however, the developer is free to add whatever other functionality is required within any one particular Case Worker subprocess.

Integration Process. This is a reusable subprocess that includes a step in which a component such as a broker performs an integration function. It can be called from the coordinator, line of business, or case-worker process, Whichever is appropriate. Examples of such subprocesses include document create and print, e-mail send, and middleware adapter call. Functions delivered with the process template are as follows:

- Step/queue for broker
- Error checking
- Branch to exception subprocess on error return code
- Retry loop

The broker, middleware adapter, EJB, COM object, or other entity that performs the integration would have to be developed by an appropriately experienced technical resource. This can, however, be done in isolation from the business process development and called upon by the process developer when required.

14.5   RESOURCING

Two broad areas of resourcing are required to support the development and implementation of a workflow/BPM solution. Such a project should primarily be business-led, with the willing help of the IT department. It should not be an IT-led project as the business is far less likely to take ownership if it feels a system is being imposed upon it. The emphasis should be given to ensuring that the right level of knowledge is made available to the project team and that the project as a whole is being given support at the highest possible level.

14.5.1   Business

The key members of a workflow/BPM project are the business analysts who form the team to establish the requirements and build the workflow applications using the Staffware toolset. The term business analyst (BA) covers a broad range of skills, and although the Staffware product is designed for the use of BAs, it cannot be considered as simple as that, as not all BAs are the best candidates for the sort of work required in using the toolset. The best description of an appropriate resource would be a technical business analyst, one who, though not a programmer or technical specialist, can show an interest and capability in using process modeling tools such as ARIS (see Chapter 6). Familiarity with scripting tools such as Excel macros and an ability to detect and correct logic problems would be a further asset. Although all

BAs need to be able to listen and apply form to business activities, experience has shown that not all BAs can grasp the concept of being an effective user of a process automation toolset. At the very least, attendance at the standard Staffware Process Definer's course will show whether an individual is a suitable candidate. Completion of this course does not, however, make the attendant an expert in the product. They will be able to use the toolset and construct workflow processes, but they will not have enough experience to build well-designed, production-ready workflow applications. They should be mentored by an experienced workflow developer over a period of at least three months, adhere to the methodology being used by the organization, and utilize a framework approach to development as described in the set of guidelines in the section on methodology above.

14.5.2 Technical

The specific skills required by the technical members of a workflow application development team will vary from project to project but, probably more normally, from site to site. Increasingly, organizations tend to embrace a particular technology platform, be it .NET, J2EE, or any other approach that, usually, is mandated by their technical architects. Interestingly, there seems to be no clear senior management involvement as to which is an appropriate approach regarding the technology platform; thus, the environment that a workflow developer works in can be arbitrary. Nevertheless, the following key attributes for technical support of a workflow/BPM development and implementation environment can best be categorized as follows, though not necessarily inclusively:

- J2EE
- .NET
- VB/C++/Delphi
- Active Server Pages (ASP)
- Web Service Technologies
- SQL (Oracle & SQL Server)
- Middleware adapters: Tibco, MQ, Tuxedo, WebMethods, and so on

The role of the technician in a workflow application environment can be separated from a project role. The technician's role is to create the technical environment that can be used and reused by the (specialist) BAs developing the workflow applications.

14.6 CONCLUSION

The key messages that need to be conveyed and understood by all organizations that currently use or plan to implement workflow/BPM solutions are as follows:

- Implementation of a workflow/BPM solution is not a trivial exercise, so provide sponsorship at a senior level.
- Involve the process owners within the business as closely as possible; they have to feel they have ownership of the resulting application.
- Manage the people for change; workflow/BPM will dramatically alter the way they work and if the effects of change are not managed, the application will fail.

The returns from the implementation of such a solution can be dramatic; for a solution comprising more traditional workflows involving people and imaged documents, a minimum of a 30% return can be expected. Where more significant automation has been achieved at the production end of the continuum, returns of up to 200–300% usually can be realized. Ensure that the organization understands the means by which it can make such project outcomes successful.

14.7 EXERCISES

Exercise 1
Explain the differences between the Process Server and the iProcess Server. With respect to the classification introduced in Chapter 1 (i.e., person-to-person, person-to-application, and application-to-application processes), which type(s) of process does each of these two engines aim at supporting?

Exercise 2
What are the basic building blocks of a procedure definition in Staffware? Compare these elements to those found in UML Activity Diagrams (Chapter 5) and EPCs (Chapter 6).

Exercise 3
What is the relationship between line of business processes and case worker processes of Staffware's "Universal Process Framework"?

Exercise 4
According to the golden rules and the resourcing framework discussed in this chapter, what should be the involvement of senior management and business analysts in a workflow design and implementation project, and why?

REFERENCES

1. M. Ader. *Workflow Comparative Study*. Workflow and Groupware Strategies, 2001.
2. Doculabs Inc. Performance Assessment of Staffware Process Suite—Executive Summary, 2002. http://www.tibco.com/resources/mk/doculabs_executive_summary.pdf
3. S. McCready. There is More than One Kind of Workflow Software. *Computerworld, 2*, 86–90, November 1992.

CHAPTER 15

The FLOWer Case-Handling Approach: Beyond Workflow Management

PAUL BERENS

15.1 OUTLINE

This chapter gives an overview of the basic principles of the FLOWer case-handling system developed by Pallas Athena in The Netherlands.

Today, high demands are made on staff regarding expertise, communication ability, and commercial skills. More and more, they have become knowledge workers. With automation, the purely routine aspects of a process can be supported more and more effectively or can even be omitted altogether. Organizations are constantly trying to improve the quantity and the quality of products and services and to lower costs. Flexibility is an essential condition for product development, process control, automation, and especially for the staff. This does not take away the fact that parts of the process can or even should be regarded as an actual production process. In this context, we speak of production workflow [4, 20] as discussed in Chapter 1. This also needs to be adequately supported.

The keyword, therefore, is flexibility [11, 16]. If we highlight this with regard to process control and management, we will see that flexibility is essential in every area:

- Process design
- Integration with information systems
- Organizational design
- Work allocation
- Execution or runtime environment
- Management and control

Furthermore, the "conceptual integrity" of these diverse areas should be maintained. FLOWer satisfies these requirements. FLOWer is based on an "information-driven" approach and takes the process as its focal point, whereas traditional workflow management systems (WfMS) [33, 34] are based on the routing of activities from work tray to work tray, leading to inflexibility and context tunneling. FLOWer focuses on handling the case as a whole, and not only on the routing of activities. This routing is merely regarded as derivative. In this chapter, it is shown that case handling is not just another paradigm, but a more powerful and more comprehensive paradigm; see also [9, 10]. This means that the FLOWer case-handling system allows you to manage effectively not only flexibly structured processes but also offers the functionality necessary for production workflow. It, therefore, allows for managing a broader set of business processes than traditional WfMS.

In Section 15.2, we give an overview of case handling and FLOWer in comparison to traditional WfMS. In Section 15.3, we will focus on the conceptual integrity of the FLOWer case-handling paradigm. In other words, is all required functionality present and is it present at the right place? We take the complete life cycle of a process as our starting point and describe how it is mapped onto the FLOWer case-handling system. This will give a detailed description of the functional components of FLOWer. In Section 15.4, we give eight golden rules of process management.

The exercises are part of the sections. They focus not only on the modeling of processes, but also on the other aspects and the relations between these aspects, ensuring that the reader understands the conceptual integrity of FLOWer.

15.2 OVERVIEW OF CASE HANDLING AND FLOWer

In this section, we describe the concept of case handling implemented by FLOWer. We do this by comparing case handling and workflow on the basis of the definitions of the workflow management coalition (WfMC) [33, 34]. Per topic, the related part of the definition is put in bold text. As our starting point we choose a simple process with 10 sequential activities, as illustrated in Figure 15.1. The activities are called activities 1 to 10 and should be handled in that order.

15.2.1 Basic Element for Control

The WfMC gives the following definition of WfM:

> **The automation of a business process, in whole or part, during which documents, information or tasks *are passed from one participant to another for action,* according to a set of procedural rules.**

It is interesting that the WfMC refers to tasks (or steps or activities) that are passed from one person to another. In other words, the basic paradigm is *routing*. A process, therefore, consists of a collection of "processing stations," that need to do "processing" in a particular order, fully determined by the routing rules. What actually happens in such a processing station, that is, which activities need to be carried

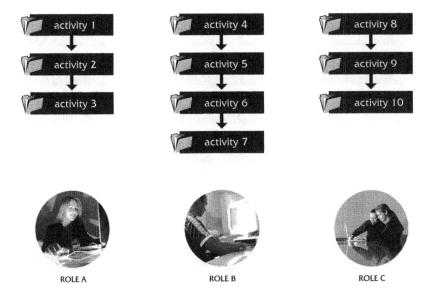

Figure 15.1 A simple process consisting of 10 activities.

out and in which order, does not appear to be relevant here, even if it forms a large part of the process. In other words, the control must be included in an application that can be invoked; thus, the control is *outside* the traditional WfMS. Therefore, nothing has changed with regard to the past; the control is still hard-coded in the applications. Assuming three processing stations, the process in Figure 15.1 would be illustrated in a traditional WfMS as in Figure 15.2.

Indeed, in practice, this process is often modeled in the three "steps." One could say this is "routing-based modeling," that is, the routing determines the process design. In a traditional WfMS, one could, of course, actually model the process as a process consisting of 10 activities. This is "activity-based modeling," that is, the individual activities determine the modeling. But that will not do. As soon as an activity is completed, for example activity 1, the traditional WfMS will calculate the next activity (activity 2) and will determine who will get it, that is, in which user's work tray the activity should be placed. It usually will be the same employee who carried out the previous activity. This can be "forced" by explicitly including in the process design instructions that this employee should get this activity. It solves the problem of other employees possibly getting the activity. In the example, this person is the employee with role A. But she still needs to go to the work tray, look up activity 2, and start it again. This, in turn, can also be solved by having activity 2 start up automatically as soon as activity 1 is closed. These changes place a considerable restriction on the behavior of the system—in *all* cases of this process after activity 1, activity 2 will be automatically started up with the same person who carried out activity 1. Exceptions are no longer possible. Furthermore, this means that activity 2 is explicitly allocated to the employee with role A. If the person with role

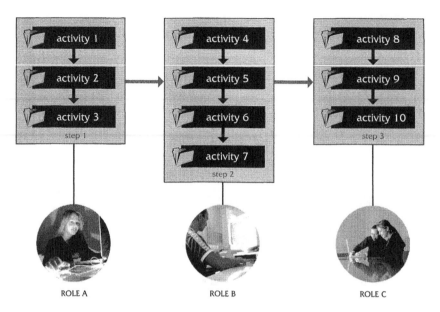

Figure 15.2 The granularity of a workflow management system, based upon routing.

A, for whatever reason, cannot carry out activity 2, this activity will have to be re-distributed.

The traditional WfMS partly acknowledges this problem. Some offer modules or have announced modules in which the control flow *within* a step can also be included in the design [12, 28]. The system behaves in exactly the same way, however, as the one described above—the next activity (for example, 2) is started up immediately for the employee who carried out activity 1, or is placed in his or her work tray only. This, therefore, also fails to provide a full solution. An employee with role A and role B, for example, is not helped by this solution. In principle, this person should be able to carry out activities 1 to 7, but these were already split into two steps. So the employee now needs to pick step 2 (involving activities 4, 5, 6, and 7) from his or her work tray before another colleague takes it. A solution for the employee with role A and role B would be to include activities 1 to 7 in a single step, but this would then cause problems for employees who only have role A or role B. The problem really boils down to the routing-based paradigm of the traditional WfMS.

Exercise 1. Suppose there are three roles, A, B, and C. Role C is higher than A and B, meaning an employee with role C can do all the work an employee with role A or an employee with role B can do. A and B are not related hierarchically. Now visualize for the three roles A, B, and C the routing-based approach for the simple process depicted in Figure 15.3. Is there any routing-based breakdown possible, other than routing between each and every activity?

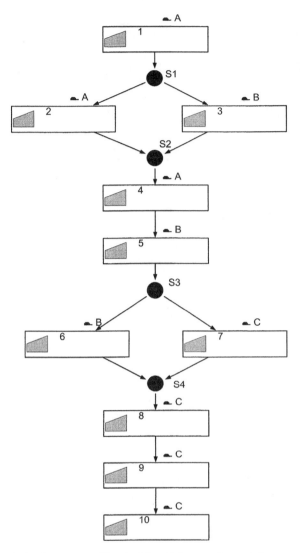

Figure 15.3   A simple process with a decision. Roles are shown at the upper right-hand corner.

The above situation is definitely not fictitious. A knowledge worker is expected to be able to carry out a large part of the process [3, 4, 16]. This growing demand for flexibility is becoming increasingly urgent. So the number of processing stations is declining while the need for process control for knowledge workers is increasing.

In case handling, we use activity based modeling, see Figure 15.4. Case handling is based on controlling all elementary activities; the routing is derived from this, as described in the following subsection.

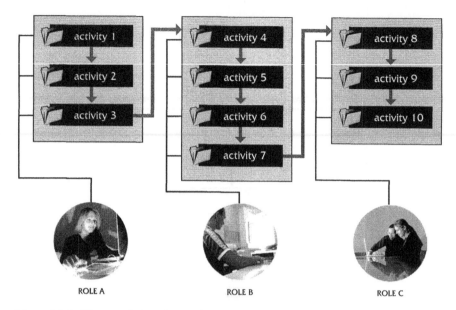

ROLE A   ROLE B   ROLE C

Figure 15.4 The granularity of a case-handling system. The activities form the basis of the system.

15.2.2 Implicit and Explicit Routing

Based on what we described in the previous subsection, we can determine the consequences of a process change. We refer yet again to the definition of the WfMC:

> *The automation of a business process, in whole or part, during which documents, information or tasks* are passed *from one participant to another for action,* according to a set of procedural rules.

This basis has serious consequences for a process change. Suppose that we have to redistribute the four activities (activities 4 to 7) in the process in Figure 15.1 due to a change in legislation or due to an organizational change like the empowerment of roles A and C. In routing terms, this would mean, "We remove the routing step (step 2) and distribute its contents over step 1 (activity 4 and activity 5) and step 3 (activity 6 and activity 7)." This would result in the following (see Figures 15.5, 15.6, and 15.7).

Figure 15.5 illustrates the starting situation. Figure 15.6 shows the new situation. In Figure 15.7, the dashed lines indicate what changes are necessary because of these apparently simple and frequently occurring adjustments. Both steps need to be adjusted, as does the sequence of activities within the step. In case handling, the change is minimal. Case handling controls the processing of all elementary activities, with the distribution being a derived, implicit function that can be adjusted sep-

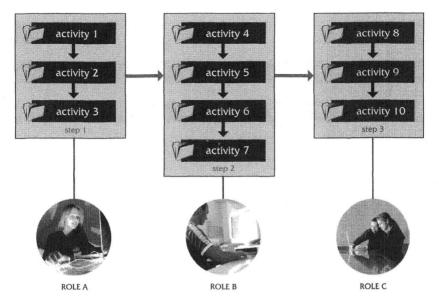

Figure 15.5 Starting situation in a traditional WfMS—three steps.

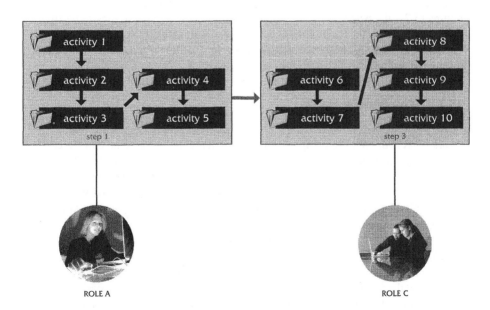

Figure 15.6 The two remaining routing points in a traditional WfMS.

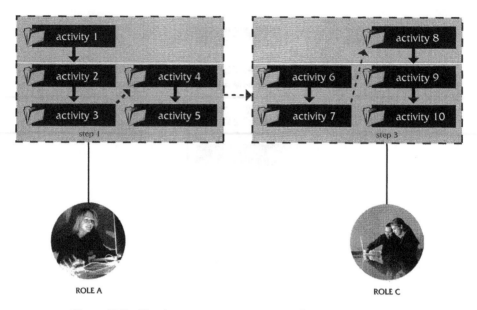

Figure 15.7 The changes necessary to remove the second routing step.

arately from the process definition. From this aspect, the process would be as shown in Figure 15.8.

It will be clear that a routing-based WfMS leads to considerable control problems. But a WfMS based on 10 activities will also encounter problems, as shown in Figure 15.2—the behavior of the system requires subsequent activities to be rigorously allocated to the same employee and also automatically started up. But this again leads to control problems and the need for additional functionality to allow for the redistribution of the activities (see Figure 15.9).

15.2.3 The Work in Hand

In the previous two subsections, the importance of the entire object, the case or the "work object" has not been emphasized. In case handling, we assume that the entire case is, in principle, available to the employee working on the case. The "work in hand" for the employee is, therefore, the entire case. Depending on the role of the employee, he or she can process and see the information. In a traditional WfMS, we find a limitation that leads to what has been called context tunneling [3]. Although most systems support the notion of "work cases," only information that is required for the current activity within the work case is available. The full context of the case is difficult to access or not present at all. The context consists of the status of the case, both in terms of the process and the available information. It includes the case history and the information gathered so far. It also includes access to possible future activities or tasks. This allows for processing information belonging to an activity

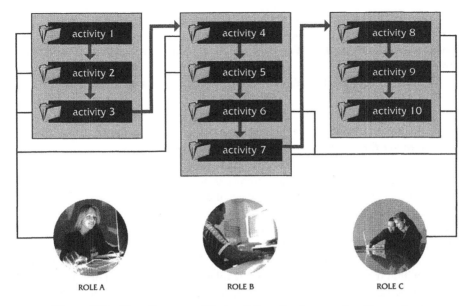

Figure 15.8 The adjustment of the implicit routing in a case-handling system.

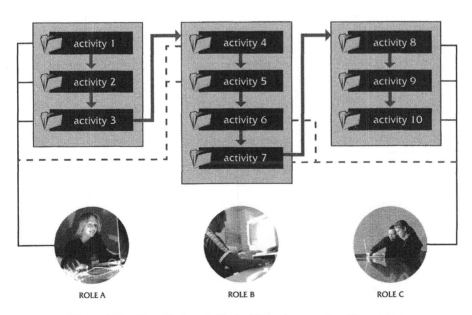

Figure 15.9 The adjustments (dashed lines) in a case-handling system.

not due yet, for example, one provided by a customer during a phone call. Of course, this should be controlled using proper authorization mechanisms, like the ones described in the FLOWer data handling model (see subsection 9.3.1 on process design).

For the WfMC, the information is regarded as being external to the process:

> **_The automation of a business process, in whole or part, during which_ documents, information, or tasks _are passed from one participant to another for action, according to a set of procedural rules._**

In Figures 15.10 and 15.11 the differences are illustrated. The information belonging to an activity is available only when the step is being processed.

The information belonging to all activities is available all the time. It will be obvious that the employee has a much better overview here of what has already happened and what still needs to happen. This means he or she can communicate with the customer more effectively. Moreover, this approach also offers the possibility [within the authorization rules of the role(s) of the employee] of carrying out activities or skipping and/or doing them again. It is also possible to enter data that is already known, even if the activity to which the data logically belongs is not yet due to be handled. In our view, information in the form of data elements and documents, for example, is of crucial importance for the status and progress of a case (see Section 15.3 on the conceptual integrity of FLOWer). This does, however, go

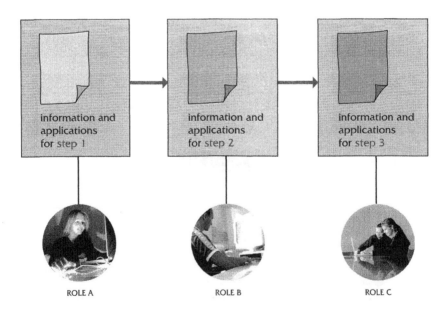

Figure 15.10   Context tunneling in workflow management (WfM). The step as the basis for the process.

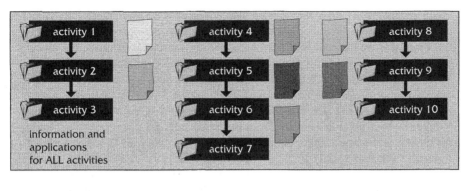

Figure 15.11   The case as the basis for the process.

against the popular view of a strict division between data and processes, but this boundary is impossible to uphold in practice. Data (information) can and most probably will play a role in the content as well as in the process. Progress made in a process instance normally is due to the gathering of information [2, 3, 8, 24, 25, 26, 27]. Using a case as the work in hand, follows the practice in administrative and service environments. Employees refer to *cases,* not to activities. A case is passed on, advice is requested from a colleague about a case, a case progresses through the organization. So we can speak of office activities as being centered around cases. It is, therefore, logical to use cases in the electronic representation as well.

15.2.4   Direct Allocation Versus Indirect Allocation

Another important difference between WfM and case handling is the way in which the system deals with work allocation. The WfMC definition says:

> *The automation of a business process, in whole or part, during which documents, information, or tasks* **are passed from one participant to another for action,** *according to a set of procedural rules.*

This definition does allow for how and when tasks are passed on to the next employee, but things are different in practice. As soon as an activity has been completed, the next activity or activities that can be carried out, as well as who can perform them, are calculated, all in accordance with the process design. The activity is then allocated to exactly those employees. All other employees do not even see the activity. From this we can conclude that there is a very strict link between the process design and the execution. It is not possible, for example, to arrange for allocation based on a department or regional office. If a person satisfies the "execution rule" attached to an activity, he will get the activity in his in-tray or work list. An employee only sees those activities that are due to be handled and that can actually be handled by him or her under the rules of the process design. This means that there is no separation between authorization aspects ("What may I do?") and distribution aspects ("What can I see?").

In fact, the traditional WfMS behaves in a rather panic-stricken way. An activity (within the case) is dropped like a hot potato directly onto the plate of employees who are authorized to carry out the activity now due to be processed. There is no way of specifying that certain employees can view an activity (or, even better, a case). Furthermore, nothing more can be changed in the method of allocation. The employees that satisfy the rules receive the activity; all the others do not. This approach is far too rigid for most organizations. In order to resolve these problems to some extent, tremendously complicated rules relating to work allocation are being created *as part of the process design, not in the organization modeling.* In these rules, authorization and distribution aspects are all jumbled together. This violates the conceptual integrity. Most traditional WfMSs offer a large range of functionalities to reverse the allocation, for example, in the event that an employee is temporarily absent. And with every application of these systems, customized solutions are developed to allow insight into the current cases, to search for a case on the basis of a client's query, and so on.

In case handling, a different paradigm is followed. Logically, the information that is relevant for the allocation is gathered for each process type. This could relate, for example, to the role needed to carry out an activity, to interesting process control data that is used to distribute the work (Zip Code, regional code, etc.), or to time-related information. By filtering and combining this information, it is possible to define a *work list* or *queue of work*. As the filter can be easily adjusted and is not part of the process design, there is optimal flexibility. The problems referred to above do not arise if the case-handling paradigm is used because:

- The link between design and allocation is not strict, but runs through a separate control environment in which the filters can be defined.
- There is a separation between authorization aspects ("which role do I need to carry this out?") and distribution aspects ("which cases does this filter provide me with?").

The flexibility does have a drawback—more design and control work is required. The extra functionality, however, more than compensates for this. Moreover, we shall see in Section 15.3 on conceptual integrity of FLOWer that, while retaining flexibility, FLOWer can automatically generate standard work trays for each employee containing exactly those cases they can carry out (i.e., cases for which there are now activities they can carry out), according to their profiles. These profiles combine what employees can carry out (authorization) and what they can view (distribution).

15.2.5 Summary

Based on the WfMC definition of workflow management, we can conclude as follows:

- WfM involves the routing of steps that, in turn, consist of a number of elementary activities. Case handling is based on processing elementary activities—the routing is derived from this.
- WfM routes external objects. With case handling, all information is accessible except those parts that an employee is not authorized to see.
- WfM has rigid routing. A distribution change results in a radical process change. With case handling, authorization and distribution are separated. Also, the distribution is set up separately from the process design.
- WfM has a strict link between design and execution. There is limited scope for different allocations for each organizational unit. Case handling allows for this difference.
- WfM causes context tunneling. Case handling allows the complete context to be seen at all times.
- Normally, WfM has few, if any ad hoc retrieval functions; only the activity that happens to be in the work tray of an employee is available. In practice, retrieval functionality is developed from scratch each time it is required. Case handling does have ad-hoc retrieval, if necessary. Again, it is restricted on the basis of organizational design.

A word of caution is required here. Each of the above-mentioned points has been addressed by workflow management systems by adding functionality to the basic paradigm, which is routing of steps. The combination of points, however, is hard to implement using that basic paradigm.

15.3 CONCEPTUAL INTEGRITY OF FLOWer

In Section 15.2, we looked at the main differences between case handling and traditional routing-based workflow management. In this section, we will focus on the conceptual integrity of the FLOWer case-handling paradigm. In other words: is all required functionality present and is it present at the right place? We shall, therefore, look at aspects in a broader framework, namely the complete life cycle of a process driven by the FLOWer case-handling system (see also [13, 17]). This brings us to the functional division of the components of FLOWer. We shall review the following (see Figure 15.12):

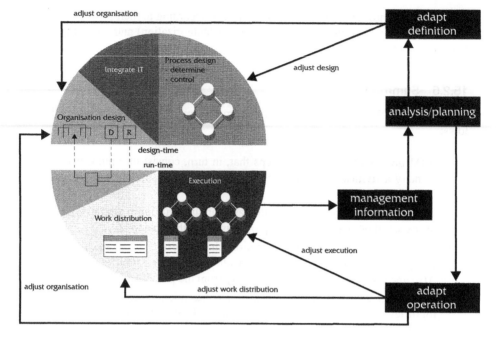

Figure 15.12   Design and operational use of processes.

- Process design
- Organizational design
- Work distribution
- Process execution

It is beyond the scope of this chapter to provide more details on the integration facilities. Today, most tools offer a wide range of integration functionality, both to the outside world and by means of API sets to the tool itself. If the above-mentioned components are present and integrated in a conceptually sound way, it will be possible, using measurement and analysis, to set in motion a cycle of continuous improvement.

We shall discuss the various parts of Figure 15.12, starting from the top right, anti-clockwise. The starting point is that an organization will develop a process based on its objectives.

The definition takes place in the upper half of the circle in Figure 15.12, the operation in the lower half. Designing the organizational aspects is found in both halves: part of this can or will be defined at the definition time, the rest can or will be designed at the operational level. In addition we provide an overview how FLOWer supports the functionality.

The FLOWer components are shown in Figure 15.13.

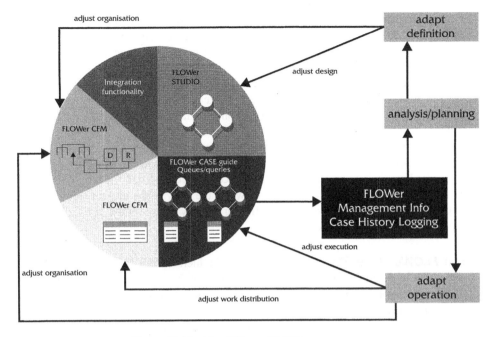

Figure 15.13 The different FLOWer components.

15.3.1 Process Design

The process is modeled in the process design component, called the Studio in FLOWer. Among other things, one defines activities, roles (authorizations) need-ed to carry out an activity, subprocesses, triggers, data for control and distribution purposes, and processing sequences. There are huge differences between the sev-eral products on the market. Three important aspects have to be taken into ac-count:

1. The supported modeling constructs
2. The capability to define exceptions
3. The robustness of the design against changes

The supported modeling constructs can be evaluated on the basis of the well-known workflow patterns (see Chapter 8) and the research carried out world-wide (see [1, 5, 6, 15, 19, 21, 30, 31]). In [6] many different WfMS products, including FLOWer, are evaluated. Please note that this research is restricted to the process modeling or design environment. This, however, does not limit its value. As far as the process modeling environment is concerned, it is currently the only in-depth evaluation on how frequently occurring workflow patterns are supported. But it ob-viously does not account for several other components, nor does it evaluate how these components interact. These other aspects are also important and should be taken into account as well. Or in other words, a high score on the supported patterns is a *necessary* but not a *sufficient* condition for successful implementations.

One of the striking elements in administrative processes is the frequent occurrence of exceptions, [7, 14, 29, 32]. The capability to define exceptions is, therefore, of utmost importance. Most processes, except for the very few extremely structured processes, contain many exceptions. In fact, exceptions are the rule. Without a proper mechanism to model exceptions on another level of abstraction than by simply drawing more arcs in the flow, one has to model each and every exception, leading to spaghetti-like flows. Or, alternatively, one has to model on an aggregate level, taking large pieces of the process together in one "activity," thereby, of course, losing the connection between the previously present activities. FLOWer has tackled this problem by offering an easy to use model to handle exceptions, namely, the role model.

The FLOWer Role Model. FLOWer uses the hierarchical relationships in the role graph to determine what a user can do based on his roles. In Figure 15.14, a simple role graph is depicted, with roles A, B, C, and D. A is a higher role than B, C, and D, but B, C, and D are not related hierarchically. If role R1 is higher than role R2, a person with role R1 is authorized to do all operations for which role R1 is required or for which role R2 is required. A person with role R2 is not authorized to do operations for which role R1 is required.

The FLOWer role model requires the process designer to define three roles for each process object (like activity, milestone, etc.):

- Execute role
- Redo role
- Skip role

The execute role is the role necessary to carry out an activity or to start a subprocess. This role is consistent with what the WfMC calls a role ([33, 34]).

The redo role of an activity is the role necessary to return the case to a status for which that activity is due. Suppose, in a case such as "Settle Motor Claim," an em-

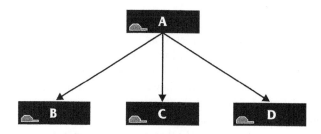

Figure 15.14 A simple role graph.

ployee wants to redo an activity carried out earlier, such as "Register Claim Data." This means that all the intermediate activities also need to be carried out again. The employee may only do this if he or she has a role that is at least equally high as all the redo roles of the intermediate activities and the "Register Claim Data" activity itself. Also, the employee must then at least have the execute role of "Register Claim Data" in order to handle this activity again.

The skip role is necessary to pass over an activity. In order to skip two consecutive activities, for example, the employee must have a role that is at least equal to the skip role of those two activities.

The three roles are a very powerful mechanism for modeling a wide range of exceptions. The redo role ensures a very dynamic (as it is dependent on the role of the employee and the status of the case) and flexible form of a *loop*. We note that FLOWer has two additional mechanisms to model loops. One is the dynamic subprocess or subplan to model multiple independent occurrences of parts of the process. The other is the sequential subplan to model parts of a process *repeating in sequence.*

The skip role takes care of a range of exceptions that would otherwise have to be modeled explicitly (drawn) in the process flow in order to pass over activities. Of course, there are ways of avoiding undesirable effects—you can define the "*no-one*" or "*nobody*" role in every process that is higher than all the other roles and that no user can perform. You can also define an "*everyone*" role that is lower than all others. An activity with the "*no-one*" redo role can never be undone again and it would then also not be possible to go back to a predecessor of that activity. This efficiently models a "point of no return." An execute role *everyone* means that the activity can be carried out by anyone who at least has a role in that process (because his role is then at least equal to the role *everyone*).

How the role model works is demonstrated in Figure 15.15. We have used the role graph from Figure 15.14 to create this figure. The case is right at the start, that is, the first activity (step 1) is due.

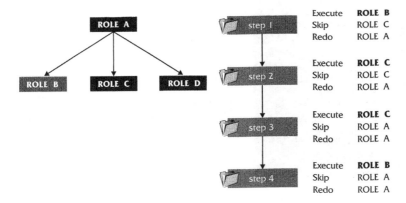

Figure 15.15 The options for an employee with role B.

Exercise 2. Give the options for an employee with:

- Role C
- Role D
- Role A
- Role C and B

Exercise 3. Suppose the FLOWer role model is not used. Draw the different arrows between the steps and provide the (role-dependent) conditions set by the redo and skip roles. In Figure 15.16, one such condition is given in pseudocode.

The example used may seem rather artificial, but does demonstrate the power of the model. It will be clear that it offers tremendous flexibility. If we want to model this relatively simple process model with only an execute role (WfMC), we have to define all role expressions as decisions and, thus, more arcs appear in the process flow. Also, the dynamic behavior whereby the employee's role determines which exceptions are allowed is lacking. In practice, we see the process models in a traditional WfMS becoming unreadable or we see much of the control hidden in the applications that are accessed. In a more complicated process model than this simple sequential process flow, with parallel paths, subprocesses, and so on, it would become even more complicated.

In the first part of this chapter, we indicated that flexibly structured processes are common. The process design functionality should support this. This means, for ex-

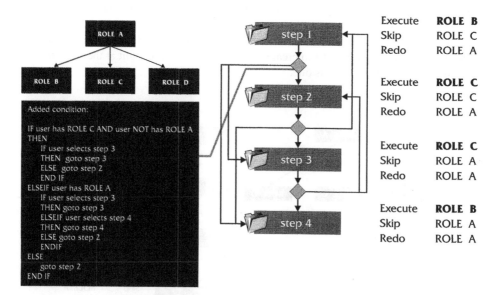

Figure 15.16 Added condition between step 1 and step 2 if no skip or redo roles are supported.

ample, that possible exceptions can be included without having to draw incomprehensible process flows.

The FLOWer Data Handling Model. Apart from the role model, FLOWer is also information driven; the information present for a case determines, to a large extent, its status. In administrative processes, the progress is determined by the information that is present (see [2, 3, 8, 24, 25, 26, 27]). Some of this information is purely process control data that can be used, for example, in decisions. In this way, for example, labeling a case as "straightforward" or "complex" determines the subsequent activities. Other data are purely content-related, such as the telephone number of a client. Some data objects include both aspects and are relevant both for the content and the control. An example is "claim amount," which determines the control (high amounts go the experienced claim handlers) and is also stored in a primary information system.

FLOWer has extensive possibilities for defining data, including *structures* and *arrays*. It is possible to define *all* data in FLOWer. But this is not necessary, of course. One could store data in the primary systems only.

Data important for process control or for management information is defined in FLOWer. For control *within* a case, FLOWer utilizes the data handling model explained below. For work distribution, in fact, part of the control *between* cases and all data objects of a process can be used. It is sufficient to label a data object for *publishing* (see the FLOWer publishing model in this subsection). This approach makes the work distribution extremely flexible. For data objects that are used for management information, it is sufficient to label them as such in the process design.

To explain the basics of the data handling model, note the following. The status of a case is (at runtime) represented by the status line (also called the time line or wave front). The status line shows all activities that are due now. A user can carry out the activities on the status line, provided he has the proper role (the execute role). These activities are visible in color; the other activities are visible in gray. Activities that have already been carried out (completed or skipped) are behind the status line, the others are in front of the line. A data object within a process can be connected to zero or more activities. The nature of this connection determines the importance of the data object for the activity. We distinguish three forms:

1. *Free.* The value of the data object can always be entered or changed, as long as the employee has written authorization for the case. An example is a client's telephone number.

2. *Mandatory* for one or more activities. In order to complete an activity, it is necessary (but not sufficient; see below) that all mandatory data objects of the activity have a value. The value of the data object can only be entered or changed if the employee has a role that is equal to or higher than the execute role in at least one of the activities for which the data object is mandatory. It is, therefore, *not* necessary that one of these activities is the next one due to be carried out at that moment, that is, it is on the status line. An example of this is entering the policy number of a client.

3. *Restricted* to one or more activities. The value of the data object can only be entered or changed if one of the activities is, in fact, the next one due to be carried out at that moment, that is, it is on the status line. An example is the "Approved?" data object that can only be carried out if the "Authorize" activity is due to be carried out.

These relationships are necessary to determine whether an activity has been completed. An activity is completed if:

- All previous activities have been completed (or skipped)
- All mandatory data objects of the activity have a value
- The so-called *completion condition* of an activity is true. This condition is normally set to "TRUE," which means that it is sufficient to give all mandatory data objects a value.

If an employee wants to give a value to a restricted data object, the accompanying activity must be due to be carried out (that is, be positioned on the status line) or the employee must move the status line by *skipping* the activities that are in between, or by *redoing* the activity again. This is only possible if he has the appropriate role, as described in the FLOWer role model.

Redoing influences the activities that have already been carried out and are now in front of the status line *again*. These now have to be carried out once more. It is not required to give all involved data objects a value again. Suppose only a small typo has to be corrected. FLOWer remembers the value entered earlier and gives the data objects the special status "to be confirmed." The user can then confirm the value for each data object.

The FLOWer data handling model in combination with the role model results in an extremely flexible model with which the user, depending on his or her role, can process a very large number of exceptions. On the other hand, by using restricted data objects and the *"nobody"* role, it is also relatively simple to force a straightforward sequential order of handling.

From the above, we can conclude that one of the striking differences between FLOWer and traditional WfMS systems is the semantics of the order of activities or plan elements. FLOWer requires that all previous plan nodes have been done (i.e., closed, skipped, or refused in the case of decisions) before a plan node can be *finished,* but one can *start* plan elements even if the predecessors have not been finished yet. This is very different from the more common notion, supported by other tools [4, 33, 34], that a plan element can only be *started* if its predecessors have all been done. In other words, FLOWer allows for "prefilling." Of course, this can easily be avoided if required by the circumstances, since this is both role dependent and data dependent. So one can easily block a decision for sending a payment check if it is made prematurely, but one can just as easily decide that certain data elements can be changed at will, irrespective of the state the case is in, as long as it is not closed yet.

The FLOWer Publishing Model. The third part of the FLOWer Studio supports the FLOWer publishing model. The FLOWer publishing model forms the main part of the interface between the process modeling environment and the organization modeling environment, FLOWer Configuration Management. In order to distribute work, it is essential to know the status of the cases. A frequently occurring status that is of interest for work distribution is, of course, that an activity is "due to be carried out"; in other words, it is on the status line. In a traditional WfMS, it also stops here—only that status is available. Moreover, the activity is only allocated to those people who have the role that belongs to that activity. FLOWer offers a more powerful solution. In FLOWer, every status and every data object can be *published.* You could publish, for example, that a specific activity has already taken place or has not yet taken place and, furthermore, is not yet due to be carried out. You can also indicate which role or roles should be published as well. This does not need to be the execute role. It can, for example, also be a manager. And, finally, you can publish a data object with its value. Also, the model is not *panic driven* (see Section 15.2.4 on direct allocation versus indirect allocation). Note that actual work distribution to users does not take place as yet; the status is published but it is only in the FLOWer configuration management that the work trays are prepared by utilizing these published statuses. How this works is explained in subsections 15.3.2 and 15.3.3. A few examples of publishing are listed below.

- The most obvious situation is publish all activities on the status line with role equal to the execute role of the activity. FLOWer allows you to fill a standard work tray for each employee based on his or her roles. Take, for example, employee Jan with the "Handler" role for the "Settle Motor Claim" process. Jan's standard work tray contains all cases in which an activity is on the status line and the execute role is the same as "Handler." With publishing, it is possible to use FLOWer to imitate the behavior of a traditional WfMS. Of course, you can restrict the standard work tray using the distribution aspects. So you can arrange to give Jan only those cases that took place in the "North" region. Simply publish the data element *"region"* from the process model.

- You can also publish those motor claims in which a specific activity, such as "Request Police Report," has been skipped and the amount payable is greater than 10,000 Euros. These activities can be published with the "Manager" role. This allows you to give the manager or another role an overview of these special cases. He or she can then determine what should be done. Of course, another role could receive these cases, or this can also be restricted using a regional code.

- It is possible to publish all data objects that are of interest for distribution.

It will be clear that this publishing approach is extremely flexible. Publishing can be compared to stocking a pool with various kinds of fish. A work tray is nothing more or less than a net to catch the right fish. You then only need to give the right employees the right net. This is done in FLOWer configuration management, as explained in the next subsection.

15.3.2   Organizational Design

The organizational design is the next item to be dealt with. As mentioned above, this aspect is on the borderline between the definition and the operation. Depending on the policy of the organization, there is a choice of placing the emphasis on the definition (for example, in centrally controlled organizations) or on the operational level (in decentrally controlled organizations). It is common practice that the main outlines of the design are generally valid and are, therefore, defined during the definition phase, and the details are defined at the operational level. It is possible, for example, to determine during the design phase which (large) organizational units there should be, and that cases should be distributed among these organizational units according to, for example, a *Zip Code* or *region*. It can also be decided during the design phase what criteria should be available for work distribution. But afterwards, the manager of an organizational unit can then take over control of the design of the unit, divide it into teams, determine the allocation, and decide on the available criteria to do this, as long as he or she stays within the boundaries set during design.

Next, the authorizations for each employee should be established. In order to guarantee the necessary flexibility, a distinction should be made here between authorization aspects and distribution aspects. An employee, for example, acts as a "Claims Handler" for the "Settle Motor Claim" process and works in the department that handles cases for region "*North.*" The authorization aspects are determined in terms of roles ("Claims Handler"). The department, in fact, forms a distribution unit; the work that satisfies certain criteria should be distributed to this department. An employee, thus, receives a work profile by being allocated a number of roles and being included in a unit to which work is allocated. This should, of course, be done in such a way that it remains easy to manage. FLOWer can do this.

We note that the above approach ensures a proper division of the various units. The process design produces a number of roles and criteria that could be of interest for the work distribution, whereas the employees are identified during the organizational design. It is then possible to compose a number of work profiles. This is the basis for the work distribution, discussed in the next subsection.

15.3.3   Work Distribution

In the previous subsection, we defined work profiles for employees. These work profiles already contain all the roles and are already connected to distribution profiles. To complete the work distribution, all we need to do now is to indicate the values of the criteria for the distribution profiles. We need to indicate, for example, that for the North department, the value of the region criterion should be "*North.*" In this way we define work profiles whereby employees are allocated work for which they are authorized and that also belongs in their organizational unit. Every employee, therefore, has a standard work tray containing the cases he or she can process. That is, cases that can be processed in that employee's organizational unit and for which he or she can carry out activities. If desired, the cases can be sorted and prioritized. Allocation can also be arranged so that it is for a single user ("push" distribution) or a group ("pull" distribution). For pure production work, the work allocation has now been arranged.

This is, however, not sufficient. The employees should be able, for example, to look up a case in the event of questions from a client. A team leader or manager must have insight into the open cases. It boils down to the fact that, besides a standard work tray, other queries should be available. This functionality is particularly important for a front-office environment or a call center.

We note yet again that this aspect is also difficult to define entirely centrally, let alone to handle in the *process modeling* environment. For example, one department divides the work differently from other departments, or provides the employees with different overviews. In one department or in one workgroup, a sudden rush of work will mean that the work allocation needs to be changed temporarily. Of course, the freedom of scope is centrally defined. This is one of the aspects of *operational flexibility,* which means that the operational environment can be adapted without having to change the design or the centrally defined organizational design. In this context, we can speak of operational flexibility in the work distribution. But operational flexibility also means that the boundaries, set out in the design or the central organizational design, may not be crossed. So operational flexibility does not mean that the employee or manager themselves can determine *everything.* See [17] for more information on dynamic work distribution in WfMS.

FLOWer Configuration Management. In FLOWer, the functionality is offered through FLOWer Configuration Management (CFM). FLOWer CFM can be used centrally and/or decentrally. This allows you to divide up the organizational design task between all those involved. In practice, the outlines are usually set out centrally first; the managers of the various units can then fill in the details for their employees.

In FLOWer CFM, the following can be defined:

- Users
- Distribution profiles
- Work profiles
- Function profiles
- Queues
- Queries

In Figure 15.17, the different elements are shown.

A user is assigned to one or more work profiles. A work profile is a combination of one or more function profiles and one or more distribution profiles. A function profile can be compared to a *function*. Roles can be grouped into a function profile.

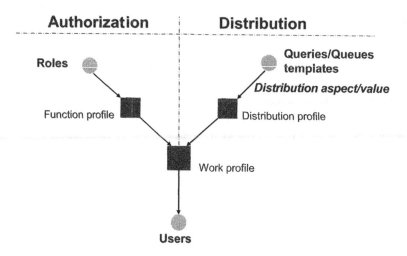

Figure 15.17   The different elements of FLOWer Configuration Management.

A role is always defined in the context of a process, as described earlier. Once the function profiles have been attached to the work profile, it is determined what the user can do, that is, what he or she is authorized to do. But it does not determine what a user can see. This is done using distribution profiles.

A queue definition or queue template contains a recipe for the selection of cases. This recipe can use data objects such as "regional code," "priority," "amount," and so on. These data objects must be published in the process design. The queue template can then be related to a distribution profile. A queue is formed by defining the value(s) of the data objects in the queue template. In this way a "North" distribution profile is created, for example, to which the queue "all motor claims with regional code North" is related. Of course, other queues can be connected to this "North" distribution profile. This distribution profile can then be related to the work profile "Claims for North."

Also, the function of "Motor Claims Correspondent" is related to this work profile, and the role of "Claims Handler" for the "Settle Motor Claim" process is related to this function. In this way, authorization and distribution are properly separated and can be set up very flexibly.

FLOWer bases the content of the queue that the employee sees on the work profile of that employee. The queue or work tray contains exactly the cases that the employee can see *and* for which he is authorized (both selected on the basis of the publishing model). The employee, therefore, can, in fact, actually work on every case in this work tray.

As you can also publish roles other than the execute role, you can also create work trays in this way containing cases that the employee can see but not necessarily work on. This could be very well suited to on-line management information.

Finally, it is also possible to define queries that also produce cases. The query can make use of all the published data. This data can also be offered as a parameter for the user of the query. For example, the *postal code* of the client could be included as a parameter in a query. This would allow a call center or a front office to find cases quickly and inform the client of its status. Queries, like queues, can also be connected to work profiles (via their relation with distribution profiles).

In both queues and queries, various setting options are possible, such as only opening read-only files, or opening the case automatically on a specific form. This also allows the production workflow elements of a process to be very effectively driven.

FLOWer Configuration Management is also available at the operational level in order to introduce further refinements such as:

- Making detailed profiles for teams or individual employees
- Defining specific queries, for example, for on-line management information
- Managing users in your own environment
- Adapting features such as queues and queries

In short, flexibility of design as well as work distribution can really be achieved at the operational level but, naturally, within the boundaries set out by the organization. The conceptual integrity of the different environments, process modeling, organizational modeling, and work distribution is preserved. Optimal flexibility is the result.

Exercise 4. Describe the effect of the following changes on the process design if no abstraction of teams/workgroups is available in the process modeling environment, so only roles are supported or roles and teams are defined using the same mechanism. In other words, suppose there is a direct coupling between the "execution condition," that is, the role expression, of an activity and the runtime environment. A user then simply gets all activities if his profile obeys the execution condition.

Suppose a process is being handled in 25 different offices, having from 5 to 50 employees. In the smaller offices, employees have many roles; in the larger offices, employees have only one or at most three roles. In addition, in some offices, independent of their size, a pull-type work distribution is in use; in other offices, a push-type distribution is used.

- Add a new office.
- Split one office into two offices.
- Allow for special treatment of cases that are due; that is, make for each office a special work tray with due cases.
- Reorganize completely. The process stays the same, but now we have 25 small front offices and one large back office.

15.3.4 Execution of Work

Finally, we come to the execution—the handling of work. Employees get work on the basis of their profiles. This work is presented in the employees' standard work trays or queues, which contain the cases the employees can process. The employees can also look up cases through the "queries" allocated to them. It could be that the respective employee can only "view" a particular case. In that case, he or she does not have the role or permission to execute an activity but can, for example, provide information about the case to a customer.

Suppose that the employee opens a case for which he or she has a role that allows processing. We are then at the process-control level within a single case. FLOWer gives the employee an overview of the entire case, not only the activity or activities that can be executed now. The information of the case is also shown. It is possible, of course, to hide certain parts of the case and the related information. An example is the "Process Medical Examination" subprocess that can *only* be viewed by employees having been assigned the role of "Medical Adviser" *directly,* that is, not assigned the role because they have a higher role than "Medical Adviser."

Previously, we described the importance of flexibly structured processes. Within the limitations set at design time (in the process modeling and organizational modeling environments), the employees should be able to make their own choices, process exceptions, apply their knowledge optimally, and so on. This is a second aspect of *operational flexibility*—the ability to adapt the processing of the case.

For more information on the process execution environment, please refer to the Pallas Athena Website, http://www.pallas-athena.com.

15.3.5   Maintenance

In the previous subsections, we described process design, organizational design, and the execution of cases. Subsequently, it should be possible change the design. This requires powerful management and maintenance functionality [7, 14, 32]. In other words, it should be possible to effectively adapt the process design, the integration with other information systems, the organizational design, the work distribution, and the execution. It cannot be overstated that conceptual integrity is crucial. So the available functionality must be complete and available at precisely the right spot. To state it bluntly, one can automate each and every process, provided one does enough programming and has a large budget, but both cost and elapsed time will be far beyond any reasonable level without proper maintenance functionality.

15.3.6   Management Information

One important aspect necessary to achieve continuous improvement is still lacking—management information. Only with good management information can one analyze how a process can be improved. In addition, logging management information may be a legal obligation or required by a company's internal auditors. An effective process management system should, therefore, record the essential basic information that shows who did what and when. Preferably, the relevant details should also be recorded. On this basis, the results can be analyzed and then changes can be identified and implemented on all levels named in the previous Subsection 15.3.5.

We make a distinction, however, between *on-line* management information and *off-line* management information. On-line management information provides direct information about the current situation and provides the opportunity for immediate intervention. A few examples are:

- An overview for a team leader of the current cases of each employee in his team, on the basis of which he can reallocate work
- An overview of high-priority cases in a department
- A departmental overview of cases that need to be completed within two weeks

With the mechanisms available in FLOWer Configuration Management, described earlier, on-line management reports can be prepared easily. The on-line property plus the representation in the form of cases allows for direct intervention, like reallocating cases.

The off-line management information is suitable for elaborate analysis, possibly and most often combined with information gathered in other systems. This makes it possible, for example, to use off-line management information for simulation purposes.

FLOWer offers a wide range of options in the area of management information. The off-line management information in FLOWer consists of two parts:

1. Management information relating to process objects, also called *case node logging*
2. Management information relating to data objects, also called *case data logging*

Case Node Logging.   Normally, FLOWer will log all changes in the status of a case. In a database, it logs which process objects have changed, when they changed, who changed them, why the change occured, the processing time since the previous status change, and the new status. FLOWer also logs when the case was created, opened, or closed and by whom. It also logs the time passed between the moment of opening of the case and the moment the user actually begins to process it, that is, starts keying in information or invoking an application. From this, the time required to make the *context switch* from one case to another can be measured.

The stored management data that can be converted into management reports using a report generator.

Case Data Logging.   As we stated earlier, data is crucial for administrative processes [3, 10, 24, 27]. FLOWer can log case data in a database. When doing this, FLOWer only logs the data objects that were labeled as *management information* during the process design (FLOWer Studio). In general, this involves data that is crucial for management information, such as the amount of a claim and results of decisions. This information, too, is stored in database tables. In combination with case node logging and/or information from the primary systems or an information warehouse one can generate reports.

Case Instance History. Apart from the management information mentioned above, FLOWer also records the entire history of a case. This information is recorded in a file and contains information about the execution of a case such as both status and data changes, timestamps, and people involved. This information can be read by those employees who are authorized to do so, using the Activity Log Viewer. So, if required the entire history of the case is available. During the design phase (FLOWer Studio) the information to be included in the case history is determined.

15.3.7   A Process of Continuous Improvement

If the elements mentioned in the previous subsections are present, a process of continuous improvement can be started. This forms an integral part of the required flexibility—processes and products must be adaptable in the shortest possible time and at the lowest possible cost. This is crucial for every organization in today's market [7, 14]. Therefore, conceptual integrity should be one of the most important discriminators in selecting a tool.

We conclude this section with an exercise involving some of the FLOWer components.

Exercise 5. Describe how the FLOWer role model and data handling model should be used to turn a FLOWer process model into a traditional workflow model. Also, describe using the FLOWer publishing model, how to define a work tray for an employee containing precisely those cases for which there are activities on the status line with an execute role equal to one of the roles of the employee.

15.4   GOLDEN RULES OF PROCESS MANAGEMENT

We conclude this chapter with eight golden rules of process management. The list is limited to what we believe are the most important ones that are at least in some way typical for process management.

1. Process management is complex. Do not oversimplify. Through the years, most existing business processes have become too complex, especially the ones that are not controlled by dedicated software. In most situations, they can be simplified. To avoid the risk of suboptimizing, one should start remodeling the business process instead of simply automating the current process. But most business processes will still turn out to be complex, even if the modeling is done properly. Some managers, however, have a tendency to downplay the complexity of the work, because the standard cases are indeed easy to handle. They also take a high-level perspective, thereby ignoring the exceptions. But, in practice, the world is more complex. To oversimplify, therefore, exposes one to real risk.

2. The impact of process management is high. Process management influences the way people work more than the data management applications (legacy, registration of data, etc.). The latter focus on automating data-handling tasks and thus influence only some aspects of the daily work. Process management influences the way processes are handled; that is, almost all other aspects of the daily work. Therefore, implementing process management should be done with great care. It is not just another application supporting a few tasks in a process but an application automating the processes themselves. Make sure the users are involved right from the beginning and appoint a dedicated and experienced change manager.

3. Don't design a production line The main reason for the failure of using the industrial production line paradigm in office processes is that this paradigm simply does not fit. The best proof for that is that it has almost never been applied successfully, even after decades of trying. Implementing a process on the basis of industrial/logistic production principles only is not good enough. Another paradigm is required. Therefore, take advantage of the lessons learned in industrial production lines and logistics; there is certainly much to be gained by applying some of the principles, but it is not the only source of knowledge.

4. Prototyping leads to higher quality in less time. As it is very difficult to fully comprehend a process specification and its impact on the organization (see the first rule), it is of utmost importance to develop prototypes. The information gathered from prototyping sessions is valuable for further development and, in addition, prototyping will help determine the right expectations for users and their management. Products allowing for rapid prototyping, like FLOWer itself or those used in combination with the Pallas Athena process modeling tool PROTOS, deliver higher quality, shorten development time, and reduce costs.

5. Develop incrementally. By developing incrementally, one can combine the benefits of prototyping and traditional development. For process management, with its heavy impact on the way people work, this is important.

6. Make the process itself a starting point, not the routing. Do not make routing the basis of a process. The activities of the process are the basis. If one makes routing the basis, one will soon conclude that the process flow within a routing step must either be supported in a separate module or must be built once again in the business applications. As shown in Subsection 9.2.1, the basic element of control—a separate module—is no real solution. The latter solution, incorporating parts of process flow in the business application, obviously is no improvement since it violates the very basis of workflow management: the separation of process and business applications.

7. Exceptions are the rule. Accept the notion that not everything can be determined beforehand. Exceptions are the rule in processes. Determine what has to be obeyed at all times, but leave other parts to the workers, allowing them to make decisions depending on their skills (and roles). Keep in mind that the mere fact that humans are involved indicates that human skills are required. What can be automated will be automated, now or in the near future; what cannot, should be left to the workers. That is, system should support these workers to make the right decision, depending on their roles and the boundaries set by process modelers and management.

The benefits from supporting exceptions, that is, not solely focusing on the straight-through processes, are real. In most processes, except for the very few production processes without exceptions, the number of cases with exceptions is high. Twenty percent or more is a good estimate. Note that these 20% of the cases take about 80% of the time, and they take even more than 80% of the cost since the exceptions are handled by experienced staff. The straightforward cases take 20% of the time. So, obviously, you can achieve high benefits by improving on the difficult cases.

8. Never compromise the conceptual integrity. The chosen product must support conceptual integrity. The cost of maintenance will be exceptionally high without it. Keep in mind that the full circle of improvement must be supported to be able to adapt rapidly to changes, not the least to changes affecting the (business) environment, like new products from competitors and new legislation. During all system development phases, regularly analyze where changes can occur and what their nature is. Ensure that these changes are indeed limited to the component they logically belong to. For example, a change in the organization like adding a second claims-handling department, adding some staff to a team, or changing the way work is distributed to certain employees, should not affect the process design. Make sure the system is as robust to changes as possible.

15.5 CONCLUSION

In this chapter, we have given an overview of the FLOWer case handling approach to process management. We have shown how FLOWer differs from traditional workflow management systems. We have provided an overview of the different aspects of process management and the way they are implemented in FLOWer. In doing so, we have demonstrated the conceptual integrity of the FLOWer case-handling paradigm for process management. Case handling is a powerful paradigm for process management. The FLOWer case handling system allows you to manage effectively not only flexibly structured processes, but straightforward production workflow. FLOWer offers organizations a process management functionality, allowing them to respond effectively and successfully to the needs of their clients in today's constantly expanding and highly competitive market. For more information on FLOWer, see [23].

ACKNOWLEDGMENT

The contents of this chapter, apart from Section 15.4, are part of the FLOWer position paper [22], copyright Pallas Athena, The Netherlands, and available on the Pallas Athena Website, http://www.pallas-athena.com.

REFERENCES

1. W. M. P. van der Aalst. The Application of Petri Nets to Workflow Management. *The Journal of Circuits, Systems and Computers, 8*(1):21–66, 1998.

2. W. M. P. van der Aalst. On the Automatic Generation of Workflow Processes Based on Product Structures. *Computers in Industry, 39*(2): 97–111, 1999.

3. W. M. P. van der Aalst and P. J. S. Berens. Beyond Worklow Management: Product-Driven Case Handling. In S. Ellis, T. Rodden, and I. Zigurs (Eds.), *International ACM SIG-GROUP Conference on Supporting Group Work (GROUP 2001)*, pp. 42–51. ACM Press, New York, 2001.

4. W. M. P. van der Aalst and K. M. van Hee. *Workflow Management: Models, Methods, and Systems*. MIT Press, Cambridge, 2002.

5. W. M. P. van der Aalst, A. H. M. ter Hofstede, B. Kiepuszewski, and A. P. Barros. Advanced Workflow Patterns. In O. Etzion and P. Scheuermann (Eds.), *Seventh International Conference on Cooperative Information Systems (CoopIS 2000)*, volume 1901 of Lecture Notes in Computer Science, pp. 18–29. Springer-Verlag, Berlin, 2000.

6. W.M.P. van der Aalst, A.H.M. ter Hofstede, B. Kiepuszewski, and A.P. Barros. Workflow Patterns. *Distributed and Parallel Databases, 14*(1):5–51, 2003.

7. W. M. P. van der Aalst and S. Jablonski. Dealing with Workflow Change: Identification of Issues and Solutions. *International Journal of Computer Systems, Science, and Engineering, 15*(5):267–276, 2000.

8. W. M. P. van der Aalst, H. A. Reijers, and S. Limam. Product-driven Workflow Design. In W. Shen et al. (Eds.), *Proceedings of the Sixth International Conference on Computer Supported Cooperative Work in Design 2001*, pp. 397–402. NRC Research Press, Ottawa, 2001.

9. W. M. P. van der Aalst, M. Stoffele, and J. W. F. Wamelink. Case Handling in Construction. *Automation in Construction, 12*(3):303–320, 2003.

10. W. M. P. van der Aalst, M. Weske, and D. Grünbauer. Case Handling: A New Paradigm for Business Process Support. *Data and Knowledge Engineering, 2004*.

11. A. Agostini and G. De Michelis. Improving Flexibility of Worklow Management Systems. In W. M. P. van der Aalst, J. Desel, and A. Oberweis (Eds.), *Business Process Management: Models, Techniques, and Empirical Studies*, volume 1806 of Lecture Notes in Computer Science, pp. 218–234. Springer-Verlag, Berlin, 2000.

12. BPi. *Activity Manager: Standard Program—Standard Forms (Version 1.2). Workflow Management Solutions*, Oosterbeek, The Netherlands, 2002.

13. F. Casati, S. Ceri, B. Pernici, and G. Pozzi. Conceptual Modeling of Workflows. In M. P. Papazoglou (Ed.), *Proceedings of the OOER'95, 14th International Object-Oriented and Entity-Relationship Modelling Conference*, volume 1021 of Lecture Notes in Computer Science, pp. 341–354. Springer-Verlag, 1995.

14. C. A. Ellis and K. Keddara. A Workflow Change Is a Workflow. In W. M. P. van der Aalst, J. Desel, and A. Oberweis (Eds.), *Business Process Management: Models, Techniques, and Empirical Studies*, volume 1806 of Lecture Notes in Computer Science, pp. 201–217. Springer Verlag, Berlin, 2000.

15. K. M. van Hee and H. A. Reijers. Using Formal Analysis Techniques in Business Process Redesign. In W. van der Aalst, J. Desel, and A. Oberweis (Eds.), *Business Process Management*, Lecture Notes in Computer Science 1806, pp. 142–160. Springer-Verlag, Berlin, 2000.

16. T. Herrmann, M. Hoffmann, K. U. Loser, and K. Moysich. Semistructured models are surprisingly useful for user-centered design. In G. De Michelis, A. Giboin, L. Karsenty, and R. Dieng (Eds.), *Designing Cooperative Systems (Coop 2000)*, pp. 159–174. IOS Press, Amsterdam, 2000.

17. S. Jablonski and C. Bussler. *Workflow Management: Modeling Concepts, Architecture and Implementation.* International Thomson Computer Press, 1996.

18. A. Kumar, W. M. P. van der Aalst, and H. M. W. Verbeek. Dynamic Work Distribution in Workflow Management Systems: How to Balance Quality and Performance? *Journal of Management Information Systems, 18*(3):157–193, 2002.

19. B. Kiepuszewski, A. H. M. ter Hofstede, and C. Bussler. On Structured Workflow Modelling. In B. Wangler and L. Bergman (Eds.), *Proceedings of the Twelfth International Conference on Advanced Information Systems Engineering (CAiSE'2000)*, volume 1789 of Lecture Notes in Computer Science, pp. 431–445, Stockholm, Sweden, June 2000. Springer-Verlag.

20. F. Leymann and D. Roller. Production *Workflow: Concepts and Techniques.* Prentice-Hall PTR, Upper Saddle River, New Jersey, 1999.

21. G. Meszaros and K. Brown. A Pattern Language for Workflow Systems. In *Proceedings of the 4th Pattern Languages of Programming Conference,* Washington University Technical Report 97-34 (WUCS-97-34), 1997.

22. Pallas Athena. *Case Handling with FLOWer: Beyond workflow management.* Pallas Athena BV, Apeldoorn, The Netherlands, 2002.

23. Pallas Athena. FLOWer Manuals. Pallas Athena BV, Apeldoorn, The Netherlands, 2001–2004. www.pallas-athena.com.

24. H. A. Reijers. Product-Based Design of Business Processes Applied within the Financial Services. *Journal of Research and Practice in Information Technology, 34*(2):34–46, 2002.

25. H.A. Reijers, S. Limam, and W.M.P. van der Aalst. Product-based Workflow Design. *Journal of Management Information Systems, 20*(1):229–262, 2003.

26. H. A. Reijers and K. Voorhoeve. Optimal Design of Process and Information Systems: A Manifesto for a Product Focus. *Informatie, 42*(12):50–57, 2000. (In Dutch)

27. N. Russell, A. H. M. ter Hofstede, D. Edmond, and W. M. P. van der Aalst. *Workflow Data Patterns CITI Technical Report, FIT-TR-2004-01, QUT,* 2004; see http://www.citi.qut.edu.au/pubs/technical.jsp.

28. Staffware. *Staffware Case Handler—White Paper.* Staffware PLC, Berkshire, UK, 2000.

29. D. M. Strong and S. M. Miller. Exceptions and Exception Handling in Computerized Information Processes. *ACM Transactions on Information Systems, 13*(2):206–233, 1995.

30. K. de Vries and O. Ommert. Advanced Workow Patterns in Practice (1): Experiences Based on Pension Processing (in Dutch). *Business Process Magazine, 7*(6):15–18, 2001.

31. K. de Vries and O. Ommert. Advanced Workflow Patterns in Practice (2): Experiences Based on Judicial Processes (in Dutch). *Business Process Magazine, 8*(1):20–23, 2002.

32. M. Weske. Formal Foundation and Conceptual Design of Dynamic Adaptations in a Workflow Management System. In R. Sprague (Ed.), *Proceedings of the Thirty-Fourth Annual Hawaii International Conference on System Science (HICSS-34).* IEEE Computer Society Press, Los Alamitos, CA, 2001.

33. WFMC. *Workflow Management Coalition Terminology & Glossary,* Document Number WFMC-TC-1011, Document Status—Issue 3.0, February. Technical report, Workflow Management Coalition, Brussels, 1999.

34. WFMC. Workflow Management Coalition, *The Workflow Reference Model,* Document Number TC00-1003, Document Status—Issue 1.1, January 1995, Author: David Hollingsworth.

Readings and Resources

This appendix includes lists of references to sources of further information regarding various aspects of process-aware information systems. It is not intended to be exhaustive. In particular, direct references to Web sites of commercial software tools are not included.

URLs are current as of October 2004. The descriptions and comments included reflect the book editors' understanding and viewpoints. The editors do not make any warranties or representations regarding the accuracy, suitability, completeness, currency, or correctness of the provided information.

BOOKS

B2B Integration by Christoph Bussler (Hardcover, 400 pages, Springer-Verlag, 2003). Presents an overview of software architectures for business-to-business integration. Emphasis is on general principles and concepts rather than products or standard-specific features. Many of the concepts and principles described are relevant to the design and implementation of application-to-application processes within and across organizational boundaries. The book constitutes an excellent source of complementary material for readers interested in the topics covered by the same author in Chapter 4 of this book.

Design and Control of Workflow Processes: Business Process Management for the Service Industry by Hajo Reijers (Paperback, 320 pages, Springer-Verlag, 2003). Addresses issues related to the design and redesign of workflow processes as discussed by the same author in Chapter 9 of this book. Focus is on performance of business processes rather than their technical realization. Among other things, the book looks at resource allocation in workflow processes (addressing Goldratt's conjecture), product-based workflow design (looking at the essential data rather than the existing way of doing things), and redesign heuristics. Most of the examples are taken from the service industry.

Production Workflow: Concepts and Techniques by Frank Leymann and Dieter Roller (Paperback, 479 pages, Prentice-Hall, 1999). Provides a relatively comprehensive presentation of workflow systems, addressing three key questions: what is a workflow system? how does one use a workflow system? and, to a lesser extent, how does one build a workflow system? Although the authors attempt to be "product-neutral" in their treatment of the topic, a significant part of the discussions and examples refer to IBM's MQSeries Workflow (later renamed IBM Websphere MQ Workflow) and its associated flow definition language, which ultimately influenced the design of the Business Process Execution Language for Web Services presented in Chapter 13 of this book. Also, the workflow system architecture presented reflects the design and technological choices incorporated in this product.

Web Services: Concepts, Architectures and Applications by Gustavo Alonso, Fabio Casati, Harumi Kuno, and Vijay Machiraju (Hardcover, 354 pages, Springer-Verlag, 2003). Provides a conceptual overview of the technology and design techniques associated with Web services. Especially relevant to the topic of process-aware information systems is the second part of the book, which, among other things, introduces the notions of Web service coordination and composition, which together constitute a paradigm for application-to-application integration. Consistent with the style of the book, the discussion on Web service composition is mainly placed at a conceptual level, and the Business Process Execution Language for Web Services (see Chapter 13 of this book) is presented as a particular realization of this notion. The book also covers transactional aspects of middleware in general, and Web services in particular, expanding on some of the considerations of Chapter 11 of this book.

Workflow Management: Modeling Concepts, Architecture and Implementation by Stefan Jablonski and Christoph Bussler (Paperback, 351 pages, International Thomson Computer Press, 1996). One of the earliest books on workflow management covering a broad range of topics including, among others, available tool support (both research prototypes and commercial products), a treatment of various identified perspectives on workflow management, and implementation aspects. As a general approach to workflow management, the MOBILE model and system presented in this book provide a context in which applicable concepts and approaches are seamlessly brought together.

Workflow Management: Models, Methods, and Techniques by Wil van der Aalst and Kees van Hee (Hardcover, 374 pages, MIT Press, 2002). Presents a model-driven approach to process-aware systems. The focus is on workflow technology and the modeling and analysis of workflow processes. As a modeling language, a variant of Petri nets (better known as "workflow nets") is used, and at the end of each chapter there are exercises. The book has a supporting Web site with interactive examples and lecture material, http://www.workflow-course.com.

Workflow-based Process Controlling: Foundation, Design, and Application of Workflow-driven Process Information Systems by Michael zur Muehlen (Paperback, 315 pages, Logos, Berlin, 2004). Focuses on the use of workflow models and data collected from workflow executions (i.e. audit trails) for the purpose of decision making. It presents a reference model for representing workflow applications and their audit trails, and provides a comprehensive overview of organizational aspects of process management. The book also includes thorough background information on workflow technology and related standards (including those discussed in Chapter 12). Among others, a historic view into the evolution workflow tools is provided, covering both research prototypes and commercial tools. More information about the book as well as an electronic version of it can be found at http://www.workflow-research.com.

PORTALS AND WEB SITES OF INTEREST

AAIM—Association for Information and Image Management (http://www.aiim. org). The main Web site of an association dedicated to the promotion of technologies for content management in general, with an emphasis on content management for process automation. Among other things, AAIM organizes and advertises professional development events (e.g., online and traditional seminars) and publishes a magazine targeted at professionals (*AAIM E-DOC Magazine*).

BP Trends (http://www.bptrends.com). This portal is dedicated primarily to managerial aspects of business processes but also covers technology-related topics, especially regarding process modeling, analysis, and simulation tools as well as standardization efforts. Among other activities, it disseminates a newsletter and maintains lists of relevant events and resources.

BPM Center (http://www.bpmcenter.org). This site is jointly maintained by the research groups with which the editors of this book are affiliated: the BPM Group of the Queensland University of Technology, Australia, and the Department of Technology Management at Eindhoven University of Technology (TU/e), The Netherlands. These associated groups conduct research across a broad spectrum of business process management topics. Among other activities, the site maintains information about ongoing activities and a series of technical reports.

BPM Institute (http://www.bpminstitute.org). This portal aims at facilitating exchanges between practitioners of business process management (BPM). Like the BPM Institute, it is mainly oriented toward managerial aspects but also covers technological issues. Among other activities, it disseminates white papers and articles, hosts discussions forums and round tables, and maintains a directory of vendors of BPM solutions.

e-Workflow (http://www.e-workflow.org). Associated with the WfMC, WARIA, and BPMI standardization initiatives (see below), this portal is dedicated to workflow technologies. It maintains a collection of case studies, a repository of white papers, and hosts a discussion forum.

Workflow Patterns (http://www.workflowpatterns.com). This site provides full descriptions, documentation, and animations for the workflow patterns introduced in Chapter 8, as well as other patterns covering different perspectives of workflow modeling. The site also features a "Vendors Corner" where commercial workflow vendors post reports describing how their tools support these workflow patterns.

Workflow Research (http://www.workflow-research.com) This portal provides information about academic activities in the area of workflow and business process management. It includes listings of research groups, universities and other institutions offering relevant courses, bibliography listings, and an open discussion forum.

STANDARDIZATION INITIATIVES

Workflow Management Coalition—WfMC (http://www.wfmc.org). This industry-driven coalition is devoted to the development of standards for workflow management, including the WfMC reference model and glossary, the XML Process Definition Language (XPDL), and the Workflow XML protocol (Wf-XML) presented in Chapter 12 of this book.

Object Management Group—OMG. This organization produces standards in a wide range of areas related to object-oriented technology. It is responsible, among others things, for the Unified Modeling Language (see Chapter 5) and has launched a number of standardization initiatives directly related to process-aware information systems (see http://www.omg.org/bp-corner/introduction.htm).

Business Process Management Initiative (BPMI). This industry consortium is mainly devoted to the development and promotion of standards in the area of Business Process Management. It promotes, among others things, the Business Process Modeling Notation (BPMN), a notation with a scope similar to UML activity diagrams (see Chapter 5) and Event Process Chains (see Chapter 6). Note that the BPMN initiative has its own Web site (http://www.bpmn.org).

RosettaNet (http://www.rosettanet.org). This consortium is dedicated to the development of standards for electronic business, with an emphasis on procurement and supply chain management processes. Among other activities, RosettaNet defines interface processes (see Chapter 5) for specific electronic business activities (e.g., placing a purchase order). These processes are called Partner Interface Protocols (PIPs).

Organization for the Advancement of Structured Information Standards (OASIS). This organization is dedicated to the development and promotion of standards in the area of information systems in general, and process-aware information systems in particular. Among other activities, it hosts the Technical Committee on Web Services Business Process Execution Language (WS-BPEL), which is responsible for the development of the language presented in Chapter 13. It also cooperates with the United Nations Centre for Trade Facilitation and Electronic Business (UN/CEFACT) in the development of the electronic business XML standard, ebXML (http://www.ebxml.org), which, among other things, includes a language for describing collaboration protocols (a notion closely related to that of interface processes discussed in Chapter 4).

World Wide Web Consortium (W3C)—Web Services Choreography Working Group (http://www.w3.org/2002/ws/chor). This group works toward defining a model and a language for describing collaboration protocols involving Web services. These collaboration protocols are called service choreographies. A service choreography captures the way a collection of services interact with each other without taking the viewpoint of any of the services involved, unlike interface processes (see Chapter 5), which would adopt the viewpoint of one of the services.

TOOLS

Call Center, Bug Tracking, and Project Management Tools for Linux (http://linas.org/linux/pm.html). This site maintains a annotated list of tools covering a large portion of the PAIS landscape (see Chapter 1), with a focus on tools that may run on top of the Linux operating system. It provides critical comments regarding the overlap and complementarity of various types of tools and briefly discusses their suitability in different settings.

jBPM (http://www.jbpm.org). This open-source initiative develops a workflow execution engine supporting a modeling language based on UML activity diagrams (see Chapter 5). In October 2004, jBPM joined the JBoss Professional Open Source Federation and was renamed "JBoss jBPM."

OBE—Open Business Engine (http://www.openbusinessengine.org). The main Web site of the open-source workflow engine, implementing the XPDL and Wf-XML standards (see Chapter 12).

ProductWatchlist—Process Integration (http://www.jenzundpartner.de/Resources/Product_Watchlist/product_watchlist.htm). Maintained by consultancy firm Jenz & Partner, this Web site provides a list of tools related to process-aware information systems development, with an emphasis on tools for process modeling and tools for application-to-application processes.

Process Modeling Tools (http://is.twi.tudelft.nl/~hommes/tools.html). List of business process modeling (and related) tools maintained at the Information Systems Algorithms Department, Delft University of Technology.

Project Management Tools (http://www.startwright.com/project1.htm). Web site maintained by StarWright, with a list of project management tools and related links.

Topicus Open Source Workflow Initiatives (http://www.topicus.nl/topwfm). Maintained by Dutch consulting company Topicus, this site provides a list of open-source initiatives in the area of workflow management systems.

Yet Another Workflow Language—YAWL (http://www.yawl-system.com). Web site of the YAWL workflow language and system briefly discussed in Chapter 8.

Index

Printed and bound by CPI Group (UK) Ltd, Croydon, CR0 4YY

27/10/2024

14580255-0003